The rights and wrongs of zebrafish: Behavioral phenotyping of zebrafish

Allan V. Kalueff
Editor

The rights and wrongs of zebrafish: Behavioral phenotyping of zebrafish

Springer

Editor
Allan V. Kalueff
The International Zebrafish Neurosceince Research Consortium (ZNRC) and ZENEREI Research Center
Slidell, LA, USA

College of Food Science and Technology
Research Institute of Marine Drugs and Nutrition
Guangdong Ocean University
Zhanjiang, Guangdong, China

Institute of Translational Biomedicine
St. Petersburg State University
St. Petersburg, Russia

Institutes of Chemical Technology and Natural Sciences
Ural Federal University
Ekaterinburg, Russia

ISBN 978-3-319-33773-9   ISBN 978-3-319-33774-6 (eBook)
DOI 10.1007/978-3-319-33774-6

Library of Congress Control Number: 2016957392

Printed on acid-free paper

This Springer imprint is published by Springer Nature
The registered company is Springer International Publishing AG
The registered company address is: Gewerbestrasse 11, 6330 Cham, Switzerland

Preface

Animal models and tests have become increasingly important for biomedical research, enabling a better understanding of pathogenic pathways involved in various human disorders. Over the last decades, zebrafish (*Danio rerio*) has become a very popular model organism in biomedical research. But, as we are often asked, why zebrafish? This small aquatic vertebrate fish species has traveled all the way from its natural habitat in India to emerge as a promising model organism in developmental biology, genetics, physiology, and toxicology. Recently, this fish has entered the waters of neuroscience and biological psychiatry, quickly becoming an indispensable model species in this field. With a high genetic homology to humans (~75 % based on coding regions), it is not surprising that humans and fish are very similar physiologically (and behaviorally)—perhaps, more than we would like to admit.

Therefore, it should not come as a surprise that zebrafish can be an excellent model of human neuropsychiatric disorders. While some classical psychiatrists may not too easily be persuaded by this generalization, the current book *The Rights and Wrongs of Zebrafish: Principles of Behavioral Phenotyping and CNS Disease Modeling* explains, in a domain-by-domain manner, how exactly zebrafish models can be used to target a wide range of human brain disorders and aberrant phenotypes. Chapter "Mutagenesis and Transgenesis in Zebrafish" discusses zebrafish genetic (mutant and transgenic) models. The book's next two chapters describe zebrafish models relevant to the two most common brain disorders—anxiety and depression. Chapter "Assessing Cognitive Phenotypes in Zebrafish" explains how to best assess zebrafish cognitive phenotypes, and chapter "Social Phenotypes in Zebrafish" comprehensively evaluates the spectrum of zebrafish social behaviors. Zebrafish models of obsessive-compulsive disorder and attention deficit/hyperactivity disorder are discussed in chapters "Modeling OCD Endophenotypes in Zebrafish" and "Zebrafish Models of Attention-Deficit/Hyperactivity Disorder (ADHD)", followed by chapters "Zebrafish Neurobehavioral Assays for Drug Addiction Research" to "Zebrafish Behavioral Models of Ageing" on zebrafish neurotoxicity, sleep, addiction, and aging. Chapter "Integrating Morphological and Behavioral Phenotypes in Developing Zebrafish" discusses neurophenotyping of developing zebrafish, and chapter "Neuroimaging Phenotypes in Zebrafish" highlights the importance of neuroimaging biomarkers in zebrafish models of CNS disorders. Finally, chapter "Illustrated Zebrafish Neurobehavioral Glossary" contains a very useful zebrafish neurobehavioral catalogue—a comprehensive updated and illustrated glossary of all major larval and adult zebrafish behaviors currently known.

The contributors to this book are leading international scholars whose work spearheads innovative zebrafish neuroscience research around the world. Written by top experts in the field, this book makes for a useful, balanced, and up-to-date reading that outlines the use of zebrafish to study the pathological mechanisms underlying neuropsychiatric disorders.

Finally, the authors want to make sure that this book is actively used in research laboratories. So go on and cover its pages with notes, question marks, and little pencil drawings in the margins—perhaps, with designs of new experiments or future, better, and more sensitive zebrafish behavioral models. The authors would not have wanted this book any other way.

Slidell, LA, USA Allan V. Kalueff

Contents

Mutagenesis and Transgenesis in Zebrafish

Fabienne E. Poulain

Abstract Over the last decades, the zebrafish has emerged as a powerful vertebrate model for studying development, diseases and behavior, and conducting high-throughput screens for therapeutic development. Large forward genetic screens have led to the generation of multiple mutant lines with developmental or behavioral defects, while transposon-based integration technologies have enabled the creation of transgenic lines essential for the functional analysis of cell and tissue movement, gene regulation, and gene function. The recent development of engineered endonucleases including ZFNs, TALENs and the CRISPR/Cas9 system has revolutionized reverse genetic approaches in zebrafish, allowing for the first time precise genome editing for targeted mutagenesis and transgenesis. In this chapter, we provide an overview of the different approaches used for mutagenesis and transgenesis in zebrafish, with an emphasis on the recent progress in targeted genetic manipulations. Examples of selected mutant and transgenic zebrafish strains are given to illustrate their growing utility for neurobehavioral phenomics and biological psychiatry.

Keywords Genome editing • CRISPR/Cas9 • TALEN • Zinc-finger nuclease • Screen • NHEJ • Homologous recombination • Tol2 transposon • Knock-in

Abbreviations

Cas	CRISPR-associated
CRISPR	Clustered regulatory interspaced short palindromic repeats
DSB	Double strand break
EENs	Engineered endonucleases
ENU	N-ethyl-N-nitrosourea
HDR	Homology-directed repair
HR	Homologous recombination
KI	Knock-in

F.E. Poulain (✉)
Department of Biological Sciences, University of South Carolina, Columbia, SC 29208, USA
e-mail: fpoulain@mailbox.sc.edu

A.V. Kalueff (ed.), *The rights and wrongs of zebrafish: Behavioral phenotyping of zebrafish*, DOI 10.1007/978-3-319-33774-6_1

NHEJ	Non homologous end joining
ssDNA	Single-stranded DNA
TALEN	Transcription-activator like effector nuclease
WT	Wild-type
ZFN	Zinc-finger nuclease

1 Introduction

In the 1960s, George Streisinger and colleagues introduced the zebrafish as a new vertebrate model for the study of developmental genetics. Thanks to its small size, ease of breeding, and external fertilization producing large clutches of transparent embryos, the zebrafish appeared especially suited for morphological observation of developmental processes. Since then, the zebrafish has become a model of choice, not only for studying vertebrate development, but also for modeling human diseases and conducting molecule screening for drug discovery. Comparison of the zebrafish and human genomes indicates that 71.4 % of human protein-coding genes have at least one zebrafish orthologue [1]. Systemic large-scale forward genetic screens combined with the annotation of the zebrafish reference genome have led to the identification of a large variety of mutations affecting embryogenesis, physiology or behavior relevant to human health [2, 3]. More recently, an explosion of new tools and techniques, in particular the development of engineered endonucleases (EENs), have open a new area for genome editing, allowing the direct manipulation of the zebrafish genome for targeted mutagenesis and transgenesis (Fig. 1). This chapter provides an overview of the different approaches used for mutagenesis and transgenesis in zebrafish, with an emphasis on the recent progress in targeted genetic manipulations, and their 'translational' applications to modeling selected brain disorders.

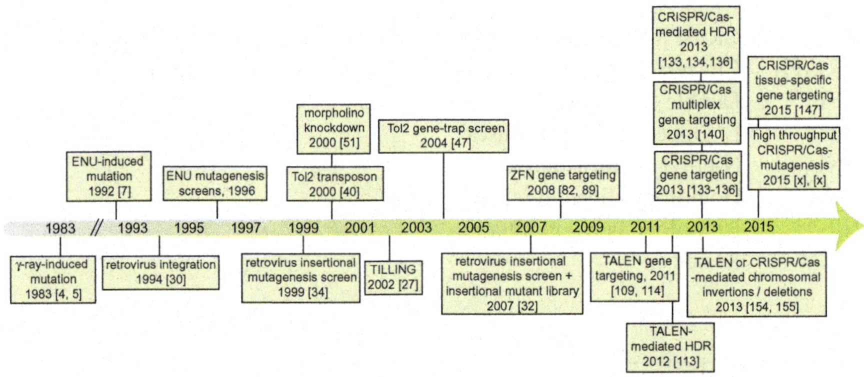

Fig. 1 Timeline recapitulating the development of mutagenesis approaches in zebrafish. This historical timeline recapitulates the major advances in the development of mutagenesis approaches in zebrafish over the last 40 years. The first ENU-based mutagenesis screens conducted in the Driever and Nüsslein-Volhard labs have generated several hundreds of mutants that were reported in a dedicated issue of *Development* in 1996. As such, no author name was reported in the corresponding box

2 Mutagenesis in Zebrafish

2.1 Chemical Genetic Screens

Forward genetic screens allow the identification of genes implicated in a specific biological pathway or process by screening a population of animals in which genome modifications have been randomly induced. Individuals that display a phenotype of interest are identified as carriers of a modified/mutated allele, and subsequent mapping of the modification/mutation responsible for the phenotype identifies the gene involved. Thanks to its external fertilization, its large clutch size, its rapid development and the transparency of its embryos, the zebrafish proved ideal for large-scale forward genetic screens. While initial work used UV light and γ-ray irradiation to trigger chromosomal breaks [4–6], N-ethyl-N-nitrosourea (ENU) has rapidly become the standard choice for chemical mutagenesis [7–9]. ENU is a DNA alkylating agent that mostly triggers point mutations, has a high mutagenic efficiency, and can be directly applied to adult male zebrafish by adding it to the water, making it very easy to use (Fig. 2). The first large-scale ENU screens performed in the Driever lab in Boston and the Nüsslein-Volhard lab in Tubingen have generated several hundreds of mutants with developmental phenotypes that were reported in a dedicated issue of *Development* in 1996 [10]. Other ENU screens have subsequently identified more genes involved in development [11–16], behavior [17–19] addiction [20], or diseases [21]. While the identification of the mutated alleles by positional cloning is often laborious, whole-genome sequencing at low coverage can now be used to map mutations rapidly [22, 23]. Two approaches using whole genome sequencing have been developed: the bulk-segregant linkage analysis (BSFseq) that involves a mapping cross, and the homozygosity mapping (HMFseq) [24]. Both rely on bioinformatic filtering for mutagenic polymorphisms, and can be analyzed with the open source computational pipeline MegaMapper available at https://wiki.med.harvard.edu/SysBio/Megason/MegaMapper. More affordable approaches using transcriptome sequencing have also been developed, such as Mutation Mapping Analysis Pipeline for Pooled RNA-seq (MMAPPR) [25] and RNA-seq-based bulk segregant analysis [26]. MMAPPR offers the advantage of identifying mutations without sequencing the parental strain or using a SNP database.

While originally used in forward screens that identify mutations after phenotypic analysis, ENU mutagenesis has also been applied in reverse genetic approaches, in which mutations are detected first and then associated with a phenotype. Targeting Induced Local Lesions in Genomes (TILLING) was first used to screen for desired mutated alleles in *Arabidopsis*, and was successfully adapted to the zebrafish 2 years later [27, 28]. In contrast to ENU-based forward genetic screens in which phenotypic analysis is conducted at the F3 generation, DNA analysis and sperm cryopreservation is performed in F1 families in TILLING (Fig. 2b). After mutations have been identified, the cryopreserved sperm is used in in vitro fertilization to generate F2 families, whose carriers are isolated by genotyping. TILLING alleles are currently being generated and distributed to the zebrafish community by the zebrafish TILLING consortium initiated by the Moens lab at the Fred Hutchinson Cancer Research Center and the Solnica-Krezel lab at Washington University

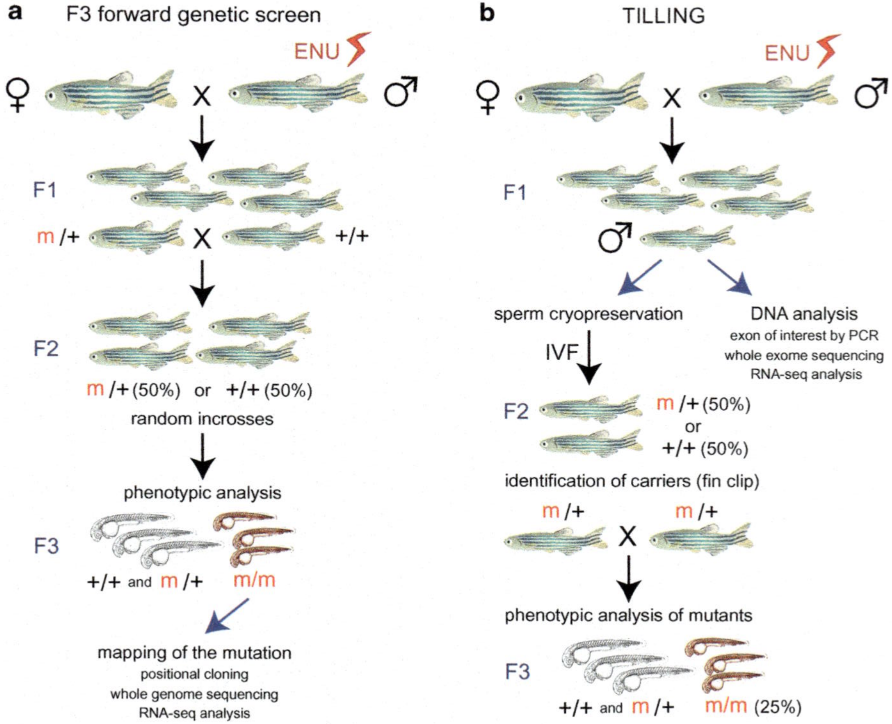

Fig. 2 Use of ENU mutagenesis in forward genetic screens and TILLING. (**a**) Overview of a classical ENU-based mutagenesis forward screen. Males are directly treated with ENU added to the water to induce random germline mutations. Mutated males are outcrossed to WT females to generate F1 families. F1 family members are outcrossed to WT to generate F2 families, whose members are randomly incrossed to produce F3 embryos. F3 embryos are examined for phenotypic defects, and their genomes are analyzed for identification of the causative mutation. (**b**) Overview of ENU-based TILLING. Males are directly treated with ENU to induce random germline mutations and are outcrossed to WT females to generate F1 families. Males from F1 families are sacrificed after cryopreservation of their sperm and sampling of their DNA for analysis. DNA from F1 males is screened for mutations in genes of interest. The cryopreserved sperm from identified carriers is then selected to generate F2 families using in vitro fertilization, and mutant embryos can be identified by genotyping

School of Medicine in St. Louis. Alleles can be requested online at http://webapps.fhcrc.org/science/tilling/index.php. Complementary to the zebrafish TILLING project, the Zebrafish Mutation Project from the Wellcome Trust Sanger Institute aims to produce a knockout allele in every protein-coding gene of the zebrafish genome, and has so far generated 26,634 alleles. Mutations are identified after whole exome enrichment and Illumina next generation sequencing, and each allele is analyzed for morphological defects [29]. A list of available lines with mutations is available online at http://www.sanger.ac.uk/sanger/Zebrafish_Zmpbrowse.

2.2 *Retroviral and Transposon-Mediated Mutagenesis*

While ENU mutagenesis is a powerful approach to generate random mutations at a high rate, it requires a significant degree of effort and commitment to identify the mutations. Insertional mutagenesis using retroviruses or transposons offers the advantage of a fast screening of carriers and a rapid identification of the mutated gene by using the sequence of the insertional element as a "tag" for mapping. Retroviruses and transposons have different insertion site preferences and generate null or hypomorphic alleles, or have no effect, depending on where they integrate in the genome.

Retroviral-mediated mutagenesis was the first insertional mutagenesis carried out in zebrafish in the early 1990s [30]. It used a pseudo-type retrovirus derived from the Moloney murine leukemia virus (MoMLV), with the envelop protein replaced by the glycoprotein from the vesicular stomatitis virus (VSV). Like in human cells, this modified retrovirus was shown to preferentially integrate in regions close to transcriptional starts in the zebrafish genome [31, 32]. It was used by the Hopkins lab at the MIT and others to carry out several large insertional forward screens that led to the generation of hundreds of lines with development defects [33–37]. Like ENU mutagenesis and TILLING, retroviral mutagenesis has also been used in reverse genetics by injecting high-titer retroviruses into embryos [32]. Sperm from F1 males is cryopreserved, and mutations are mapped by identifying the genomic sequences flanking the insertion site, or by high-throughput Illumina sequencing to generate a proviral insertion library. Most retroviral insertions have been located in introns, with insertions into the first intron of genes often leading to a decrease in gene expression. Using this approach, the Lin lab at UCLA and the Burgess lab at NHGRI/NIH have generated the Zebrafish Insertion Collection (ZInC), in which 3054 mutations in genes have been isolated from 6144 F1 fish [38, 39]. Mutant lines can be searched for with the ZInC database (http://research.nhgri.nih.gov/ZInC/?mode=search) and requested through the Zebrafish international Resource Center (ZIRC) (http://zebrafish.org/home/guide.php).

Insertional mutagenesis using transposons has also been used for gene inactivation. Transposons, or "jumping genes", are mobile DNA sequences that can change their position within the genome, thereby altering it and creating mutations. Insertions and excisions require the activity of the transposase enzyme. Several transposable elements including *Sleeping beauty*, *Ac/Ds* and *Tol2* have been used in zebrafish for both mutagenesis and transgenesis, *Tol2* being the most common [40–42].

Transposons have been favored over retroviruses due to their ease of use and their ability to integrate large transgenes. Transposon-based gene-breaking constructs have been improved over the years to simultaneously inactivate a gene ("gene trap") and insert transgenes of interest such as fluorescent proteins ("protein trap") (Fig. 3a and b), or to inactivate a gene in a conditional manner. The FlipTrap cassette, for instance, allows conditional mutagenesis thanks to the insertion of loxP and FRT sites for Cre-mediated and Flp-mediated recombination, respectively [43]. When integrated into an intron of a gene, the FlipTrap cassette forms a fusion protein of citrine and the endogenous protein, thereby revealing the expression profile of the targeted gene when it is expressed. Exposure to Cre recombinase removes the citrine sequence and a splice donor sequence associated to it, thereby inducing a truncation of the gene. Flp-mediated recombination allows the exchange of the cassette with any DNA sequence after the integration has occurred. Several FlipTrap lines have been made available through the FlipTrap database (http://www.fliptrap.org/static/anatomies.html). Other efficient gene-breaking constructs such as the RP2 cassette [44], the FlEx cassette [45] or a recently developed bipartite Gal4-containing vector [46] also function as conditional alleles thanks to the presence of loxP and/or FRT sites flanking the mutagenic cassette. Transposon-based cassettes have been used to perform several mutagenesis screens [32, 47–49]. A list of gene trap fish lines obtained from the Kawakami lab is provided through the zTrap database (http://kawakami.lab.nig.ac.jp/ztrap/). The zTrap database allows the search for gene trap insertions located within or near genes of interest [50].

2.3 *Targeted Mutagenesis*

While ENU, retroviruses and transposon-based constructs are powerful mutagenesis tools for forward genetic screens, the genome modifications they generate are random, making it effortful and time-consuming to isolate a mutant for a gene of interest. For many years, methods for engineering specific loci in the genome were restricted to organisms like the mouse, in which embryonic stem cells can be manipulated in a precise way through homologous recombination (HR). In zebrafish, gene knockdown was transiently achieved by injecting antisense morpholino oligonucleotides (MOs) designed to block the splicing or translation of a targeted mRNA [51]. MOs have been widely used to test gene function, but have recently raised some concerns regarding their specificity [52, 53]. A comparative study looking at more than 80 genes notably reported that around 80 % of the phenotypes observed in MO-injected embryos ("morphants") could not be detected in the corresponding mutants [54]. These differences have led to the assumption that MO-induced phenotypes often result from off-target effects, and that mutants should become the standard model to describe gene function. On the other hand, deleterious mutations have recently been shown to activate genetic compensatory mechanisms [55]. Further investigation will likely be required to explain the discrepancies between morphant and mutant phenotypes for a specific gene.

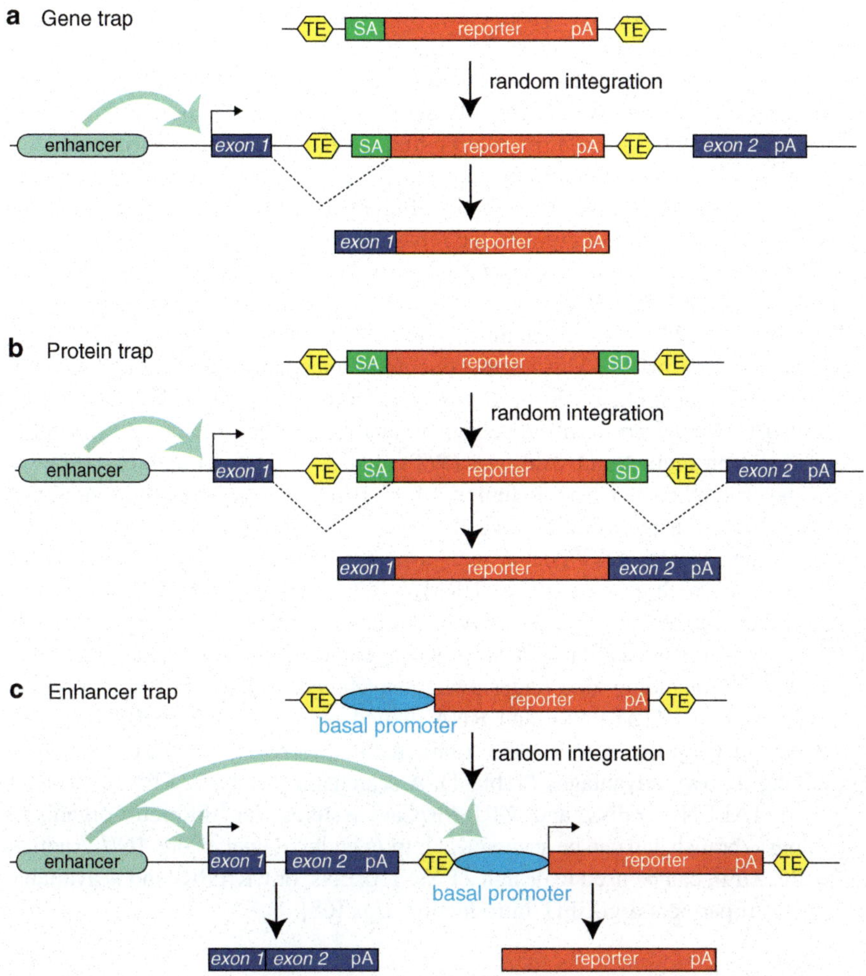

Fig. 3 Overview of transposon-based gene-trap, protein trap and enhancer trap approaches. All transposon-based gene trap, protein trap and enhancer trap vectors contain transposable elements (TE) that mediate random integration of the vector in the genome. (**a**) In a gene trap approach, the vector contains a splice acceptor site (SA) upstream of a reporter sequence with a stop codon and a polyA (pA) signal at its 3′ end. Because the reporter does not have any start codon, its transcription depends on the regulation of the endogenous gene by the upstream regulatory element (enhancer). Proper splicing of the SA to the 5′ exon of the gene integrates the reporter into the transcript and generates a truncated protein. (**b**) In a protein trap approach, the vector contains both a SA and a splice donor site (SD) flanking the reporter sequence. The reporter is devoid of start and stop codons, allowing the fusion between the reporter and the endogenous transcript when integration in an intron is in the correct orientation and proper reading frame. (**c**) In an enhancer trap, the vector contains a basal promoter with minimal activity upstream of a reporter sequence with a start codon, a stop codon and a pA signal. When the vector integrates near an endogenous transcriptional enhancer, its basal promoter becomes regulated by it and drives the expression of the reporter without any mutagenic effect

The ability to precisely manipulate the zebrafish genome has remained a long-standing quest that was only recently resolved by the discovery of sequence-specific endonucleases and their engineering as genome editing tools. All engineered endonucleases (EENs) consist of a sequence-specific DNA targeting component (protein domain or RNA) and a double-stranded DNA cleaving endonuclease (catalytic domain) that introduces double-strand breaks (DSBs) in the genome. DSBs can be repaired by two different pathways: non-homologous end joining (NHEJ) and homology-directed repair (HDR). NHEJ can ligate the cleaved DNA double strands without any template but introduces insertions or deletions (indels) at the cut site. HDR, on the other hand, uses a homologous template of DNA to repair DSBs. NHEJ is ten times more active than HDR or HR during zebrafish development [56–58]. This error-prone repair mechanism is exploited to introduce a frameshift mutation leading to a non-functional protein. Indel mutations generated by NHEJ are easily detected by analyzing the formation of heteroduplexes between mutant and wild-type (WT) alleles, either by a mobility assay, in which heteroduplexes and homoduplexes have different electrophoretic migration profiles [59], by using enzymes like the endonucleases *Surveyor* or *Cel-I* or the bacteriophage resolvase *T7E1* that recognize and cut mismatches [60–62], or by high resolution melt curve analysis (HRMA) [63–65]. The nature of indels can be further characterized by directly analyzing Sanger sequencing data with the poly peak parser software available at http://yost.genetics.utah.edu/software.php [66]. Several EENs including zinc finger nucleases (ZFNs), transcription activator-like effector nucleases (TALENs), and RNA-guided nucleases (CRISPR/Cas9) have been used successfully in zebrafish for targeted mutagenesis (Fig. 4), each of them presenting its own advantages (Table 1). A searchable database, EENdb, collects reported TALENs, ZFNs and CRISPR/Cas systems for different organisms including zebrafish and can be accessed at http://eendb.zfgenetics.org/ [67]. Another software, ZiFit, can be used to design ZFNs, TALENs, or CRIPSRs and is available at http://zifit.partners.org/ZiFiT/Introduction.aspx [68].

Fig. 4 (continued) active upon dimerization, ZFNs work in pairs, cleaving DNA only after each of them has bound to its target sequence. (**b**) TALENs are constructed by fusing the catalytic domain of *FokI* to the DNA-binding transcription activator-like effector (TALE) proteins. Each TALE contains an N-terminal translocation domain that recognizes a 5′-T (in red in the DNA sequence), a DNA-binding central repeat domain, and a C-terminal sequence. The central domain contains repeat units composed of 33–35 conserved amino acids, with differences at amino acids 12 and 13 that form the repeat variable di-residue (RVD). Each RVD recognizes and binds to a single specific nucleotide and is therefore responsible for the DNA binding specificity of each repeat unit. Like ZFNs, TALENs function by pairs to cleave DNA. (**c**) In the CRISPR/Cas9 system, a single guide RNA (sgRNA) recruits the endonuclease Cas9 to the genomic sequence it complements. The sgRNA is composed of 20 nt sequence that directly matches the DNA target sequence, followed by 72–80 nt of the bacterial crRNA/tracrRNA sequence that are required for the formation of hairpin loops stabilizing the sgRNA. Cas9 has two catalytic domains, RuvC and HNH, that each cleaves a DNA strand. The presence of NGG as a protospacer adjacent motif (PAM) is required in 3′ of the target sequence for DNA recognition by Cas9

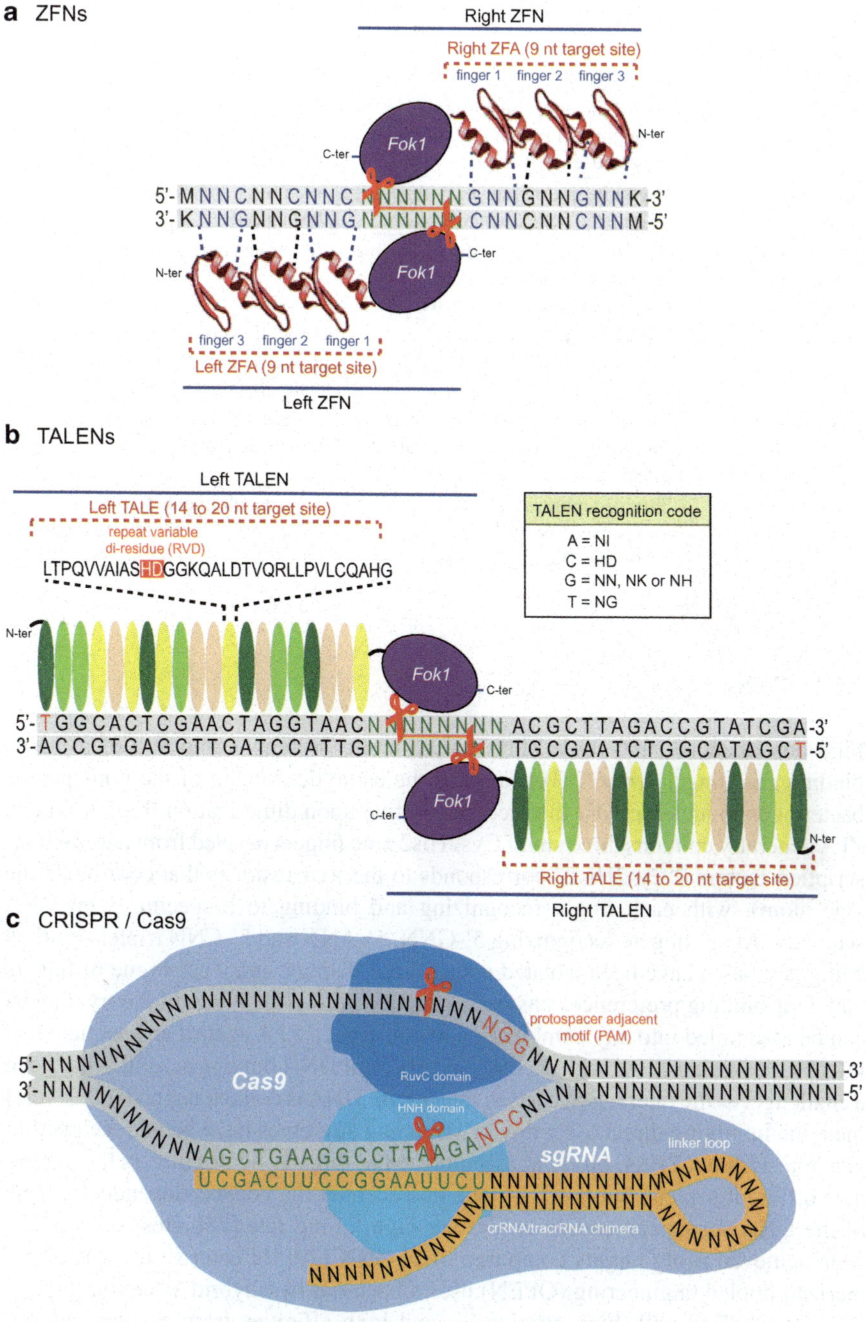

Fig. 4 Engineered endonucleases for targeted mutagenesis. (**a**) ZFNs are composed of zinc finger arrays (ZFAs) fused to the catalytic domain of the *FokI* endonuclease. Each ZFA generally consists in three fingers that each recognizes and binds to a specific 3 nt DNA sequence. Since *FokI* becomes

Table 1 Comparison of mutagenesis approaches available in zebrafish

Mutagenesis approach	Advantages	Disadvantages
ENU mutagenesis	High efficiency, high mutagenic throughput, random mutations, easy to use	Mapping of mutations work-intensive, high rate of background mutations
Retroviral mutagenesis	High integration rate, easy mapping of mutated genes, scalable	Integration not random, low throughput
Transposon-based mutagenesis	Scalable, easy to use, easy mapping of mutated genes, integration of large transgenes, inexpensive	Low transgenesis rate
ZFNs	Targeted mutagenesis	Difficult assembly, target not present in every gene, not scalable, some off-target effects
TALENs	Targeted mutagenesis, no constraint for target selection, rare off-target effects	Not easy to design, not scalable, quite expensive
CRISPR/Cas9	Targeted mutagenesis, easy to use, inexpensive, multiplexing, tissue-specific mutagenesis	Some requirements for target selection, off-target effects more likely

2.3.1 ZFNs

First described in 1996 [69], ZFNs are chimeric proteins composed of a DNA-binding zinc finger array (ZFA) fused to the catalytic domain of the non-specific bacterial endonuclease *FokI* that becomes active upon dimerization (Fig. 4a). Each ZFA generally contains three small Cys2His2 zinc fingers derived from natural transcription factors ("Cys2His2" corresponds to the four residues that coordinate the zinc atom), with each finger recognizing and binding to a specific 3 bp DNA sequence. Many fingers recognizing 5′-GNN, 5′-ANN and 5′-CNN triplets (with N being any base) have been isolated using phage display, and a catalogue of fingers and their binding preferences has been generated [70–77]. While in theory, fingers can be assembled into any combination to construct a ZFA against any sequence of interest, designing ZFAs with specific and efficient DNA binding activities has been a challenge, as the interaction of each finger with DNA is context dependent. Several methods involving direct assembly or screening strategies have been developed to generate efficient ZFAs. Modular assembly (MA) directly ligates fingers that recognize different triplets, but does not take into account the context-dependent effects of the DNA sequence, leading to a rather high failure rate [78]. Best success has been achieved using targets composed of 5′-GNN [79]. In contrast to MA, oligomerized pooled engineering (OPEN) uses a bacterial two-hybrid selection method to identify ZFAs with high efficiencies and high affinities from a combinatorial library of multi-finger arrays recognizing 9 bp sequences [80, 81]. A similar approach with a one-hybrid selection system has also been used [82]. While more efficient, these approaches require expertise in constructing libraries and are quite labor-intensive. A more recent and easier method named Context-dependent assembly (CoDA) assembles three-finger arrays by selecting N- and C-terminal fingers from known ZFAs containing a common middle finger, thereby accounting for context-dependent effects between adjacent fingers [83].

Since the *FokI* endonuclease domain must dimerize to be active [84], ZFNs function by pairs, cleaving DNA only after each of them has bound to its target sequence (Fig. 4a). Obligate heterodimer modifications have been introduced in the *FokI* catalytic domain to increase ZFN efficacy and reduce off target cleavages [85, 86]. The spacer that separates the ZFA target sequences is relatively short, of variable size and has no sequence requirement. Two to five amino acids can be introduced between the ZFA and the *FokI* as an inter-domain linker to accommodate the variable size of the spacer in the DNA sequence [87, 88]. While the requirement of two ZFNs to target a sequence provides good specificity and limits off target effects, finding a target sequence in the 5′ region of a gene to generate a null mutation can be limited by the context-dependent affinity of each zinc finger within a ZFA. Nonetheless, ZFNs have been successfully employed for gene targeting in zebrafish since the first reports of their use [82, 89, 90]. ZFN target sites can be identified in several organisms including zebrafish with ZFNgenome, a comprehensive open source accessible at http://bindr.gdcb.iastate.edu/ZFNGenome/ [91]. mRNAs encoding ZFNs are then injected at one-cell stage after ZFNs have been assembled.

2.3.2 TALENs

While useful for targeted mutagenesis, ZFNs have rapidly been challenged by the development of TALENs, which appear to be more mutagenic in zebrafish [92, 93]. TALENs are chimera proteins obtained by fusing the DNA-binding transcription activator-like effectors (TALEs) to the catalytic domain of *FokI* (Fig. 4b). TALEs were originally identified in the bacterial plant pathogen *Xanthomonas* and were named for their ability to trigger the expression of genes promoting infection in the host cell [94].

Each TALE is composed of an N-terminal translocation domain that recognizes a 5′-T, a DNA-binding central repeat domain, and a C-terminal sequence. The central domain contains 15.5–19.5 repeat units composed of 33–35 conserved amino acids, with differences at amino acids 12 and 13 forming the repeat variable di-residue (RVD) (the last repeat unit contains only 20 amino acids and is referred to as a half repeat). Each RVD recognizes and binds to a single specific nucleotide and is therefore responsible for the DNA binding specificity of each repeat unit. The RVDs NI, HD and NG are commonly used to target the nucleotides A, C and T, respectively, while NN, NK and NH can be employed for targeting a guanine, with NK and NH binding more specifically but with a weaker affinity [95–97]. In contrast to ZFNs, whose efficiency is context-dependent, TALENs do not have much requirement in terms of the targeted sequence besides a 5′-T and a minimum length of 11 RDVs for the binding domain [98]. Several online tools such as TALE-NT [99] (https://tale-nt.cac.cornell.edu/node/add/talen), Mojo Hand [100] (http://www.talendesign.org/), or idTALE [101] (http://omictools.com/idtale-s5415.html) can be used to identify the optimal TALEN target sequence within a gene of interest.

Like ZFNs, TALENs are engineered with the *FokI* endonuclease catalytic domain as an obligate heterodimer and must therefore work by pairs to cleave DNA. The optimal spacer length seems to depend on the scaffold of the TALEN, with TALENs containing short C-terminal lengths (17–28 amino acids) being more efficient with shorter spacers (12–14 bp). Several methods have been developed for constructing TALENs and dictate the scaffold used. The Golden Gate cloning strategy uses restriction digest of a TALE plasmid library by type II endonucleases followed by ligation [102]. This approach is theoretically a one-step assembly that can construct TALE repeats in one digest and ligation reaction. While several Golden Gate derived methods have been generated and some commercially available, they do not always use the same TALE scaffold and are not always compatible [103–107]. Other approaches such as the Unit assembly method [108, 109] and the restriction enzyme and ligation (REAL) method [110] rely on standard molecular cloning using hierarchical restriction digests and ligations. While effective, these methods are labor-intensive and do not allow the construction of TALENs in a large scale. High-throughput can be achieved with the fast ligation-based automatable solid-phase high-throughput (FLASH) or the iterative capped assembly (ICA) methods that use solid-phase ligation on magnetic beads instead of the time-consuming transformation and growing of bacteria [111, 112].

Due to their higher mutation frequencies, the rarity of off-target effects, and the presence of target sequences in almost every gene, TALENs have quickly become the method of choice for mutagenesis in zebrafish since their first application in 2011–2012 [64, 109, 113–115]. As for ZFNs, mRNAs encoding TALENs are injected at one-cell stage. TALEN efficiency can be assessed the day after by analyzing heteroduplex formation in injected embryos.

2.3.3 CRISPR/Cas9 System

The CRISPR (clustered regularly interspaced short palindromic repeats) / Cas (CRISPR associated proteins) system was originally identified as a defense mechanism used by bacteria and archae against the introduction of foreign nucleotides form bacteriophage and exogenous plasmids [116–118]. Invading nucleic acids are first recognized as foreign and integrated as spacers between short DNA repeats (the CRISPR locus) in the host genome, thereby forming CRISPR arrays. Transcription of these CRISPR arrays generates primary transcripts, or pre CRISPR RNAs (pre-crRNA) that are subsequently cleaved into small CRISPR RNAs (crRNAs). Upon infection, the crRNAs whose spacers have a sequence close to the invading nucleic acids bind to them, and recruit a second non-coding RNA with partial complementarity to the crRNA named auxiliary trans-activating crRNA (tracrRNA). The complex tracrRNA/crRNA in turn recruits nucleases associated with the CRISPR locus named Cas to degrade the intruder nucleic acids and prevent pathogen invasion. Of particular interest is Cas9, an endonuclease that introduces DSBs in the target DNA thanks to its two nuclease active sites, RuvC and HNH, that each cleaves a DNA strand (Fig. 4c). Several groups saw the genome-editing possibilities offered by the CRISPR/Cas9 system and adapted it for its use in eukaryotic cells. The crRNA and tracrRNA of *Streptococcus pyogenes* were fused into a single guide RNA (sgRNA) named for its ability to recruit and activate Cas9 [119]. On the other hand, the sequence of the *Streptococcus pyogenes* Cas9 has been modified by codon optimization and the introduction of nuclear localization signals to promote its use in eukaryotic cells [119–121].

Since the targeting properties of the CRISPR/Cas9 system only rely on the sequence of the sgRNA, it has become very easy to target any sequence of interest in the genome. Each sgRNA is composed of a 20 nt sequence that directly matches the target sequence, followed by 72–80 nt of the 3′ crRNA/tracrRNA sequence that are required for the formation of hairpin structures stabilizing the sgRNA [122, 123]. The only constraint for the design of sgRNAs is the presence of NGG at the 3′ end of the target site that acts as a protospacer adjacent motif (PAM) required for DNA recognition by Cas9 and Cas9 subsequent activation (Fig. 4c) [124–127]. The requirement of NGG as a PAM currently limits the number of sequences recognized by Cas9, but a recent study has successfully engineered efficient Cas9 derivatives with altered PAM specificities, thereby expanding the repertoire of PAMs needed [128]. Several servers and online softwares have been specifically developed for the design of sgRNAs, including CRISPRdirect (http://crispr.dbcls.jp/) [129], the Optimized CRISPR Design (http://crispr.mit.edu/) from the Zhang lab, the Cas9 Online Designer (http://cas9.wicp.net/) developed by Dayong Guo, Cas-Designer (http://rgenome.net/cas-designer/) [130], sgRNACas9 (http://www.biootools.com/) [131], and CHOPCHOP (https://chopchop.rc.fas.harvard.edu/) [132].

Because of its ease of use and affordability, the CRISPR/cas9 system has rapidly been applied in zebrafish, with mutagenesis rates comparable or superior to those obtained with TALENs [133–136]. sgRNAs are obtained by in vitro transcription from plasmids or oligos and co-injected with Cas9 mRNA at one-cell stage to induce DSBs in the target sequence. Some studies have also directly injected the Cas9 protein with sgRNAs to increase mutagenic activity [137–139]. A major advantage of the CRISPR/Cas9 system is that its high efficiency is sometimes sufficient to introduce extensive biallelic mutations causing phenotypes in injected embryos, a feature not often seen using TALENs [140]. In addition, it offers the possibility to simultaneously target multiple sequences at once by co-injecting several sgRNAs, or by using a plasmid with multiple sgRNA cassettes under the control of U6 or H1 promoters, a process named CRISPR multiplexing [140–143]. Multiplexing has recently been employed in a high-throughput mutagenesis set-up to successfully generate mutations in 83 different genes in the zebrafish genome [144]. At lower scale, multiplexing can be very useful to generate double or triple mutants in related genes for which single mutants would lack a phenotype due to compensatory mechanisms. It can also be employed to study the role of non-coding RNA genes that are not affected by changing the frame of translation. The identification of optimal targets in multiple locus has been facilitated by the recent development of specialized softwares such as CRISPRseek (http://www.bioconductor.org/packages/release/bioc/html/CRISPRseek.html) [145] or CRISPR MultiTargeter (http://www.multicrispr.net/) [146]. Finally, a last advantage of the CRISPR/Cas9 system is the possibility of disrupting gene function in a spatially controlled manner by injecting a modular vector that contains an sgRNA cassette under the control of a U6 promoter and a Cas9 cassette under the control of a tissue- or cell-specific promoter [147].

Altogether, the CRISPR/Cas9 system offers so many advantages that it has quickly become the method of choice for targeted mutagenesis in zebrafish. A major drawback, though, is the rather high frequency of off-target effects observed in various models [122, 123, 148–150]. The short length of the target sequence, and the tolerance of Cas9 for mismatches between the target sequence and the sgRNA, can lead to the mutation of secondary targets in the genome that would need repeated outcrossing to be eliminated. Choosing unique target sequences using the specific softwares mentioned above is thus important. Specificity can be further improved by using truncated sgRNAs (17 nt) that have a decreased mutagenesis rate at off-target sites [151]. Finally, Cas9 variants possessing only one nuclease catalytic site instead of two can be used [122, 152, 153]. These Cas9 "nickases" introduce nicks in one DNA strand only, and must be used in pairs with two sgRNAs to introduce DSBs. This system is thus analogous to TALENs or ZFNs by requiring a dual recognition of the targeted DNA sequence.

2.4 Chromosomal Deletions and Inversions

In addition to generating small indel mutations in a gene of interest, large genomic deletions or inversions can be introduced by injecting several TALEN pairs or multiple sgRNAs. DSBs are introduced simultaneously at two separate sites, leading to the loss, or more rarely the inversion, of the DNA fragment in between. In zebrafish, genomic deletions with sizes ranging from several hundred bases to 1 Mb have been reported [141, 154, 155]. Introducing large deletions has proved useful in different systems to study the role of cis-regulatory sequences [156, 157], or to recreate translocations similar to those found in human diseases [158–160].

3 Transgenesis in Zebrafish

Transgenesis is defined as the introduction of exogenous genes, or "transgenes", into the genome of a living organism. The first zebrafish transgenic lines were obtained by the random integration of transgenes and regulatory promoters in the genome using retroviruses or transposon-based systems. By expressing transgenes such as fluorescent proteins or genes with dominant negative mutations, and by providing a spatial and/or temporal control of gene activation, these transgenic lines proved to be powerful tools for observing the fate and behavior of cells and tissues, studying gene regulation, and testing gene function in development, behavior and diseases. More recently, TALENs and the CRISPR/Cas9 system have revolutionized zebrafish research by allowing the insertion of sequences into specific loci in the genome, making the generation of knock-ins (KIs) finally possible.

3.1 Transposon-Based Transgenesis with Random Insertion

3.1.1 Enhancer Traps and Protein Traps

As mentioned previously for transposon-based mutagenesis, transposon-based constructs have been used in gene trap, enhancer trap or protein trap configurations (Fig. 3) in the context of high throughput screens. All constructs possess transposable elements (TE) derived from *Sleeping beauty* or *Tol2* transposons that allow random integration in the genome. While gene traps are used for insertional mutagenesis (discussed earlier in section 2.2.), enhancer traps have a limited mutagenic effect and are designed to report the transcriptional activity of enhancers located nearby their site of integration (Fig. 3c). Proteins traps are constructed to create a fusion between the full-length trapped gene and the reporter, allowing the visualization of protein expression in the embryo (Fig. 3b).

Numerous enhancer trap screens have been conducted in zebrafish and have led to the creation of a large library of transgenic lines expressing fluorescent reporters or drivers in specific cells and tissues. Several basal promoters have been employed, including *keratin 4* (*krt4*) and *keratin 8* (*krt8*), *gata2*, *hsp70*, *c-fos*, *Eb1*, *ef1a*, *thymidine kinase*, the *carp β-actin promoter* (TKBA), and the medaka *edar* locus [161–170]. These basal promoters have various trapping efficiencies and can drive different expression profiles based on their sensitivity to the genomic enhancer regulating them [163]. Although all basal promoters have been useful to reveal specific patterns of expression during development, some have a bias for traps with expression in specific structures (for instance, the *E1b* promoter has a strong bias for cranial ganglia), while others can drive non-specific background expression is tissues such as the muscles or the dermis. Several reporters have also been used, the most common being fluorescent reporters like EGFP to monitor transcriptional activity and follow the movement or differentiation of the cells labeled, and Gal4, to drive effector gene expression where Gal4 is expressed using the Gal4/UAS system. Several Gal4 enhancer trap lines have been generated in combination with a UAS:EGFP or UAS:Kaede reporter, where the photo-convertible Kaede fluorescent protein can be used for mapping neural circuits or cell lineages [163, 164]. A collection of enhancer trap lines is described in the ZETRAP 2.0 database available at http://plover.imcb.a-star.edu.sg/webpages/home.html [171].

In parallel to enhancer traps, protein traps have been developed to generate an in-frame fusion between the full-length trapped gene and the reporter (Fig. 3b). By retaining all the regulatory sequences of the endogenous genes, this approach allows detailed studies on the expression of the protein trapped and its regulation as well as its localization within cells. As mentioned previously, protein traps have mostly been combined with gene-traps to allow simultaneous gene inactivation and protein inactivation [43].

3.1.2 The Gateway System for Easy Transgenesis

While enhancer and protein trap constructs have been instrumental for visualizing developing tissues or protein localization, they are not particularly suited for over-expressing a gene of interest in a temporally or spatially controlled manner. Transgenesis is an essential tool for testing gene and cell function, but has been historically laborious in zebrafish due to technical limitations such as laborious conventional cloning and low rates of germline transmission when using supercoiled or linear DNA [172, 173]. To overcome these limitations, the Tol2Kit system (that uses the recombination-based cloning of multiple DNA fragments) was designed to easily generate expression constructs for transgenesis [174]. This multisite Gateway technology relies on the *att* site specific recombination system from the λ phage [175], and uses different engineered *att* sites that recombine specifically to assemble up to five DNA fragments in a directional manner. Three different "entry" clones containing a promoter, a coding sequence of interest, and a polyA or a tag, respectively, are recombined into a "destination" vector that also possesses Tol2 recombination elements for integration in the genome with high efficiency. The plasmid hence generated is co-injected with *transposase* mRNA at one-cell stage for transient or stable transgenesis. Carriers of the transgene are usually easily identified by the expression of a reporter gene. A main advantage of this approach is its modularity that allows the generation of libraries of entry clones with promoters or genes of interest. For instance, entry clones with the promoter element from the *hsp70* gene [176] or a UAS promoter have been generated for conditional expression. A list of essential Tol2Kit clones can be requested online at http://tol2kit.genetics.utah.edu/index.php/Main_Page. By providing a simple, affordable and flexible system to generate transgenesis constructs, the Tol2Kit has largely facilitated zebrafish research, promoting the sharing of clones within the zebrafish community and making transgenesis available for any lab. Several labs have expanded the number of clones using the gateway technology and made their resources available (http://lawsonlab.umassmed.edu/gateway.html). To date, a list of 14,524 transgenic lines generated with either "trap" or Gateway constructs can be viewed on Zfin at http://zfin.org/action/fish/search.

3.2 *Targeted Transgenesis and the Generation of Knock-Ins*

By providing an efficient approach to integrate DNA constructs into the genome, transposon-derived elements have been instrumental for the study of gene function and tissue morphogenesis in zebrafish. However, transposon-mediated integrations occur randomly, precluding precise genome editing. TALENs and CRISPR/Cas9, on the other hand, allow targeted engineering and have been recently employed to insert small or large sequences at precise loci into the genome. Several methods involving the NHEJ or the HDR pathways have successfully led to the generation of the first KIs in zebrafish.

3.2.1 Integration via the HDR Pathway

DNA integration mediated by HDR has been achieved with both TALENs and the CRISPR/Cas9 system. Several templates including linearized plasmid and single-stranded DNA (ssDNA) have been used with various integration efficiencies.

One of the first reports of gene targeting via HR in zebrafish used TALENs and a linearized DNA vector containing the cassette to be inserted flanked by homologous sequences to the genomic target of around 800 and 900 bp on each side [177]. Several cassettes with loxP, eGFP, or eGFP-stop sequences were used to modify three different loci in the genome. Authors co-injected the linearized donor plasmids with TALEN mRNAs at one cell stage and were able to detect HR between the donor plasmids and the endogenous loci, with transmission to the germline in one case, albeit with low efficiency (about 1.5 %). Subsequent studies demonstrated that the length of homology arms as well as the configuration of the targeting construct have a significant impact on the efficiency of HDR [178]. In particular, increasing the length of the left and right arms to 1 and 2 kb, and introducing a DSB in the shorter homology arm, were shown to greatly improve efficient HR and germline transmission (over 10 %).

HDR has also been achieved using ssDNA with short homology arms as a template together with TALENs [113] or the CRISPR/Cas9 system [133, 134, 136]. Short fragments encoding restriction sites (6 bp) or loxP sites (34 bp) have been successfully integrated after co-injecting TALEN mRNAs and ssDNA oligonucleotides with short homology arms of 20 and 18 bp [113]. Interestingly in that case, increasing the length of homology arms seemed to reduce the frequency of HDR integrations. While germline transmission of the integrated DNA could be observed in 10 % of the cases, a major drawback of this approach was the frequent imprecise integration of the donor DNA with additional indel mutations. Similar results were obtained after co-injecting ssDNA oligonucleotides with sgRNAs and Cas9 mRNA [133, 134, 136]. All studies reported so far achieved precise integration of the template in the targeted genome location with various efficiencies. However, in all cases, imprecise repair events were frequently detected as a probable result of NHEJ. Inhibit the NHEJ pathway by blocking the activity of endogenous DNA ligase IV with the Scr7 inhibitor has recently been shown to increase the efficiency of HDR-mediated genome editing in mammalian cells and mice [179, 180], and might lead to similar improvement in zebrafish in the future.

3.2.2 Integration via the NHEJ Pathway

Considering the prevalence of NHEJ in zebrafish during development [56–58], recent studies have exploited the NHEJ pathway to elicit targeted integration of large donor DNAs [181, 182]. In this approach, the donor vector contains a short sequence bearing the TALEN or the CRISPR target site upstream the cassette to be integrated. Co-injection of this donor plasmid with sgRNA and *Cas9* mRNA (or alternatively TALEN mRNAs) lead to the concurrent cleavage of the plasmid and the genomic target, and the subsequent integration of the linearized plasmid at the genomic target by an NHEJ repair mechanism. Alternatively, two different sgRNAs can be used, one for cleaving the genome target, and the other for cleaving the donor plasmid. This method has proved to be very efficient for generating KIs allele, with rates of germline transmission over 30 %. However, since the integration of the donor plasmid can occur in both forward or reverse orientation and three different frames, screening efforts are necessary to isolate the appropriate lines. Introducing short homologous sequences flanking the sgRNA target site into the donor seems to improve the precision of integration by involving both NHEJ and HDR mechanisms [183]. Alternatively, selecting an sgRNA target in the intron of the gene sequence, and using a donor vector with a homologous arm spanning from that sgRNA site to the 3′ region of the targeted gene, can be employed to circumvent the requirement of in-frame insertions and increase KI efficiency [184].

4 Conclusions and Future Directions

Overall, the last 40 years have witnessed major advances in zebrafish research, from the first mutagenesis screens to the use of EENs for targeted genome editing. With precise genomic manipulations now available, the zebrafish has caught-up with other vertebrate organisms, combining genomic approaches previously restricted to the mouse with screening and high-resolution imaging techniques only possible in fish. This progress has promoted the development and use of mutant and transgenic lines in a wide number of research areas, and in neurobehavioral phenotyping research in particular. The following examples illustrate the wide spectrum of neural phenotypes studied, and their relevance to selected human brain disorders (see also Tables 2 and 3 for additional examples).

The Allan–Herndon–Dudley syndrome (AHDS) syndrome is a rare developmental nervous system disorder characterized by severe intellectual disability, muscle hypotonia and spastic paraplegia. It is caused by mutations in the *mct8* (*slc16a2*) gene located on the X chromosome that encodes a thyroid hormone receptor. Impaired Mct8 function is thought to prevent the entry of the active T3 hormone into neurons, leading to abnormal neurological development. While the MCT8 knockout mouse recapitulates the metabolic and endocrine defects seen in patients, they do not have any neurological or behavioral phenotype. In order to determine the functions of Mct8 in AHDS, Zada and colleagues used ZFNs to generate an *mct8* zebrafish mutant line [185]. Video-tracking behavioral imaging as well as time-lapse imaging of neuronal circuits showed that *Mct8* zebrafish mutants had a reduced locomotor activity that correlated with defects in synaptic density of motoneuron arbors and abnormal axonal branching in sensory neurons. Additional behavioral defects were observed, including increased and more fragmented sleep, and altered responses to light variations. Thus, the use of ZFNs to induce targeted mutation in *Mct8* in zebrafish lead to the development of the first vertebrate model of AHDS that recapitulates the full spectrum of defects seen in patients. As zebrafish is particularly suited for large-throughput approaches, this *mct8* mutant line could further be used in pharmacological screens for therapeutic development.

Table 2 Selected examples of aberrant neurobehavioral phenotypes demonstrated by mutant zebrafish lines

Gene	Biological function	Mutagenesis approach	Abnormal zebrafish phenotype	Relevance to human disorder	Ref.
tpp1	Lysosomal serine protease	ENU	Abnormal axon tract formation, early onset neuronal degeneration (apoptosis), reduced cell proliferation in the CNS, motor defects, reduced survival	CNL2 disease	[194]
sod1	Superoxide dismutase	TILLING	Altered neuromuscular junctions, reduced number of motoneurons, adult-onset motor defects	Familial ALS	[195]
dbh	Dopamine β-hydroxylase	ZFNs	Lower overall activity, increased sleep, reduced arousal threshold	Sleep disorders, depression, ADHD	[196]
nptx2a	Pentraxin, synaptic protein	TALENs	Decreased synapse density in motoneuron axons and reduced locomotor response to stimuli	Learning disorders, memory, seizures	[197]
kcnh4a	Voltage-gated potassium channel	CRISPR/Cas9	Locomotor activity slightly increased, decreased number and length of sleep episodes	Sleep disorders	[198]

ADHD attention deficit hyperactivity disorder, *ALS* amyotrophic lateral sclerosis, *CNL* ceroid lipofuscinosis neuronal

In addition to modeling neurological developmental diseases, zebrafish mutant lines have been used to study complex behaviors. Sleep, for instance, is an evolutionary conserved state that is essential to all organisms, but whose regulation remains poorly understood. In particular, the factors that transmit circadian information to regulate sleep are largely unknown. Using TALENs as a mutagenesis approach, Gandhi and colleagues generated a new zebrafish line harboring a null mutation in the *aanat2* gene that encodes an enzyme essential for melatonin synthesis [186]. Videotracking assays revealed that *aanat2* mutants had a normal sleep pattern during daytime but a reduced sleep and a longer sleep latency during night. Importantly, circadian rhythms were not disrupted in mutants, revealing for the first time that melatonin is not required to initiate or maintain the circadian clock. Altogether, the use of TALENs is this example allowed the generation of the first genetic loss-of-function model for melatonin in a diurnal vertebrate, and lead to the discovery of the endogenous functions of melatonin in sleep regulation.

Complementary to mutants, transgenic lines have proved very helpful to decipher the mechanisms underlying complex behaviors and neurological disorders (see Table 3). Narcolepsy, for example, is a rare chronic sleep disorder involving excessive daytime sleepiness, sleep fragmentation and paralysis at night, hypnagogic

Table 3 Selected examples of aberrant neurobehavioral phenotypes demonstrated by transgenic zebrafish lines

Transgene	Transgene function	Zebrafish phenotype	Relevance to human disorder	Ref.
dat:CFP-NTR	Expression of CFP-nitroreductase fusion protein in dopaminergic neurons	Metronidazole-induced loss of dopaminergic neurons, locomotor defects	Parkinson's disease	[199]
UAS:hTAU-P301L; HuC:Gal4	Expression of mutated human Tau in neurons (UAS/Gal4 system)[a]	Neuronal degeneration, altered motoneuron morphology, locomotor defects	Tauopathies, FTD	[200]
UAS:TeTXlc-CFP;Gal4^{s1019t} (Gal4^{s1019t}: enhancer trap line)	Expression of GFP-tetanus toxin fusion protein in the dorsal habenula (UAS/Gal4 system)	Deficit in avoidance learning	Anxiety disorders	[201]
tph2:nfsB-mCherry	Expression of nitroreductase–mCherry fusion protein in dorsal raphe neurons	Increase in visual sensitivity during arousal abolished	Attention disorders	[202]
UAS:Arch3-GFP; Gal4^{y252} (Gal4^{y252}: enhancer trap line)	Expression of CFP-Arch3 (light-activated proton pump) in *Gsx1*-expressing neurons	Optogenetic inhibition of Gsx1-expressing neurons; reduced prepulse inhibition of the startle reflex, increased startle response	Schizophrenia	[203]

Lines mentioned in this table have been generated using Tol1- or Tol2-transposon mediated transgenesis
ALS amyotrophic lateral sclerosis, *CFP* cyan fluorescent protein, *FTD* frontotemporal dementia, *GFP* green fluorescent protein
[a]TAU-P301L has been linked to frontotemporal dementia

hallucinations and cataplexy. It is caused by the selective degeneration of hypothalamic hypocretin/orexin (HCRT) neurons, whose activity is known to regulate several other behaviors including food intake, reward or drug addiction. To generate a zebrafish model of narcolepsy, Elbaz and colleagues used Tol2-mediated transgenesis to establish a stable transgenic line expressing the nitroreductase *nfsB* gene under the control of the *hcrt* promoter [187]. Exposing transgenic embryos to the drug metrodinazole induces the apoptosis of cells expressing *nfsB*, providing an inducible method to selectively ablate HCRT neurons at specific times. As expected, transgenic embryos lacking HCRT neurons recapitulated the defects seen in narcoleptic patients, including increased sleep time and transitions between wake and sleep states. They further had altered locomotor responses to light and sound, suggesting a broader function of HCRT neurons in mediating behavioral responses to external stimuli.

Transgenesis not only allows the ablation or silencing of a specific class of neurons regulating complex behaviors, but can also be used to express specific mutations in genes causing human disorders. Amyotrophic lateral sclerosis (ALS) is an adult-onset lethal neurodegenerative disease characterized by the progressive loss of motor neurons. While the majority of ALS is sporadic, around 10 % of cases are familial and caused by mutations in certain genes. Among them, mutations in the superoxide dismutase *Sod1* gene have been associated with 20 % of familial ALS (fALS). Several zebrafish transgenic lines expressing the SOD1-G93A mutation have been generated using Tol2-mediated transgenesis [188, 189]. They all showed defects associated with fALS including abnormal motor neuron outgrowth and branching, loss of neuromuscular junctions, muscle atrophy and motor neuron cell loss leading to premature death. These new transgenic lines thus provide an additional system for observing the progression of ALS directly in vivo in an intact organism, and isolating new effective compounds in therapeutics screens.

Overall, targeted mutagenesis and transgenesis have broaden the field of zebrafish research in many areas. The development of conditional mutant and targeted transgenic lines is now under way and will expand the repertoire of lines and resources currently available. For instance, one recent study has introduced *attP* site at specific loci into the genome for future recombination-mediated site-specific transgenesis [190]. Other new approaches that could be adapted to zebrafish research include Cas9 engineering to regulate transcription [191–193]. The zebrafish model is looking at a bright future, one that George Streisinger would be proud of.

Acknowledgements We apologize to authors whose work was omitted from this article due to space limitations. Work in the F.E.P. laboratory is supported in part by R00NS083714 from the National Institute of Neurological Disorders and Stroke at the National Institute of Health.

References

1. Howe K, Clark MD, Torroja CF, Torrance J, Berthelot C, Muffato M, et al. The zebrafish reference genome sequence and its relationship to the human genome. Nature. 2013;496:498–503.
2. Driever W, Solnica-Krezel L, Schier AF, Neuhauss SC, Malicki J, Stemple DL, et al. A genetic screen for mutations affecting embryogenesis in zebrafish. Development. 1996;123:37–46.
3. Haffter P, Granato M, Brand M, Mullins MC, Hammerschmidt M, Kane DA, et al. The identification of genes with unique and essential functions in the development of the zebrafish, *Danio rerio*. Development. 1996;123:1–36.
4. Chakrabarti S, Streisinger G, Singer F, Walker C. Frequency of gamma-ray induced specific locus and recessive lethal mutations in mature germ cells of the zebrafish, Brachydanio rerio. Genetics. 1983;103:109–23.
5. Walker C, Streisinger G. Induction of mutations by gamma-rays in pregonial germ cells of zebrafish embryos. Genetics. 1983;103:125–36.
6. Grunwald DJ, Streisinger G. Induction of mutations in the zebrafish with ultraviolet light. Genet Res. 1992;59:93–101.

7. Grunwald DJ, Streisinger G. Induction of recessive lethal and specific locus mutations in the zebrafish with ethyl nitrosourea. Genet Res. 1992;59:103–16.
8. Mullins MC, Hammerschmidt M, Haffter P, Nusslein-Volhard C. Large-scale mutagenesis in the zebrafish: in search of genes controlling development in a vertebrate. Curr Biol. 1994;4:189–202.
9. Solnica-Krezel L, Schier AF, Driever W. Efficient recovery of ENU-induced mutations from the zebrafish germline. Genetics. 1994;136:1401–20.
10. Nusslein-Volhard C. The zebrafish issue of development. Development. 2012;139: 4099–103.
11. Xiao T, Roeser T, Staub W, Baier H. A GFP-based genetic screen reveals mutations that disrupt the architecture of the zebrafish retinotectal projection. Development. 2005;132: 2955–67.
12. Pogoda HM, Sternheim N, Lyons DA, Diamond B, Hawkins TA, Woods IG, et al. A genetic screen identifies genes essential for development of myelinated axons in zebrafish. Dev Biol. 2006;298:118–31.
13. Kuhlman J, Eisen JS. Genetic screen for mutations affecting development and function of the enteric nervous system. Dev Dyn. 2007;236:118–27.
14. Covassin LD, Siekmann AF, Kacergis MC, Laver E, Moore JC, Villefranc JA, et al. A genetic screen for vascular mutants in zebrafish reveals dynamic roles for Vegf/Plcg1 signaling during artery development. Dev Biol. 2009;329:212–26.
15. Lee J, Cox BD, Daly CM, Lee C, Nuckels RJ, Tittle RK, et al. An ENU mutagenesis screen in zebrafish for visual system mutants identifies a novel splice-acceptor site mutation in patched2 that results in Colobomas. Invest Ophthalmol Vis Sci. 2012;53:8214–21.
16. Dai ZX, Yan G, Chen YH, Liu W, Huo ZJ, Wen ZH, et al. Forward genetic screening for zebrafish mutants defective in myelopoiesis. Nan Fang Yi Ke Da Xue Xue Bao. 2010;30: 1230–3.
17. Muto A, Orger MB, Wehman AM, Smear MC, Kay JN, Page-McCaw PS, et al. Forward genetic analysis of visual behavior in zebrafish. PLoS Genet. 2005;1:e66.
18. Baraban SC, Dinday MT, Castro PA, Chege S, Guyenet S, Taylor MR. A large-scale mutagenesis screen to identify seizure-resistant zebrafish. Epilepsia. 2007;48:1151–7.
19. Wolman MA, Jain RA, Marsden KC, Bell H, Skinner J, Hayer KE, et al. A genome-wide screen identifies PAPP-AA-mediated IGFR signaling as a novel regulator of habituation learning. Neuron. 2015;85:1200–11.
20. Darland T, Dowling JE. Behavioral screening for cocaine sensitivity in mutagenized zebrafish. Proc Natl Acad Sci U S A. 2001;98:11691–6.
21. Kim SH, Wu SY, Baek JI, Choi SY, Su Y, Flynn CR, et al. A post-developmental genetic screen for zebrafish models of inherited liver disease. PLoS One. 2015;10:e0125980.
22. Voz ML, Coppieters W, Manfroid I, Baudhuin A, Von Berg V, Charlier C, et al. Fast homozygosity mapping and identification of a zebrafish ENU-induced mutation by whole-genome sequencing. PLoS One. 2012;7:e34671.
23. Bowen ME, Henke K, Siegfried KR, Warman ML, Harris MP. Efficient mapping and cloning of mutations in zebrafish by low-coverage whole-genome sequencing. Genetics. 2012;190:1017–24.
24. Obholzer N, Swinburne IA, Schwab E, Nechiporuk AV, Nicolson T, Megason SG. Rapid positional cloning of zebrafish mutations by linkage and homozygosity mapping using whole-genome sequencing. Development. 2012;139:4280–90.
25. Hill JT, Demarest BL, Bisgrove BW, Gorsi B, Su YC, Yost HJ. MMAPPR: mutation mapping analysis pipeline for pooled RNA-seq. Genome Res. 2013;23:687–97.
26. Miller AC, Obholzer ND, Shah AN, Megason SG, Moens CB. RNA-seq-based mapping and candidate identification of mutations from forward genetic screens. Genome Res. 2013;23:679–86.
27. Wienholds E, Schulte-Merker S, Walderich B, Plasterk RH. Target-selected inactivation of the zebrafish rag1 gene. Science. 2002;297:99–102.
28. Wienholds E, van Eeden F, Kosters M, Mudde J, Plasterk RH, Cuppen E. Efficient target-selected mutagenesis in zebrafish. Genome Res. 2003;13:2700–7.

29. Kettleborough RN, Busch-Nentwich EM, Harvey SA, Dooley CM, de Bruijn E, van Eeden F, et al. A systematic genome-wide analysis of zebrafish protein-coding gene function. Nature. 2013;496:494–7.
30. Lin S, Gaiano N, Culp P, Burns JC, Friedmann T, Yee JK, et al. Integration and germ-line transmission of a pseudotyped retroviral vector in zebrafish. Science. 1994;265:666–9.
31. Wu X, Li Y, Crise B, Burgess SM. Transcription start regions in the human genome are favored targets for MLV integration. Science. 2003;300:1749–51.
32. Wang D, Jao LE, Zheng N, Dolan K, Ivey J, Zonies S, et al. Efficient genome-wide mutagenesis of zebrafish genes by retroviral insertions. Proc Natl Acad Sci U S A. 2007;104: 12428–33.
33. Gaiano N, Amsterdam A, Kawakami K, Allende M, Becker T, Hopkins N. Insertional mutagenesis and rapid cloning of essential genes in zebrafish. Nature. 1996;383:829–32.
34. Amsterdam A, Burgess S, Golling G, Chen W, Sun Z, Townsend K, et al. A large-scale insertional mutagenesis screen in zebrafish. Genes Dev. 1999;13:2713–24.
35. Golling G, Amsterdam A, Sun Z, Antonelli M, Maldonado E, Chen W, et al. Insertional mutagenesis in zebrafish rapidly identifies genes essential for early vertebrate development. Nat Genet. 2002;31:135–40.
36. Amsterdam A, Nissen RM, Sun Z, Swindell EC, Farrington S, Hopkins N. Identification of 315 genes essential for early zebrafish development. Proc Natl Acad Sci U S A. 2004;101: 12792–7.
37. Sun Z, Amsterdam A, Pazour GJ, Cole DG, Miller MS, Hopkins N. A genetic screen in zebrafish identifies cilia genes as a principal cause of cystic kidney. Development. 2004;131: 4085–93.
38. Varshney GK, Huang H, Zhang S, Lu J, Gildea DE, Yang Z, et al. The Zebrafish Insertion Collection (ZInC): a web based, searchable collection of zebrafish mutations generated by DNA insertion. Nucleic Acids Res. 2013;41:D861–4.
39. Varshney GK, Lu J, Gildea DE, Huang H, Pei W, Yang Z, et al. A large-scale zebrafish gene knockout resource for the genome-wide study of gene function. Genome Res. 2013;23: 727–35.
40. Kawakami K, Shima A, Kawakami N. Identification of a functional transposase of the Tol2 element, an Ac-like element from the Japanese medaka fish, and its transposition in the zebrafish germ lineage. Proc Natl Acad Sci U S A. 2000;97:11403–8.
41. Davidson AE, Balciunas D, Mohn D, Shaffer J, Hermanson S, Sivasubbu S, et al. Efficient gene delivery and gene expression in zebrafish using the Sleeping Beauty transposon. Dev Biol. 2003;263:191–202.
42. Emelyanov A, Gao Y, Naqvi NI, Parinov S. Trans-kingdom transposition of the maize dissociation element. Genetics. 2006;174:1095–104.
43. le Trinh A, Hochgreb T, Graham M, Wu D, Ruf-Zamojski F, Jayasena CS, et al. A versatile gene trap to visualize and interrogate the function of the vertebrate proteome. Genes Dev. 2011;25:2306–20.
44. Clark KJ, Balciunas D, Pogoda HM, Ding Y, Westcot SE, Bedell VM, et al. In vivo protein trapping produces a functional expression codex of the vertebrate proteome. Nat Methods. 2011;8:506–15.
45. Ni TT, Lu J, Zhu M, Maddison LA, Boyd KL, Huskey L, et al. Conditional control of gene function by an invertible gene trap in zebrafish. Proc Natl Acad Sci U S A. 2012;109: 15389–94.
46. Balciuniene J, Nagelberg D, Walsh KT, Camerota D, Georlette D, Biemar F, et al. Efficient disruption of Zebrafish genes using a Gal4-containing gene trap. BMC Genomics. 2013;14:619.
47. Kawakami K, Takeda H, Kawakami N, Kobayashi M, Matsuda N, Mishina M. A transposon-mediated gene trap approach identifies developmentally regulated genes in zebrafish. Dev Cell. 2004;7:133–44.
48. Petzold AM, Balciunas D, Sivasubbu S, Clark KJ, Bedell VM, Westcot SE, et al. Nicotine response genetics in the zebrafish. Proc Natl Acad Sci U S A. 2009;106:18662–7.

49. Seiler C, Gebhart N, Zhang Y, Shinton SA, Li YS, Ross NL, et al. Mutagenesis screen identifies agtpbp1 and eps15L1 as essential for T lymphocyte development in zebrafish. PLoS One. 2015;10:e0131908.
50. Kawakami K, Abe G, Asada T, Asakawa K, Fukuda R, Ito A, et al. zTrap: zebrafish gene trap and enhancer trap database. BMC Dev Biol. 2010;10:105.
51. Nasevicius A, Ekker SC. Effective targeted gene 'knockdown' in zebrafish. Nat Genet. 2000;26:216–20.
52. Schulte-Merker S, Stainier DY. Out with the old, in with the new: reassessing morpholino knockdowns in light of genome editing technology. Development. 2014;141:3103–4.
53. Stainier DY, Kontarakis Z, Rossi A. Making sense of anti-sense data. Dev Cell. 2015;32:7–8.
54. Kok FO, Shin M, Ni CW, Gupta A, Grosse AS, van Impel A, et al. Reverse genetic screening reveals poor correlation between morpholino-induced and mutant phenotypes in zebrafish. Dev Cell. 2015;32:97–108.
55. Rossi A, Kontarakis Z, Gerri C, Nolte H, Holper S, Kruger M, et al. Genetic compensation induced by deleterious mutations but not gene knockdowns. Nature. 2015;524:230–3.
56. Hagmann M, Bruggmann R, Xue L, Georgiev O, Schaffner W, Rungger D, et al. Homologous recombination and DNA-end joining reactions in zygotes and early embryos of zebrafish (*Danio rerio*) and *Drosophila melanogaster*. Biol Chem. 1998;379:673–81.
57. Dai J, Cui X, Zhu Z, Hu W. Non-homologous end joining plays a key role in transgene concatemer formation in transgenic zebrafish embryos. Int J Biol Sci. 2010;6:756–68.
58. Liu J, Gong L, Chang C, Liu C, Peng J, Chen J. Development of novel visual-plus quantitative analysis systems for studying DNA double-strand break repairs in zebrafish. J Genet Genomics. 2012;39:489–502.
59. Ota S, Hisano Y, Muraki M, Hoshijima K, Dahlem TJ, Grunwald DJ, et al. Efficient identification of TALEN-mediated genome modifications using heteroduplex mobility assays. Genes Cells. 2013;18:450–8.
60. Oleykowski CA, Bronson Mullins CR, Godwin AK, Yeung AT. Mutation detection using a novel plant endonuclease. Nucleic Acids Res. 1998;26:4597–602.
61. Qiu P, Shandilya H, D'Alessio JM, O'Connor K, Durocher J, Gerard GF. Mutation detection using surveyor nuclease. Biotechniques. 2004;36:702–7.
62. Vouillot L, Thelie A, Pollet N. Comparison of T7E1 and surveyor mismatch cleavage assays to detect mutations triggered by engineered nucleases. G3 (Bethesda). 2015;5:407–15.
63. Parant JM, George SA, Pryor R, Wittwer CT, Yost HJ. A rapid and efficient method of genotyping zebrafish mutants. Dev Dyn. 2009;238:3168–74.
64. Dahlem TJ, Hoshijima K, Jurynec MJ, Gunther D, Starker CG, Locke AS, et al. Simple methods for generating and detecting locus-specific mutations induced with TALENs in the zebrafish genome. PLoS Genet. 2012;8:e1002861.
65. Xing L, Quist TS, Stevenson TJ, Dahlem TJ, Bonkowsky JL. Rapid and efficient zebrafish genotyping using PCR with high-resolution melt analysis. J Vis Exp. 2014:e51138.
66. Hill JT, Demarest BL, Bisgrove BW, Su YC, Smith M, Yost HJ. Poly peak parser: method and software for identification of unknown indels using sanger sequencing of polymerase chain reaction products. Dev Dyn. 2014;243:1632–6.
67. Xiao A, Wu Y, Yang Z, Hu Y, Wang W, Zhang Y, et al. EENdb: a database and knowledge base of ZFNs and TALENs for endonuclease engineering. Nucleic Acids Res. 2013;41:D415–22.
68. Sander JD, Maeder ML, Reyon D, Voytas DF, Joung JK, Dobbs D. ZiFiT (Zinc Finger Targeter): an updated zinc finger engineering tool. Nucleic Acids Res. 2010;38:W462–8.
69. Kim YG, Cha J, Chandrasegaran S. Hybrid restriction enzymes: zinc finger fusions to Fok I cleavage domain. Proc Natl Acad Sci U S A. 1996;93:1156–60.
70. Choo Y, Klug A. Selection of DNA binding sites for zinc fingers using rationally randomized DNA reveals coded interactions. Proc Natl Acad Sci U S A. 1994;91:11168–72.
71. Choo Y, Klug A. Toward a code for the interactions of zinc fingers with DNA: selection of randomized fingers displayed on phage. Proc Natl Acad Sci U S A. 1994;91:11163–7.

72. Greisman HA, Pabo CO. A general strategy for selecting high-affinity zinc finger proteins for diverse DNA target sites. Science. 1997;275:657–61.
73. Segal DJ, Dreier B, Beerli RR, Barbas 3rd CF. Toward controlling gene expression at will: selection and design of zinc finger domains recognizing each of the 5′-GNN-3′ DNA target sequences. Proc Natl Acad Sci U S A. 1999;96:2758–63.
74. Dreier B, Beerli RR, Segal DJ, Flippin JD, Barbas 3rd CF. Development of zinc finger domains for recognition of the 5′-ANN-3′ family of DNA sequences and their use in the construction of artificial transcription factors. J Biol Chem. 2001;276:29466–78.
75. Dreier B, Fuller RP, Segal DJ, Lund CV, Blancafort P, Huber A, et al. Development of zinc finger domains for recognition of the 5′-CNN-3′ family DNA sequences and their use in the construction of artificial transcription factors. J Biol Chem. 2005;280:35588–97.
76. Liu Q, Xia Z, Zhong X, Case CC. Validated zinc finger protein designs for all 16 GNN DNA triplet targets. J Biol Chem. 2002;277:3850–6.
77. Kim HJ, Lee HJ, Kim H, Cho SW, Kim JS. Targeted genome editing in human cells with zinc finger nucleases constructed via modular assembly. Genome Res. 2009;19:1279–88.
78. Ramirez CL, Foley JE, Wright DA, Muller-Lerch F, Rahman SH, Cornu TI, et al. Unexpected failure rates for modular assembly of engineered zinc fingers. Nat Methods. 2008;5:374–5.
79. Carroll D, Morton JJ, Beumer KJ, Segal DJ. Design, construction and in vitro testing of zinc finger nucleases. Nat Protoc. 2006;1:1329–41.
80. Maeder ML, Thibodeau-Beganny S, Osiak A, Wright DA, Anthony RM, Eichtinger M, et al. Rapid "open-source" engineering of customized zinc-finger nucleases for highly efficient gene modification. Mol Cell. 2008;31:294–301.
81. Maeder ML, Thibodeau-Beganny S, Sander JD, Voytas DF, Joung JK. Oligomerized pool engineering (OPEN): an 'open-source' protocol for making customized zinc-finger arrays. Nat Protoc. 2009;4:1471–501.
82. Meng X, Noyes MB, Zhu LJ, Lawson ND, Wolfe SA. Targeted gene inactivation in zebrafish using engineered zinc-finger nucleases. Nat Biotechnol. 2008;26:695–701.
83. Sander JD, Dahlborg EJ, Goodwin MJ, Cade L, Zhang F, Cifuentes D, et al. Selection-free zinc-finger-nuclease engineering by context-dependent assembly (CoDA). Nat Methods. 2011;8:67–9.
84. Smith J, Bibikova M, Whitby FG, Reddy AR, Chandrasegaran S, Carroll D. Requirements for double-strand cleavage by chimeric restriction enzymes with zinc finger DNA-recognition domains. Nucleic Acids Res. 2000;28:3361–9.
85. Miller JC, Holmes MC, Wang J, Guschin DY, Lee YL, Rupniewski I, et al. An improved zinc-finger nuclease architecture for highly specific genome editing. Nat Biotechnol. 2007;25: 778–85.
86. Szczepek M, Brondani V, Buchel J, Serrano L, Segal DJ, Cathomen T. Structure-based redesign of the dimerization interface reduces the toxicity of zinc-finger nucleases. Nat Biotechnol. 2007;25:786–93.
87. Handel EM, Alwin S, Cathomen T. Expanding or restricting the target site repertoire of zinc-finger nucleases: the inter-domain linker as a major determinant of target site selectivity. Mol Ther. 2009;17:104–11.
88. Wilson KA, McEwen AE, Pruett-Miller SM, Zhang J, Kildebeck EJ, Porteus MH. Expanding the repertoire of target sites for zinc finger nuclease-mediated genome modification. Mol Ther Nucleic Acids. 2013;2:e88.
89. Doyon Y, McCammon JM, Miller JC, Faraji F, Ngo C, Katibah GE, et al. Heritable targeted gene disruption in zebrafish using designed zinc-finger nucleases. Nat Biotechnol. 2008;26:702–8.
90. Foley JE, Yeh JR, Maeder ML, Reyon D, Sander JD, Peterson RT, et al. Rapid mutation of endogenous zebrafish genes using zinc finger nucleases made by Oligomerized Pool ENgineering (OPEN). PLoS One. 2009;4:e4348.
91. Reyon D, Kirkpatrick JR, Sander JD, Zhang F, Voytas DF, Joung JK, et al. ZFNGenome: a comprehensive resource for locating zinc finger nuclease target sites in model organisms. BMC Genomics. 2011;12:83.

92. Chen S, Oikonomou G, Chiu CN, Niles BJ, Liu J, Lee DA, et al. A large-scale in vivo analysis reveals that TALENs are significantly more mutagenic than ZFNs generated using context-dependent assembly. Nucleic Acids Res. 2013;41:2769–78.
93. Moore FE, Reyon D, Sander JD, Martinez SA, Blackburn JS, Khayter C, et al. Improved somatic mutagenesis in zebrafish using transcription activator-like effector nucleases (TALENs). PLoS One. 2012;7:e37877.
94. Bogdanove AJ, Schornack S, Lahaye T. TAL effectors: finding plant genes for disease and defense. Curr Opin Plant Biol. 2010;13:394–401.
95. Christian ML, Demorest ZL, Starker CG, Osborn MJ, Nyquist MD, Zhang Y, et al. Targeting G with TAL effectors: a comparison of activities of TALENs constructed with NN and NK repeat variable di-residues. PLoS One. 2012;7:e45383.
96. Cong L, Zhou R, Kuo YC, Cunniff M, Zhang F. Comprehensive interrogation of natural TALE DNA-binding modules and transcriptional repressor domains. Nat Commun. 2012;3:968.
97. Streubel J, Blucher C, Landgraf A, Boch J. TAL effector RVD specificities and efficiencies. Nat Biotechnol. 2012;30:593–5.
98. Boch J, Scholze H, Schornack S, Landgraf A, Hahn S, Kay S, et al. Breaking the code of DNA binding specificity of TAL-type III effectors. Science. 2009;326:1509–12.
99. Doyle EL, Booher NJ, Standage DS, Voytas DF, Brendel VP, Vandyk JK, et al. TAL Effector-Nucleotide Targeter (TALE-NT) 2.0: tools for TAL effector design and target prediction. Nucleic Acids Res. 2012;40:W117–22.
100. Neff KL, Argue DP, Ma AC, Lee HB, Clark KJ, Ekker SC. Mojo Hand, a TALEN design tool for genome editing applications. BMC Bioinf. 2013;14:1.
101. Li L, Piatek MJ, Atef A, Piatek A, Wibowo A, Fang X, et al. Rapid and highly efficient construction of TALE-based transcriptional regulators and nucleases for genome modification. Plant Mol Biol. 2012;78:407–16.
102. Cermak T, Doyle EL, Christian M, Wang L, Zhang Y, Schmidt C, et al. Efficient design and assembly of custom TALEN and other TAL effector-based constructs for DNA targeting. Nucleic Acids Res. 2011;39:e82.
103. Morbitzer R, Elsaesser J, Hausner J, Lahaye T. Assembly of custom TALE-type DNA binding domains by modular cloning. Nucleic Acids Res. 2011;39:5790–9.
104. Sakuma T, Hosoi S, Woltjen K, Suzuki K, Kashiwagi K, Wada H, et al. Efficient TALEN construction and evaluation methods for human cell and animal applications. Genes Cells. 2013;18:315–26.
105. Ding Q, Lee YK, Schaefer EA, Peters DT, Veres A, Kim K, et al. A TALEN genome-editing system for generating human stem cell-based disease models. Cell Stem Cell. 2013;12:238–51.
106. Ma AC, Lee HB, Clark KJ, Ekker SC. High efficiency In Vivo genome engineering with a simplified 15-RVD GoldyTALEN design. PLoS One. 2013;8:e65259.
107. Uhde-Stone C, Gor N, Chin T, Huang J, Lu B. A do-it-yourself protocol for simple transcription activator-like effector assembly. Biol Proced Online. 2013;15:3.
108. Huang P, Xiao A, Tong X, Zu Y, Wang Z, Zhang B. TALEN construction via "Unit Assembly" method and targeted genome modifications in zebrafish. Methods. 2014;69:67–75.
109. Huang P, Xiao A, Zhou M, Zhu Z, Lin S, Zhang B. Heritable gene targeting in zebrafish using customized TALENs. Nat Biotechnol. 2011;29:699–700.
110. Reyon D, Khayter C, Regan MR, Joung JK, Sander JD. Engineering designer transcription activator-like effector nucleases (TALENs) by REAL or REAL-Fast assembly. Curr Protoc Mol Biol. 2012;Chapter 12:Unit 12 5.
111. Reyon D, Tsai SQ, Khayter C, Foden JA, Sander JD, Joung JK. FLASH assembly of TALENs for high-throughput genome editing. Nat Biotechnol. 2012;30:460–5.
112. Briggs AW, Rios X, Chari R, Yang L, Zhang F, Mali P, et al. Iterative capped assembly: rapid and scalable synthesis of repeat-module DNA such as TAL effectors from individual monomers. Nucleic Acids Res. 2012;40:e117.

113. Bedell VM, Wang Y, Campbell JM, Poshusta TL, Starker CG, Krug 2nd RG, et al. In vivo genome editing using a high-efficiency TALEN system. Nature. 2012;491:114–8.
114. Sander JD, Cade L, Khayter C, Reyon D, Peterson RT, Joung JK, et al. Targeted gene disruption in somatic zebrafish cells using engineered TALENs. Nat Biotechnol. 2011;29:697–8.
115. Cade L, Reyon D, Hwang WY, Tsai SQ, Patel S, Khayter C, et al. Highly efficient generation of heritable zebrafish gene mutations using homo- and heterodimeric TALENs. Nucleic Acids Res. 2012;40:8001–10.
116. Barrangou R, Fremaux C, Deveau H, Richards M, Boyaval P, Moineau S, et al. CRISPR provides acquired resistance against viruses in prokaryotes. Science. 2007;315:1709–12.
117. Marraffini LA, Sontheimer EJ. CRISPR interference limits horizontal gene transfer in staphylococci by targeting DNA. Science. 2008;322:1843–5.
118. Bhaya D, Davison M, Barrangou R. CRISPR-Cas systems in bacteria and archaea: versatile small RNAs for adaptive defense and regulation. Annu Rev Genet. 2011;45:273–97.
119. Jinek M, Chylinski K, Fonfara I, Hauer M, Doudna JA, Charpentier E. A programmable dual-RNA-guided DNA endonuclease in adaptive bacterial immunity. Science. 2012;337: 816–21.
120. Mali P, Yang L, Esvelt KM, Aach J, Guell M, DiCarlo JE, et al. RNA-guided human genome engineering via Cas9. Science. 2013;339:823–6.
121. Cong L, Ran FA, Cox D, Lin S, Barretto R, Habib N, et al. Multiplex genome engineering using CRISPR/Cas systems. Science. 2013;339:819–23.
122. Mali P, Aach J, Stranges PB, Esvelt KM, Moosburner M, Kosuri S, et al. CAS9 transcriptional activators for target specificity screening and paired nickases for cooperative genome engineering. Nat Biotechnol. 2013;31:833–8.
123. Hsu PD, Scott DA, Weinstein JA, Ran FA, Konermann S, Agarwala V, et al. DNA targeting specificity of RNA-guided Cas9 nucleases. Nat Biotechnol. 2013;31:827–32.
124. Shah SA, Erdmann S, Mojica FJ, Garrett RA. Protospacer recognition motifs: mixed identities and functional diversity. RNA Biol. 2013;10:891–9.
125. Sternberg SH, Redding S, Jinek M, Greene EC, Doudna JA. DNA interrogation by the CRISPR RNA-guided endonuclease Cas9. Nature. 2014;507:62–7.
126. Jinek M, Jiang F, Taylor DW, Sternberg SH, Kaya E, Ma E, et al. Structures of Cas9 endonucleases reveal RNA-mediated conformational activation. Science. 2014;343:1247997.
127. Nishimasu H, Ran FA, Hsu PD, Konermann S, Shehata SI, Dohmae N, et al. Crystal structure of Cas9 in complex with guide RNA and target DNA. Cell. 2014;156:935–49.
128. Kleinstiver BP, Prew MS, Tsai SQ, Topkar VV, Nguyen NT, Zheng Z, et al. Engineered CRISPR-Cas9 nucleases with altered PAM specificities. Nature. 2015;523:481–5.
129. Naito Y, Hino K, Bono H, Ui-Tei K. CRISPRdirect: software for designing CRISPR/Cas guide RNA with reduced off-target sites. Bioinformatics. 2015;31:1120–3.
130. Park J, Bae S, Kim JS. Cas-Designer: a web-based tool for choice of CRISPR-Cas9 target sites. Bioinformatics. 2015;31:4014–6.
131. Xie S, Shen B, Zhang C, Huang X, Zhang Y. sgRNAcas9: a software package for designing CRISPR sgRNA and evaluating potential off-target cleavage sites. PLoS One. 2014;9:e100448.
132. Montague TG, Cruz JM, Gagnon JA, Church GM, Valen E. CHOPCHOP: a CRISPR/Cas9 and TALEN web tool for genome editing. Nucleic Acids Res. 2014;42:W401–7.
133. Hruscha A, Krawitz P, Rechenberg A, Heinrich V, Hecht J, Haass C, et al. Efficient CRISPR/Cas9 genome editing with low off-target effects in zebrafish. Development. 2013;140: 4982–7.
134. Hwang WY, Fu Y, Reyon D, Maeder ML, Kaini P, Sander JD, et al. Heritable and precise zebrafish genome editing using a CRISPR-Cas system. PLoS One. 2013;8:e68708.
135. Hwang WY, Fu Y, Reyon D, Maeder ML, Tsai SQ, Sander JD, et al. Efficient genome editing in zebrafish using a CRISPR-Cas system. Nat Biotechnol. 2013;31:227–9.
136. Chang N, Sun C, Gao L, Zhu D, Xu X, Zhu X, et al. Genome editing with RNA-guided Cas9 nuclease in zebrafish embryos. Cell Res. 2013;23:465–72.
137. Gagnon JA, Valen E, Thyme SB, Huang P, Akhmetova L, Pauli A, et al. Efficient mutagenesis by Cas9 protein-mediated oligonucleotide insertion and large-scale assessment of single-guide RNAs. PLoS One. 2014;9:e98186.

138. Sung YH, Kim JM, Kim HT, Lee J, Jeon J, Jin Y, et al. Highly efficient gene knockout in mice and zebrafish with RNA-guided endonucleases. Genome Res. 2014;24:125–31.
139. Kotani H, Taimatsu K, Ohga R, Ota S, Kawahara A. Efficient multiple genome modifications induced by the crRNAs, tracrRNA and Cas9 protein complex in zebrafish. PLoS One. 2015;10:e0128319.
140. Jao LE, Wente SR, Chen W. Efficient multiplex biallelic zebrafish genome editing using a CRISPR nuclease system. Proc Natl Acad Sci U S A. 2013;110:13904–9.
141. Ota S, Hisano Y, Ikawa Y, Kawahara A. Multiple genome modifications by the CRISPR/Cas9 system in zebrafish. Genes Cells. 2014;19:555–64.
142. Ranganathan V, Wahlin K, Maruotti J, Zack DJ. Expansion of the CRISPR-Cas9 genome targeting space through the use of H1 promoter-expressed guide RNAs. Nat Commun. 2014;5:4516.
143. Yin L, Maddison LA, Li M, Kara N, LaFave MC, Varshney GK, et al. Multiplex conditional mutagenesis using transgenic expression of Cas9 and sgRNAs. Genetics. 2015;200:431–41.
144. Varshney GK, Pei W, LaFave MC, Idol J, Xu L, Gallardo V, et al. High-throughput gene targeting and phenotyping in zebrafish using CRISPR/Cas9. Genome Res. 2015;25:1030–42.
145. Zhu LJ, Holmes BR, Aronin N, Brodsky MH. CRISPRseek: a bioconductor package to identify target-specific guide RNAs for CRISPR-Cas9 genome-editing systems. PLoS One. 2014;9:e108424.
146. Prykhozhij SV, Rajan V, Gaston D, Berman JN. CRISPR multitargeter: a web tool to find common and unique CRISPR single guide RNA targets in a set of similar sequences. PLoS One. 2015;10:e0119372.
147. Ablain J, Durand EM, Yang S, Zhou Y, Zon LI. A CRISPR/Cas9 vector system for tissue-specific gene disruption in zebrafish. Dev Cell. 2015;32:756–64.
148. Fu Y, Foden JA, Khayter C, Maeder ML, Reyon D, Joung JK, et al. High-frequency off-target mutagenesis induced by CRISPR-Cas nucleases in human cells. Nat Biotechnol. 2013;31: 822–6.
149. Pattanayak V, Lin S, Guilinger JP, Ma E, Doudna JA, Liu DR. High-throughput profiling of off-target DNA cleavage reveals RNA-programmed Cas9 nuclease specificity. Nat Biotechnol. 2013;31:839–43.
150. Liu D, Wang Z, Xiao A, Zhang Y, Li W, Zu Y, et al. Efficient gene targeting in zebrafish mediated by a zebrafish-codon-optimized cas9 and evaluation of off-targeting effect. J Genet Genomics. 2014;41:43–6.
151. Fu Y, Sander JD, Reyon D, Cascio VM, Joung JK. Improving CRISPR-Cas nuclease specificity using truncated guide RNAs. Nat Biotechnol. 2014;32:279–84.
152. Cho SW, Kim S, Kim Y, Kweon J, Kim HS, Bae S, et al. Analysis of off-target effects of CRISPR/Cas-derived RNA-guided endonucleases and nickases. Genome Res. 2014;24: 132–41.
153. Ran FA, Hsu PD, Lin CY, Gootenberg JS, Konermann S, Trevino AE, et al. Double nicking by RNA-guided CRISPR Cas9 for enhanced genome editing specificity. Cell. 2013;154: 1380–9.
154. Gupta A, Hall VL, Kok FO, Shin M, McNulty JC, Lawson ND, et al. Targeted chromosomal deletions and inversions in zebrafish. Genome Res. 2013;23:1008–17.
155. Xiao A, Wang Z, Hu Y, Wu Y, Luo Z, Yang Z, et al. Chromosomal deletions and inversions mediated by TALENs and CRISPR/Cas in zebrafish. Nucleic Acids Res. 2013;41:e141.
156. Li Y, Rivera CM, Ishii H, Jin F, Selvaraj S, Lee AY, et al. CRISPR reveals a distal super-enhancer required for Sox2 expression in mouse embryonic stem cells. PLoS One. 2014;9:e114485.
157. Wang S, Sengel C, Emerson MM, Cepko CL. A gene regulatory network controls the binary fate decision of rod and bipolar cells in the vertebrate retina. Dev Cell. 2014;30:513–27.
158. Piganeau M, Ghezraoui H, De Cian A, Guittat L, Tomishima M, Perrouault L, et al. Cancer translocations in human cells induced by zinc finger and TALE nucleases. Genome Res. 2013;23:1182–93.

159. Blasco RB, Karaca E, Ambrogio C, Cheong TC, Karayol E, Minero VG, et al. Simple and rapid in vivo generation of chromosomal rearrangements using CRISPR/Cas9 technology. Cell Rep. 2014;9:1219–27.
160. Maddalo D, Manchado E, Concepcion CP, Bonetti C, Vidigal JA, Han YC, et al. In vivo engineering of oncogenic chromosomal rearrangements with the CRISPR/Cas9 system. Nature. 2014;516:423–7.
161. Balciunas D, Davidson AE, Sivasubbu S, Hermanson SB, Welle Z, Ekker SC. Enhancer trapping in zebrafish using the Sleeping Beauty transposon. BMC Genomics. 2004;5:62.
162. Parinov S, Kondrichin I, Korzh V, Emelyanov A. Tol2 transposon-mediated enhancer trap to identify developmentally regulated zebrafish genes in vivo. Dev Dyn. 2004;231:449–59.
163. Scott EK, Baier H. The cellular architecture of the larval zebrafish tectum, as revealed by gal4 enhancer trap lines. Front Neural Circuits. 2009;3:13.
164. Scott EK, Mason L, Arrenberg AB, Ziv L, Gosse NJ, Xiao T, et al. Targeting neural circuitry in zebrafish using GAL4 enhancer trapping. Nat Methods. 2007;4:323–6.
165. Asakawa K, Kawakami K. The Tol2-mediated Gal4-UAS method for gene and enhancer trapping in zebrafish. Methods. 2009;49:275–81.
166. Asakawa K, Suster ML, Mizusawa K, Nagayoshi S, Kotani T, Urasaki A, et al. Genetic dissection of neural circuits by Tol2 transposon-mediated Gal4 gene and enhancer trapping in zebrafish. Proc Natl Acad Sci U S A. 2008;105:1255–60.
167. Distel M, Wullimann MF, Koster RW. Optimized Gal4 genetics for permanent gene expression mapping in zebrafish. Proc Natl Acad Sci U S A. 2009;106:13365–70.
168. Poon KL, Liebling M, Kondrychyn I, Garcia-Lecea M, Korzh V. Zebrafish cardiac enhancer trap lines: new tools for in vivo studies of cardiovascular development and disease. Dev Dyn. 2010;239:914–26.
169. Levesque MP, Krauss J, Koehler C, Boden C, Harris MP. New tools for the identification of developmentally regulated enhancer regions in embryonic and adult zebrafish. Zebrafish. 2013;10:21–9.
170. Otsuna H, Hutcheson DA, Duncan RN, McPherson AD, Scoresby AN, Gaynes BF, et al. High-resolution analysis of central nervous system expression patterns in zebrafish Gal4 enhancer-trap lines. Dev Dyn. 2015;244:785–96.
171. Kondrychyn I, Teh C, Garcia-Lecea M, Guan Y, Kang A, Korzh V. Zebrafish enhancer TRAP transgenic line database ZETRAP 2.0. Zebrafish. 2011;8:181–2.
172. Stuart GW, McMurray JV, Westerfield M. Replication, integration and stable germ-line transmission of foreign sequences injected into early zebrafish embryos. Development. 1988;103:403–12.
173. Stuart GW, Vielkind JR, McMurray JV, Westerfield M. Stable lines of transgenic zebrafish exhibit reproducible patterns of transgene expression. Development. 1990;109:577–84.
174. Kwan KM, Fujimoto E, Grabher C, Mangum BD, Hardy ME, Campbell DS, et al. The Tol2kit: a multisite gateway-based construction kit for Tol2 transposon transgenesis constructs. Dev Dyn. 2007;236:3088–99.
175. Hartley JL, Temple GF, Brasch MA. DNA cloning using in vitro site-specific recombination. Genome Res. 2000;10:1788–95.
176. Halloran MC, Sato-Maeda M, Warren JT, Su F, Lele Z, Krone PH, et al. Laser-induced gene expression in specific cells of transgenic zebrafish. Development. 2000;127:1953–60.
177. Zu Y, Tong X, Wang Z, Liu D, Pan R, Li Z, et al. TALEN-mediated precise genome modification by homologous recombination in zebrafish. Nat Methods. 2013;10:329–31.
178. Shin J, Chen J, Solnica-Krezel L. Efficient homologous recombination-mediated genome engineering in zebrafish using TALE nucleases. Development. 2014;141:3807–18.
179. Chu VT, Weber T, Wefers B, Wurst W, Sander S, Rajewsky K, et al. Increasing the efficiency of homology-directed repair for CRISPR-Cas9-induced precise gene editing in mammalian cells. Nat Biotechnol. 2015;33:543–8.
180. Maruyama T, Dougan SK, Truttmann MC, Bilate AM, Ingram JR, Ploegh HL. Increasing the efficiency of precise genome editing with CRISPR-Cas9 by inhibition of nonhomologous end joining. Nat Biotechnol. 2015;33:538–42.

181. Auer TO, Duroure K, De Cian A, Concordet JP, Del Bene F. Highly efficient CRISPR/Cas9-mediated knock-in in zebrafish by homology-independent DNA repair. Genome Res. 2014;24:142–53.
182. Kimura Y, Hisano Y, Kawahara A, Higashijima S. Efficient generation of knock-in transgenic zebrafish carrying reporter/driver genes by CRISPR/Cas9-mediated genome engineering. Sci Rep. 2014;4:6545.
183. Hisano Y, Sakuma T, Nakade S, Ohga R, Ota S, Okamoto H, et al. Precise in-frame integration of exogenous DNA mediated by CRISPR/Cas9 system in zebrafish. Sci Rep. 2015;5:8841.
184. Li J, Zhang BB, Ren YG, Gu SY, Xiang YH, Du JL. Intron targeting-mediated and endogenous gene integrity-maintaining knockin in zebrafish using the CRISPR/Cas9 system. Cell Res. 2015;25:634–7.
185. Zada D, Tovin A, Lerer-Goldshtein T, Vatine GD, Appelbaum L. Altered behavioral performance and live imaging of circuit-specific neural deficiencies in a zebrafish model for psychomotor retardation. PLoS Genet. 2014;10:e1004615.
186. Gandhi AV, Mosser EA, Oikonomou G, Prober DA. Melatonin is required for the circadian regulation of sleep. Neuron. 2015;85:1193–9.
187. Elbaz I, Yelin-Bekerman L, Nicenboim J, Vatine G, Appelbaum L. Genetic ablation of hypocretin neurons alters behavioral state transitions in zebrafish. J Neurosci. 2012;32: 12961–72.
188. Sakowski SA, Lunn JS, Busta AS, Oh SS, Zamora-Berridi G, Palmer M, et al. Neuromuscular effects of G93A-SOD1 expression in zebrafish. Mol Neurodegener. 2012;7:44.
189. Ramesh T, Lyon AN, Pineda RH, Wang C, Janssen PM, Canan BD, et al. A genetic model of amyotrophic lateral sclerosis in zebrafish displays phenotypic hallmarks of motoneuron disease. Dis Model Mech. 2010;3:652–62.
190. Mosimann C, Puller AC, Lawson KL, Tschopp P, Amsterdam A, Zon LI. Site-directed zebrafish transgenesis into single landing sites with the phiC31 integrase system. Dev Dyn. 2013;242:949–63.
191. Perez-Pinera P, Kocak DD, Vockley CM, Adler AF, Kabadi AM, Polstein LR, et al. RNA-guided gene activation by CRISPR-Cas9-based transcription factors. Nat Methods. 2013;10:973–6.
192. Cheng AW, Wang H, Yang H, Shi L, Katz Y, Theunissen TW, et al. Multiplexed activation of endogenous genes by CRISPR-on, an RNA-guided transcriptional activator system. Cell Res. 2013;23:1163–71.
193. Qi LS, Larson MH, Gilbert LA, Doudna JA, Weissman JS, Arkin AP, et al. Repurposing CRISPR as an RNA-guided platform for sequence-specific control of gene expression. Cell. 2013;152:1173–83.
194. Mahmood F, Fu S, Cooke J, Wilson SW, Cooper JD, Russell C. A zebrafish model of CLN2 disease is deficient in tripeptidyl peptidase 1 and displays progressive neurodegeneration accompanied by a reduction in proliferation. Brain. 2013;136:1488–507.
195. Da Costa MM, Allen CE, Higginbottom A, Ramesh T, Shaw PJ, McDermott CJ. A new zebrafish model produced by TILLING of SOD1-related amyotrophic lateral sclerosis replicates key features of the disease and represents a tool for in vivo therapeutic screening. Dis Model Mech. 2014;7:73–81.
196. Singh C, Oikonomou G, Prober DA. Norepinephrine is required to promote wakefulness and for hypocretin-induced arousal in zebrafish. Elife. 2015;4:e07000.
197. Elbaz I, Lerer-Goldshtein T, Okamoto H, Appelbaum L. Reduced synaptic density and deficient locomotor response in neuronal activity-regulated pentraxin 2a mutant zebrafish. FASEB J. 2015;29:1220–34.
198. Yelin-Bekerman L, Elbaz I, Diber A, Dahary D, Gibbs-Bar L, Alon S, et al. Hypocretin neuron-specific transcriptome profiling identifies the sleep modulator Kcnh4a. Elife. 2015;4: e08638.
199. Godoy R, Noble S, Yoon K, Anisman H, Ekker M. Chemogenetic ablation of dopaminergic neurons leads to transient locomotor impairments in zebrafish larvae. J Neurochem. 2015;135:249–60.

200. Paquet D, Bhat R, Sydow A, Mandelkow EM, Berg S, Hellberg S, et al. A zebrafish model of tauopathy allows in vivo imaging of neuronal cell death and drug evaluation. J Clin Invest. 2009;119:1382–95.
201. Lee A, Mathuru AS, Teh C, Kibat C, Korzh V, Penney TB, et al. The habenula prevents helpless behavior in larval zebrafish. Curr Biol. 2010;20:2211–6.
202. Yokogawa T, Hannan MC, Burgess HA. The dorsal raphe modulates sensory responsiveness during arousal in zebrafish. J Neurosci. 2012;32:15205–15.
203. Bergeron SA, Carrier N, Li GH, Ahn S, Burgess HA. Gsx1 expression defines neurons required for prepulse inhibition. Mol Psychiatry. 2015;20:974–85.

Developing Zebrafish Depression-Related Models

Julian Pittman and Angelo Piato

Abstract Animal models of disease are ultimately only as strong as the clinical phenotype(s) upon which they are based. Many obstacles impede our ability to design animal models of complex mental illnesses, such as depression. An animal model that attempts to re-create any disease strives to maximize construct, face, and predictive validities. Strategies to model depression in representative animals have largely focused on one or more symptoms of depression, which have left many knowledge gaps open. In approaching these knowledge gaps, there are three primary areas that we feel need to be focused on: development of translational animal models, identification of genetic determinants, and discovery of novel targets/biomarkers of depression. Here, we discuss how zebrafish may be utilized in the modeling and analysis of the mechanisms of depression. Furthermore, this chapter also provides a detailed description of the behavioral responses and makes recommendations for further development of these methods, and how they may be employed in forward genetic screening for mutations involved in depression-related phenotypes.

Keywords Depression • Animal model development • Behavioral tests • Endophenotypes • Pharmacological analysis

1 Introduction

To translate basic science lessons learned from animal models of depression to clinical acumen, animal models of depression must be considered side-by-side with human presentation of symptoms of illness. Modeling human depression (see further) in animals poses unique challenges given contributions from higher-order functions such as

J. Pittman (✉)
Department of Biological and Environmental Sciences, Troy University, McCall Hall Troy, AL 36083, USA
e-mail: jtpittman@troy.edu

A. Piato
Department of Pharmacology, Universidade Federal do Rio Grande do Sul, Porto Alegre, Brazil
e-mail: angelopiato@ufrgs.br

A.V. Kalueff (ed.), *The rights and wrongs of zebrafish: Behavioral phenotyping of zebrafish*, DOI 10.1007/978-3-319-33774-6_2

emotions and cognitions to symptom presentations that are difficult, if not impossible to pinpoint and study in animals. The foundation of research into the mechanisms of depression must involve the development of novel behavioral paradigms, as they allow the quantification of functional changes in the brain induced by mutations or drugs, and will facilitate the discovery of underlying mechanisms and drug targets.

Depression is a common, serious and debilitating brain disorder [1]. Numerous studies have examined the biological mechanisms of depression, and a considerable amount of effort has been invested in the development of pharmacological treatments [2–9]. For preclinical research, most of these studies have used rodents. Since a large amount of data has been accumulated on rodent species, it may seem logical to think that building upon this well-laid foundation is the only way to proceed. The abandonment of rodent research is certainly not likely or recommended; however, utilization of another vertebrate, zebrafish, appears to be a fruitful direction to pursue namely because they are robust, small, reproduce quickly and possess evolutionarily conserved traits.

Zebrafish are showing promise as a model organism for experimental studies of affective disorders [5, 10, 11]. This species is demonstrating the potential to be an "exceptional" animal for investigating experimental, genetic, and pharmacological models of neurobehavioral disorders, such as depression [5, 8, 12–18]. As a result of the past three decades of intensive investigation with zebrafish, this species has become geneticists' favorite model organisms [16]. Zebrafish models strike an optimal balance between system complexity and practical simplicity, possessing brain anatomy, physiology, and genome very similar to those of other vertebrates including mammals [19–25]. Furthermore, they are small, easy and cheap to maintain in the laboratory, and are highly amenable to high-throughput screening (e.g., forward genetic or drug screens). The latter is particularly noteworthy for the purposes of unraveling the genetic, and in general the biological, mechanisms of complex brain functions and the disorders of these functions. High-throughput screens may have the ability to identify a significant proportion of the potentially large number of molecular players involved in these functions [17, 26].

2 Pathogenesis of Depression and Model development

Depression remains a common disorder that affects approximately 15 million Americans, despite the increasing knowledge on its pathophysiology and treatment [19]. One of the obstacles is the lack of validated diagnostic tests based on biological markers, which would allow us to predict treatment response in depressed patients. Also, biomarkers that correlate to treatment response to antidepressants or psychotherapy have not been identified so far. While imbalances in neurotransmitter levels are certainly involved in the pathophysiology of depression, no single neurotransmitter system is considered to be exclusively responsible. This is expected considering the range of symptoms included in the depressive syndrome: depressed mood, disinterest in usual activities, inability to feel pleasure, attention deficits, sleep disturbances, appetite alterations, and suicidal ideation. A novel conceptual approach is to consider depression as a systems-level 'spectrum' disorder that concerns several critical brain regions and connecting pathways. In order to enable the development of scientifically-based rationales for innovative treatments, a comprehensive understanding of the neurobiology of depression and its genetic and environmental underpinnings is required.

The etiology of depression is currently viewed as a result of gene-environment interactions that ultimately impact the three major monoamines—serotonin (5-hydroxytryptamine, 5HT), norepinephrine (NE), and dopamine (DA). Recently developed tools in molecular biology and brain imaging have provided further evidence for the involvement of these neurotransmitter systems. Contrary to earlier views [21], recent observations now support a preeminent role for central dopaminergic circuits [27], which could explain the now well-reported suboptimal response to selective serotonin reuptake inhibitors (SSRIs) and selective serotonin-norepinephrine reuptake inhibitors (SNRIs).

Animal models cannot replicate the symptoms of depression in a complete manner, since core symptoms of the disorder such as depressed mood, low self-esteem, or suicidality are not possible to access in non-humans [25, 26, 28]. On the other hand, there are depression endophenotypes that can be individually reproduced and evaluated in animals [29]. Ideally, an animal model should represent a means to understand the molecular, genetic, and epigenetic factors involved in the etiology of depression. Animal models also afford insight into the pathology of depression by allowing us to examine underlying molecular alterations and the causal relationship between genetic or environmental factors, which are indispensable to develop novel therapies with greater efficacy. The attempt to model a single symptom or endophenotype of a disorder, rather than to recapitulate its full phenotypic expression, is especially relevant for medical disorders of unclear pathophysiology or genetic etiology, such as depression. For behavioral measures to be used as novel models they should meet reliability, predictive, construct and face validity criteria as much as possible [23, 30].

3 Novel vs. Familiar, Open Field, Social Isolation Tests

Various methods have already been developed to induce and study depression-related behaviors. Novelty is classically recognized as an anxiety-inducing factor in several species, including humans. For instance, in the "open field test" rodents [31, 32] and other animals, including fish [22, 33], are exposed to an unfamiliar (thus potentially threatening) environment. The response to this novel environment is thought to be the resultant of two opposing and conflicting tendencies: exploration, an active response associated with the natural drive to explore unfamiliar places and objects, and anxiety, a passive response associated with harm-avoidance Both behaviors are considered adaptive, as exploratory activity may reveal food resources, mates and escape routes, while passive anxiety-induced responses (immobility/freezing) may reduce predation risk [32]. This interpretation may seem speculative, but quantitative genetic studies point towards ambidirectional selection forces as the basis for open field behavior. Thus, natural selection in rodents favored individuals that displayed intermediate behaviors (not too active but not too passive either) [32], an observation that extends to other vertebrates including fish [34]. This represents a particularly valuable application for measuring depressive behavior in zebrafish and for identifying new genetic lines.

The evolutionary past of zebrafish is likely similar to that of mice and rats considering that zebrafish has also been under ambidirectional selection with regard to behavioral responses induced by novelty. Therefore, when exposed to a novel environment, zebrafish are expected to display moderate levels of anxiety-like behavior. Importantly, behavioral experimentation generally includes animal handling by humans, which also induces some level of anxiety. Analysis of novelty-induced anxiety responses in zebrafish [35], demonstrate initially low levels of exploratory activity that progressively increase across time. A typical "diving" response is observed, i.e., increased amount of time spent on the bottom of the test tank, which slowly decreases as the fish habituates to the novel environment [35] (see [22] for similar findings). Nicotine was shown to have anxiolytic properties as this drug reduced fear responses induced by novelty [35].

Decreased serotonergic activity is associated with depression and may be experimentally induced by social isolation [36]. Specifically, rodents display hyperactive and aggressive behavior following long-term social isolation, and anti-depressant treatment is able to block these consequences [37]. Such isolation paradigms based on serotonin deficits are used as experimental depression models in rodents [25], and may be similarly employed with zebrafish.

4 Stress Models

Several protocols of unpredictable chronic stress (UCS) were reported to induce depression-like behavior in rodents [38–40]. These UCS models, however, are expensive, time-consuming, long lasting (at least 4 weeks), and require a large physical infrastructure, besides presenting problems of reproducibility among laboratories [39, 41].

Although other labs [42, 43] investigated some aspects of stress in zebrafish, ref. [44] was the first report to describe an experimental protocol to study the effects of UCS in zebrafish. Compared to the most often used rodent protocols, a number of advantages can be highlighted, such as low cost, ease of maintenance and manipulation without the need for complex physical structure. In addition, while UCS protocols are usually conducted over at least 4 weeks in rodents [39, 45], zebrafish stressed during 7 or 14 days already showed behavioral, physiological and cellular responses consistent with those observed in rodents and chronically stressed humans [44]. The stress protocol induced anxiety, cognitive impairment and neuroendocrine dysfunction, as measured by increased cortisol and CRF levels and decreased GR expression. These results suggest that this model has adequate construct validity.

Subsequently, Chakravarty et al. [46] exposed zebrafish to a similar stress model for 15 days. This protocol induced anxiety-like behavior and decreased neurogenesis. The molecular markers corticotropin-releasing factor, calcineurin and phosphocyclic AMP were altered. Moreover, using proteomics analyses, 18 proteins were found to be modified in stressed-zebrafish, four of them (PHB2, SLC25A5, VDAC3 and IDH2) related with mitochondrial viability.

Another study [47] used a milder UCS protocol to study the effects of daytime and nighttime stress on inhibitory avoidance learning, cortisol levels and gene expression in Tuebingen zebrafish strain. Fish submitted to UCS displayed weaker inhibitory avoidance learning compared to the control group. Regarding cortisol, while fish submitted to 7 nights of UCS had higher levels of cortisol, no difference was observed after 7 or 14 days of UCS. Important changes in bdnf, grα, grβ, grβ/grα ratio, and mr genes were also observed after the 7-night UCS protocol.

In [48], the effects of a modified UCS protocol on molecular and physiologic parameters related to stress response were assessed. Zebrafish submitted to UCS protocol showed increase in cortisol levels and pro-opiomelanocortin, glucocorticoid and mineralocorticoid receptors, prolactin, brain-derived neurotrophic factor, hypocretin/orexin, and c-fos expression.

A recent study [49] also evaluated the effects of UCS on purinergic system in zebrafish. UCS induced decrease in ecto-ADA (adenosine deaminase) and increases in adenosine levels in zebrafish brain, without affect any ADA gene (*ada1*, *ada2.1*, *ada2.2*, *adaL*, and *adaasi*) expression using quantitative reverse transcription. The authors suggested that this increase in adenosine levels could help zebrafish to achieve homeostasis during UCS. The UCS model in zebrafish remains to be more fully pharmacologically validated, since its predictive validity was not assessed thus far. Given the rich behavioral repertoire and the complex social interactions of individuals in a group, this model may contribute to a better understanding of the effects of drugs modulating the stress axis (Table 1, Fig. 1).

Table 1 Main results of depression-related models in zebrafish

	Model	Main results	References
Genetic	Mutant gr^{s357}	↑ HPA axis	[50]
		Blunted suppression of cortisol by dexamethasone	
		↓ Spontaneous activity	[51]
Stress	Chronic stress	↑ Time in the tank bottom	[44]
		↓ GR expression	
		Impaired memory	
		↑ Cortisol and CRF expression	
		↑ Time in the tank bottom	[46]
		↓ Neurogenesis	
		Mitochondrial toxicity	
		↑ Cortisol	[47]
		Altered BDNF, grα, grβ, grβ/grα ratio, and mr genes	
		Impaired memory	
		↑ Cortisol levels	[48]
		↑ POMC, GR, MR, prolactin, BDNF, hypocretin/orexin, and c-fos expression	
		↑ Adenosine	[49]
Drug	Reserpine	Impaired locomotion	[52]

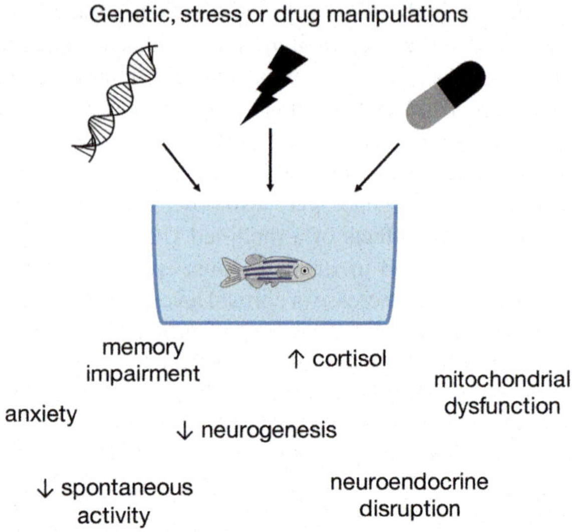

Fig. 1 Effects of different manipulations on behavioral, physiological and molecular parameters relevant to depression in zebrafish

5 Pharmacological Models for Depression-Like Responses

The motivation for the continued search for improved drugs to treat depression is not only to improve the quality of life of those suffering from it, but also to aid in our understanding of how depression develops, and what biological mechanisms may underlie this disorder cluster. Another reason is that the currently available, however numerous, drugs are often not efficacious or do not work for all patients. One way zebrafish may be beneficial for such research is by speeding up the discovery of the biological mechanisms responsible for the symptoms of depression. This may be achieved using, for example, forward genetic screens that identify mutations leading to the isolation of underlying genes. Another completely different approach has been to search for compounds, or "small molecules", which may alter expression-like symptoms. It is thus important to consider what is known about the psychopharmacological properties of zebrafish in the context of depression. For example, can one consistently detect the efficacy of "gold standard" drugs for depression using zebrafish? That is to say, does the zebrafish model have predictive validity? Predictive validity is an important question for the use of novel model organisms. The principal theme with regard to the translational relevance of laboratory model organisms concerns the notion "evolutionary homology", i.e., conservation of biological function across previously utilized species (e.g., rodents), the novel laboratory species (e.g., zebrafish), and humans.

Many different pharmacological approaches can be employed to model depression [53]. An example is the administration of psychostimulants, such as amphetamine, which leads to hyperactivity and may be reversed by the administration of anti-manic treatments, such as valproate. Additionally, repeated administration of psychostimulants induces a process of behavioral sensitization and may be used to model bipolar disorder [24]. Considering that repeated exposure to cocaine can lead to "cycling" in many neurochemical and physiological systems [54], bipolar-like behavior could be replicated in zebrafish, for instance, by combining cocaine with antipsychotic drugs. Another possibility is the induction of depressive-like behavior due to withdrawal of an anxiolytic agent, such as ethanol; this protocol requires chronic administration (minimum 3 weeks) of high doses of ethanol (1–3 %), and at least 7 days post-withdrawal before behavioral symptoms are manifest. The SSRI fluoxetine is able to reverse these depressive-like behaviors. In addition, quantitative changes in immunoreactive neurons are observed following this protocol of ethanol administration, mirroring many of the neurochemical findings of clinical depression [53].

There is a great number of studies reporting the effects of ethanol exposure across development in zebrafish. Findings comprise, for example, the strain-dependent effect of developmental alcohol exposure [55], the long-term effects of early embryonic ethanol exposure in adult animals [56], the development of adaptation (tolerance) and withdrawal symptoms following chronic ethanol exposure [34, 57, 58], and numerous alterations induced by acute ethanol administration [58]. Importantly, the behavioral effects of ethanol depend on concentration and administration regime, since lower doses of ethanol were shown to induce anxiolytic effects (see [58] and [22]), while prolonged exposure and withdrawal was associated with anxiogenic properties (see [57] and [22]). The behavioral effects induced by other drugs of abuse have also been documented for zebrafish. Cocaine, for example, has rewarding properties, and forward genetic screens have already been identified zebrafish mutants with altered cocaine reinforced place preference in [59]. Similarly to ethanol, also lead to anxiety/depression-related behaviors depending on drug concentration and administration schedules [60, 61].

Classical anti-anxiety drugs have been shown to exhibit an anxiolytic profile in zebrafish, such as fluromethylhistidine [62], benzodiazepines like diazepam, and the widely prescribed SSRI fluoxetine, that decreases bottom-dwelling, erratic movements, and whole-body cortisol levels [22], paralleling the responses observed in rodents [63]. On the other hand, acute administration of drugs known to induce anxiety in humans [64] and rodents [65], such as the benzodiazepine inverse agonist FG-7142 [61] and caffeine [22], led to increased anxiety responses in zebrafish, demonstrated by increased bottom-dwelling and erratic movements. Investigations of stress hormone levels in zebrafish have revealed numerous similarities when compared to the human stress response [5], strengthening the translational relevance of zebrafish as a model organism in depression research. The sight of a predator, for example, was shown to elevate cortisol levels in zebrafish [66]. It is important to note that cortisol is the primary stress hormone of the hypothalamic-pituitary-adrenal (HPA) axis in both human and zebrafish, but not in rodents, which use corticosterone instead. At the Society for Neuroscience meeting in San Diego (2010), Baier and his team demonstrated the generation of behavioral phenotypes resembling depression by disrupting the zebrafish stress response [67]. Another study [50] found a mutation in the glucocorticoid receptor gene in zebrafish that displayed depression-like behaviors, suggesting that depression could be connected to an individual's capability to cope with stress. Furthermore, the SSRI fluoxetine (Prozac) ameliorated depression-like behaviors is animals carrying the mutation. Molecules targeting the glucocorticoid receptor and enhancing its activity instead of blocking it may lead to promising novel therapies for the treatment of depression.

Also, depression-like motor retardation and social withdrawal have been reported in adult zebrafish several days after exposure to reserpine [3]—a dopamine-depleting drug known to elicit depression-like responses in rodents and clinical depression in humans. However, with the use of all the above pharmacological treatments, one must exercise extreme care and ensure there is some ability to provide a dissection between anxiety and depression endpoints, especially given a high degree of comorbidity of anxiety with depression clinically. This may be achieved through careful selection of pharmacological agents and behavioral tests (much development is needed in this area), and confirmation of quantitative changes in neuronal circuits involved in depression.

6 Model Limitations and Future Directions of Research

A significant difficulty with using zebrafish in depression research is the fact that only recently the behavioral repertoire of this species has begun to be explored. Although the number of behavioral studies published on zebrafish is on the rise compared to classical laboratory species such as rat, mouse, or even the fruit fly, zebrafish behavioral research is still in its infancy [28]. With the lack of reliable behavioral tests and a thorough understanding of zebrafish behavioral features, the behavioral and neurochemical consequences of gene mutation or drug exposure will remain exceedingly difficult to study.

Given the complex mechanisms involved in the pathophysiology of depression, one may assume the necessity of identifying a considerable number of molecular players, i.e., genes and their protein products and the biochemical interactions between the proteins. A possibility to tackle this complexity may be, at least initially, to employ large scale screenings for mutations and drugs. This may result in the identification of potential targets and leads that may be followed up on by more targeted hypothesis-driven analyses. We are not, however, advocating large screening as the only fruitful approach. A large number of mechanisms is awaiting to be revealed, and "blind", i.e., unbiased, screening applications may facilitate their discovery. This is where zebrafish poses a major advantage over the classical laboratory organisms.

The development of novel behavioral endpoints and observational methodologies, such as automated video-tracking systems, is important to reinforce the utility of zebrafish as a model organism for depression research. The use of biomolecular markers, such as gene and protein expression, to parallel zebrafish physiology with behavioral data represents another critical research direction to pursue.

References

1. Best JD, Alderton WK. Zebrafish: an in vivo model for the study of neurological diseases. Neuropsychiatr Dis Treat. 2008;4(3):567–76.
2. Flint JSS. Animal models of psychiatric disease. Curr Opin Genet Dev. 2008;18:235–40.
3. Kyzar ER, Roth A, Gaikwad S, Green J, Collins C, El-Ounsi M, Davis A, Pham M, Stewart AM, Cachat J, Zukowska Z, Kalueff AV. On making zebrafish sad and anxious: developing novel aquatic models of affective disorders. IBNS Abstract, 2012.
4. Mathur P, Guo S. Use of zebrafish as a model to understand mechanisms of addiction and complex neurobehaviroal phenotypes. Neurobiol Dis. 2010;40:66–72.
5. Alsop D, Vijayan MM. Development of the corticosteroid stress axis and receptor expression in zebrafish. Am J Physiol Regul Integr Comp Physiol. 2008;294:711–9.
6. Alsop D, Vijayan M. The zebrafish stress axis: molecular fallout from the teleost specific genome duplication event. Gen Comp Endocrinol. 2008;161:62–6.
7. Blaser R, Gerlai R. Behavioral phenotyping in zebrafish: comparison of three behavioral quantification methods. Behav Res Methods. 2006;38:456–69.
8. Dooley K, Zon LI. Zebrafish: a model system for the study of human disease. Curr Opin Genet Dev. 2000;10:252–6.
9. Norton W, Bally-Cuif L. Adult zebrafish as a model organism for behavioral genetics. BMC Neurosci. 2010;11:90.
10. Sprague J, Doerry E, Douglas S, Westerfield M. The zebrafish information network (ZFIN): a resource for genetic, genomic and developmental research. Nucleic Acids Res. 2001;29: 87–90.
11. Zon L, Peterson R. In vivo drug discovery in the zebrafish. Nat Rev Drug Discov. 2005;4:35–44.
12. Wullimann MF, Knipp S. Proliferation pattern changes in the zebrafish brain from embryonic through early postembryonic stages. Anat Embryol. 2000;202:385–400.
13. Ward A, Lieschke G. The zebrafish as a model system for human disease. Front Biosci. 2002;7:d827–33.
14. Shin J, Fishman M. From zebrafish to human: modular medical models. Annu Rev Genomics Hum Genet. 2002;3:311–40.
15. Moorman S. Development of sensory systems in zebrafish (*Danio rerio*). ILAR J. 2001;42: 292–8.
16. McGrath P, Li CQ. Zebrafish: a predictive model for assessing drug-induced toxicity. Drug Discov Today. 2008;13:394–401.
17. Lele Z, Krone PH. The zebrafish as a model system in developmental, toxicological and transgenic research. Biotechnol Adv. 1996;14:57–72.
18. Blackburn J, Liu S, Raimondi A, Ignatius M, Salthouse C, Langenau D. High-throughput imaging of adult fluorescent zebrafish with an LED fluorescence macroscope. Nat Protoc. 2011;6:229–41.
19. Kessler R, Chiu W, Demler O, Walters E. Prevalence, severity, and comorbidity of twelve-month DSM-IV disorders in the National Comorbidity Survey Replication (NCS-R). Arch Gen Psychiatry. 2005;62(6):617–27.
20. Willner P. The validity of animal models of depression. Psychopharmacology (Berl). 1984;83(1):1–16.
21. Nemeroff C. The neurobiology of depression. Sci Am. 1998;278:42–9.
22. Egan R, Bergner C, Hart P, Cachat J, Canavello P, Elegante M, Elkhayat S, Bartels B, Tien A, Tien D, Mohnot S, Beeson E, Glasgow E, Amri H, Zukowska Z, Kalueff A. Understanding behavioral and physiological phenotypes of stress and anxiety in zebrafish. Behav Brain Res. 2009;205:38–44.
23. Geyer M, Markou A. The role of preclinical models in the development of psychotropic drugs. In: Davis K, Charney D, Coyle J, Nemeroff C, editors. Neuropsychopharmacology: the fifth generation of progress. Philadelphia: Lippincott Williams & Wilkins; 2002. p. 446–55.

24. Kato T, Kubota M, Kasahara T. Animal models of bipolar disorder. Neurosci Biobehav Rev. 2007;31:832–42.
25. Leonard B. Animal models of depression. In: Briley M, Montgomery S, editors. Antidepressant therapy. London: Martin Dunitz Ltd; 1998. p. 87–109.
26. Willner P. Animal models of depression: an overview. Pharmacol Ther. 1990;45:425–55.
27. Dunlop B, Nemeroff C. The role of dopamine in the pathophysiology of depression. Arch Gen Psychiatry. 2007;64:327–37.
28. Weiss J, Kilts C. Animal models of depression and schizophrenia. Washington, DC: American Psychiatric Press; 1995. p. 89–131.
29. Hasler G. Discovering endophenotypes for major depression. Neuropsychopharmacology. 2004;29:1765–81.
30. Geyer M, Markou A. Animal models of psychiatric disorders. In: Bloom F, Kupfer D, editors. Psychopharmacology: the fourth generation of progress. New York: Raven; 1995. p. 787–98.
31. Prut L, Belzung C. The open field as a paradigm to measure the effects of drugs on anxiety-like behaviors: a review. Eur J Pharmacol. 2003;463:3–33.
32. Crusio E, van Abeelen JH. The genetic architecture of behavioral responses to novelty in mice. Heredity. 1986;56:55–63.
33. Csányi V, Gerlai R. Open-field behavior and the behavior-genetic analysis of the paradise fish (*Macropodus opercularis*). J Comp Psychol. 1988;102:326–36.
34. Gerlai R, Csányi V. Genotype environment interaction and the correlation structure of behavioral elements in paradise fish (*Macropodus opercularis*). Physiol Behav. 1990;47:343–56.
35. Levin E, Bencan Z, Cerutti D. Anxiolytic effects of nicotine in zebrafish. Physiol Behav. 2007;90:54–8.
36. Garattini S, Giacalone E, Valzelli L. Isolation, aggressiveness and brain 5-hydroxytriptamine turnover. J Pharm Pharmacol. 1967;19:338–9.
37. Garzon J, del Rio J. Hypersensitivity induced in rats by long term isolation: further studies on a new animal model for the detection of antidepressants. Eur J Pharmacol. 1981;74:287–94.
38. Mineur YS, Belzung C, Crusio WE. Effects of unpredictable chronic mild stress on anxiety and depression-like behavior in mice. Behav Brain Res. 2006;175(1):43–50.
39. Willner P. Chronic mild stress (CMS) revisited: consistency and behavioural-neurobiological concordance in the effects of CMS. Neuropsychobiology. 2005;52(2):90–110.
40. Yalcin I, Belzung C, Surget A. Mouse strain differences in the unpredictable chronic mild stress: a four-antidepressant survey. Behav Brain Res. 2008;193(1):140–3.
41. Willner P. Validity, reliability and utility of the chronic mild stress model of depression: a 10-year review and evaluation. Psychopharmacology (Berl). 1997;134(4):319–29.
42. Cachat J, et al. Measuring behavioral and endocrine responses to novelty stress in adult zebrafish. Nat Protoc. 2010;5(11):1786–99.
43. Champagne DL, et al. Translating rodent behavioral repertoire to zebrafish (*Danio rerio*): relevance for stress research. Behav Brain Res. 2010;214(2):332–42.
44. Piato AL, et al. Unpredictable chronic stress model in zebrafish (*Danio rerio*): behavioral and physiological responses. Prog Neuropsychopharmacol Biol Psychiatry. 2011;35(2):561–7.
45. Surget A, et al. Antidepressants recruit new neurons to improve stress response regulation. Mol Psychiatry. 2011;16(12):1177–88.
46. Chakravarty S, et al. Chronic unpredictable stress (CUS)-induced anxiety and related mood disorders in a zebrafish model: altered brain proteome profile implicates mitochondrial dysfunction. PLoS One. 2013;8(5):e63302.
47. Manuel R, et al. Unpredictable chronic stress decreases inhibitory avoidance learning in Tuebingen long-fin zebrafish: stronger effects in the resting phase than in the active phase. J Exp Biol. 2014;217(Pt 21):3919–28.
48. Pavlidis M, Theodoridi A, Tsalafouta A. Neuroendocrine regulation of the stress response in adult zebrafish, *Danio rerio*. Prog Neuropsychopharmacol Biol Psychiatry. 2015;60:121–31.
49. Zimmermann FF, et al. Unpredictable chronic stress alters adenosine metabolism in zebrafish brain. Mol Neurobiol. 2015;53:2518–28.

50. Ziv L, et al. An affective disorder in zebrafish with mutation of the glucocorticoid receptor. Mol Psychiatry. 2013;18(6):681–91.
51. Griffiths BB, et al. A zebrafish model of glucocorticoid resistance shows serotonergic modulation of the stress response. Front Behav Neurosci. 2012;6:68.
52. Kyzar E, et al. Behavioral effects of bidirectional modulators of brain monoamines reserpine and d-amphetamine in zebrafish. Brain Res. 2013;1527:108–16.
53. Pittman JT, Ichikawa KM. iPhone(R) applications as versatile video tracking tools to analyze behavior in zebrafish (*Danio rerio*). Pharmacol Biochem Behav. 2013;106:137–42.
54. Antelman S, Caggiula A, Kucinski B, Fowler H, Gerhon S, Edwards D. The effects of lithium on a potential cycling model of bipolar disorder. Prog Neuropsychopharmacol Biol Psychiatry. 1998;22:495–510.
55. Loucks E, Carvan 3rd MJ. Strain-dependent effects of developmental ethanol exposure in zebrafish. Neurotoxicol Teratol. 2004;26:745–55.
56. Fernandes Y, Gerlai R. Long-term behavioral changes in response to early developmental exposure to ethanol in zebrafish. Clin Exp Res. 2009;33:601–9.
57. Gerlai R, Chatterjee D, Pereira T, Sawashima T, Krishnannair R. Acute and chronic alcohol dose: population differences in behavior and neurochemistry of zebrafish. Genes Brain Behav. 2009;8:586–99.
58. Gerlai R, Lahav M, Guo S, Rosenthal A. Drinks like a fish: Zebra fish (*Danio rerio*) as a behavior genetic model to study alcohol effects. Pharmacol Biochem Behav. 2000;67:773–82.
59. Darland T, Dowling J. Behavioral screening for cocaine sensitivity in mutagenized zebrafish. Proc Natl Acad Sci U S A. 2001;98:11691–6.
60. Ninkovic J, Bally-Cuif L. The zebrafish as a model system for assessing the reinforcing properties of drugs of abuse. Methods. 2006;39:262–74.
61. Lopez-Patino M, Yu L, Cabral H, Zhdanova I. Anxiogenic effects of cocaine withdrawal in zebrafish. Physiol Behav. 2008;93:160–71.
62. Peitsaro N, Kaslin J, Anichtchik O, Panula P. Modulation of the histaminergic system and behaviour by alpha-fluoromethylhistidine in zebrafish. J Neurochem. 2003;86:432–41.
63. Dulawa S, Holick K, Gundersen B, Hen R. Effects of chronic fluoxetine in animal models of anxiety and depression. Neuropsychopharmacology. 2004;29:1321–30.
64. Childs E, Hohoff C, Deckert J, Xu K, Badner J, de Wit H. Association between ADORA2A and DRD2 polymorphisms and caffeine-induced anxiety. Neuropsychopharmacology. 2008;33:2791–800.
65. El Yacoubi M, Ledent C, Parmentier M, Costentin J, Vaugeois J. The anxiogenic-like effect of caffeine in two experimental procedures measuring anxiety in the mouse is not shared by selective A(2A) adenosine receptor antagonists. Psychopharmacology (Berl). 2000;148:153–63.
66. Barcellos G, Ritter F, Kreutz C, Quevedo M, Bolognesi da Silva L, Bedin C. Whole-body cortisol increases after direct and visual contact with a predator in zebrafish, *Danio rerio*. Aquaculture. 2007;272:774–8.
67. Baier H. Depression-like behavior in zebrafish mutants with disruption of the glucocorticoid receptor. Society for Neuroscience Annual Meeting, 2010. Abstract 884.1.

Zebrafish Models of Anxiety-Like Behaviors

Adam D. Collier, Allan V. Kalueff, and David J. Echevarria

Abstract Anxiety disorders are widespread psychiatric illnesses affecting approximately 7–10 % of the global population. Zebrafish are a particularly useful animal model for studying anxiety-related phenotypes. They are increasingly utilized for studying neurobiological, physiological and genetic mechanisms of anxiety, as well as for screening various anxiolytic drugs. Summarized here, accumulating evidence supports the utility of zebrafish neurobehavioral phenotyping in studying anxiety and stress neurobiology. For example, zebrafish are highly sensitive to various anxiety-evoking environmental stressors, including novelty, predator exposure, alarm pheromone, anxiogenic drugs, and drug withdrawal. Zebrafish also show high sensitivity to anxiolytic manipulations. Zebrafish anxiety-related neuroendocrine responses are also robust, sensitive, and correlate strongly (and bi-directionally) with behavioral endpoints. Finally, zebrafish are also amenable to genetic manipulations, and differences in baseline and experimentally-evoked anxiety levels can be observed in different strains of zebrafish. Collectively, this supports the validity and efficiency of both larval and adult zebrafish model for studying acute and chronic anxiety-like states.

Keywords Zebrafish • Anxiety • Stress • Endocrine response • Cortisol • Novelty • Novel tank • Open field • Light dark • Predator stress • Alarm pheromone • Strain differences • Behavioral phenotyping

---

A.D. Collier • D.J. Echevarria
Department of Psychology, University of Southern Mississippi, Hattiesburg, MS, USA

The International Zebrafish Neuroscience Research Consortium (ZNRC), Slidell, LA, USA

A.V. Kalueff (✉)
The International Zebrafish Neurosceince Research Consortium (ZNRC) and ZENEREI Research Center, Slidell, LA, USA

Research Institute of Marine Drugs and Nutrition, College of Food Science and Technology, Guangdong Ocean University, Zhanjiang, Guangdong, China

Institute of Translational Biomedicine, St. Petersburg State University, St. Petersburg, Russia

Institutes of Chemical Technology and Natural Sciences, Ural Federal University, Ekaterinburg, Russia
e-mail: avkalueff@gmail.com

A.V. Kalueff (ed.), *The rights and wrongs of zebrafish: Behavioral phenotyping of zebrafish*, DOI 10.1007/978-3-319-33774-6_3

1 Introduction

Anxiety disorders are common psychiatric illnesses that involve multifaceted interactions between behavior, neural circuitry, physiology, genetics and experience [1]. Anxiety disorders are particularly widespread, affecting approximately 7–10 % of the general population [2]. Clinical manifestations of anxiety-related disorders are summarized in Table 1 [3].

An important strategy to elucidate neural underpinnings and develop novel treatments is to study animal models which have highly conserved neural circuitry related to anxiety [4]. A variety of behavioral paradigms, pharmacological screens, and genetic manipulations in animal models especially rodents, have long been employed to study anxiety disorder pathogenesis [5–7]. However, experimental rodent models are often low-throughput, costly and time-consuming [8]. The zebrafish (*Danio rerio*) has emerged as a new advantageous in-vivo preclinical model organism used in biomedical and translational neuroscience research to study the behavioral and molecular mechanisms underlying brain disorders, including anxiety. As will be shown in this and other chapters of this book, zebrafish display complex and well-defined behavioral phenotypes [9] (also see chapter "Illustrated Zebrafish Neurobehavioral Glossary" in this Book), and are amenable to high-throughput screening due to low-cost and small size [10, 11]. Video tracking technologies are also readily available to be coupled with zebrafish behavioral assays, providing data-rich endpoints (e.g., velocity, distance traveled) and 'big data'-level analyses (e.g., three-dimensional spatial and spatiotemporal swim path reconstructions, behavioral barcoding approaches) that are impossible to generate manually [12–14]. Zebrafish also possess all major neurotransmitter systems, transporters, receptors and hormones [15–17], and have a fully sequenced genome with ~70–75 % of human genes having at least one zebrafish orthologue [18].

Zebrafish are rapidly becoming a promising model organism for anxiety and stress research, especially due to a robust and easily quantifiable cortisol stress response [15, 19], clear-cut drug-evoked phenotypes with high predictive validity [20, 21], and fish sensitivity to a wide range of experimental stressors. For example, like with rodents, stressors that can trigger zebrafish anxiety-like behaviors include novelty exposure [21], social isolation and confinement [8], predator exposure [22] and alarm substance [23]. Furthermore, a number of genetic strains that show behavioral differences are now available [24], with multiple cutting-edge genome editing tools that can be applied to zebrafish [25–27]. Here, we introduce several behavioral paradigms and outline aspects of zebrafish phenotyping related to anxiety-like states. Focusing mainly on adult zebrafish models with established neural and physiological systems, this chapter also briefly mentions conceptually similar approaches to model anxiety-like states in larval zebrafish.

Table 1 Summary of clinical symptoms of anxiety-related disorders and diagnostic criteria from DSM, Diagnostic and Statistical Manual of Mental Disorders, 5th edition [3]

Anxiety disorders	Potentially relevant zebrafish phenotypes
Generalized anxiety disorder (GAD)	
Excessive anxiety/worry about events and activities	Anxiety-like behaviors
Difficulty controlling worry	–
Restlessness	Hyper-arrousal, erratic movements
Fatigue	Fatigue
Difficulty concentrating	Poor performance in cognitive tests
Irritability	Increased social aggression
Muscle tension	–
Sleep disturbance	Sleep deficits
The anxiety, worry, or physical symptoms causes clinically significant distress and impairs general functioning (e.g., social, occupational)	Anxiety behaviors in various contexts, aversive conditioning
Specific phobia	
Marked and persistent fear or anxiety about a specific object or situation	Anxiety behaviors in various contexts
Avoidance of phobic object or situation	Neophobia
Fear or anxiety out of proportion to actual danger posed by specific object or situation	Anxiety behaviors in various contexts, aversive conditioning
Fear and anxiety of specific object or situation causes clinically significant distress and impairs general functioning (e.g., social, occupational)	Anxiety behaviors in various contexts, aversive conditioning
Social anxiety disorder	
Marked and persistent fear or anxiety about social situations	Social deficits
Fear of showing anxiety symptoms that will be negatively evaluated	Increased anxiety in social interactions or contexts/tests
Avoidance of social situations or endured with intense fear and anxiety	Anxiety behaviors in various contexts, aversive conditioning, social avoidance
Fear and anxiety out of proportion to actual danger posed by social situation	Anxiety behaviors in various contexts, aversive conditioning, social avoidance
Fear and anxiety of social situation causes clinically significant distress and impairs general functioning (e.g., social, occupational)	Anxiety behaviors in various contexts, aversive conditioning, social avoidance
Panic disorder	
Recurrent unexpected panic attacks	Increased anxiety
Accelerated heart rate	Accelerated heart rate

(continued)

Table 1 (continued)

Anxiety disorders	Potentially relevant zebrafish phenotypes
Sweating	–
Trembling or shaking	Trembling or shaking
Shortness of breath	Shortness of breath
Feelings of choking	–
Chest pain or discomfort	–
Nausea or abdominal distress	Nausea or abdominal distress
Feeling dizzy/faint	–
Chills or heat sensations	–
Paresthesias	–
Derealization/Depersonalization	–
Fear of losing control	–
Fear of dying	–
Agoraphobia	
Marked and persistent fear and anxiety about agoraphobic situations (e.g., public transportation, being in open spaces, being in enclosed places, being in a crowd, being outside of the home alone)	Increased thigmotaxis and aversive conditioning
Avoidance of agoraphobic situations	–
Fear and anxiety of agoraphobic situation is out of proportion to actual danger	–
Agoraphobic situation causes clinically significant distress and impairs general functioning (e.g., social, occupational)	–
Substance/medication-induced anxiety disorder	
Panic attacks or anxiety	Increased anxiety behaviors
Evidence that panic attacks or anxiety developed during or soon after substance intoxication or withdrawal	Pharmacogenic or withdrawal-evoked anxiety
Evidence that the substance is capable of producing panic attacks or anxiety	Pharmacogenic or withdrawal-evoked anxiety
The disturbance is not better explained by another anxiety disorder	–
The disturbance causes clinically significant distress and impairs general functioning (e.g., social, occupational)	

2 Novelty-Based Behavioral Paradigms: Open Field, Light Dark, and Novel Tank Tests

Traditionally, animal models of anxiety are often based on behavioral responses to novel environments [28, 29]. In many taxa, exposure to a novel (and, therefore, potentially dangerous) environment often triggers the expression of avoidance-related behaviors in animals that likely serve evolutionarily conserved 'anti-predatory' functions [30, 31]. Novelty exploration is believed to underlie behavioral organization in a new environment and reflect the emotional state of animals [32–34]. Typical 'spatial' behaviors include total distance traveled, average velocity, and spatial distribution of exploratory activity. Initial exploratory behaviors tend to attenuate over the testing session as animals habituate to novel environments, a form of non-associative learning and an important cognitive phenotype, the impairment of which may be associated with increased anxiety [35, 36]. Like in rodents, zebrafish novelty-based paradigms and associated phenotypes are highly sensitive to exposure to acute and chronic stressors and pharmacolological manipulations (Tables 2, 3, 4, and 5), and can therefore can be used to screen drug effects [51, 58]. Accordingly, a number of novelty-based paradigms traditionally developed and used for rodents, have been applied to zebrafish neurophenotyping.

One of the most popular animal behavioral paradigms is the open-field test, which since its invention in 1932 is most commonly used in rodents to evaluate their novelty-evoked anxiety-like behaviors [29, 59–61]. Recently, this paradigm has been adapted to neurobehavioral phenotyping of both larval [62] and adult zebrafish [34], and its typical exploratory-based behavioral endpoints include the time in/entries to the center, time in/entries to the periphery (i.e., thigmotaxis), distance traveled and average velocity (Table 2). The center and periphery of the open field apparatus may be defined differently across laboratories. For example, one group could visually divide the open field into 16 equally sized squares and define the center as the middle 4 squares and the periphery as the remaining outer squares [36]. In other studies, center can be defined arbitrarily as area within 5 cm from the walls of the apparatus [37]. There are no specific standards regarding how to best define

Table 2 Adult zebrafish anxiety-related behavioral phenotypes: the open field test

Phenotype	Treatment + effect	↑ Value indicates	References
Entries to center	Acute LSD ↓	↑ entries to center indicates ↓ anxiety	[36, 37]
Time in center	Acute LSD Ø α-fluoro-methylhistidine ↑	↑ time in center indicates ↓ anxiety	[37]
Entries to periphery	Acute LSD Ø	↑ entries to periphery indicates ↑ anxiety	[37, 38]
Time in periphery	Acute LSD Ø	↑ time in periphery indicates ↑ anxiety	[37]
Distance traveled in periphery	Cocaine withdrawal ↑ FG-7142 ↑	↑ distance traveled in periphery indicates ↑ anxiety	[39]
Distance traveled (total)	Acute LSD Ø Acute ibogaine Ø FG-7142 ↑	↑ total distance traveled indicates hyperactivity	[13, 37, 40]
Average velocity	Acute LSD Ø Acute ibogaine Ø	↑ average velocity indicates motor aspects of zebrafish swimming	[13, 37]

↑ increased/activated, ↓ reduced/inhibited/impaired, Ø no effect, *LSD* lysergic acid diethylamide, *FG-7142* a benzodiazepine antagonist

Table 3 Adult zebrafish anxiety-related behavioral phenotypes: the light dark test

Phenotype	Treatment + effect	↑ Value indicates	References
Latency to dark side	CUS ↓	↑ latency to enter the dark side indicates ↓ anxiety	[36, 37, 41, 42]
	Acute LSD Ø		
	Acute ibogaine ↑		
	Acute ketamine ↑		
Time in dark side	CUS ↑	↑ time in the dark side indicates ↑ anxiety	[36, 37, 41–44]
	Acute LSD ↓		
	Restraint stress ↑		
	Acute ibogaine ↓		
	Acute caffeine ↑		
	Acute ZM241385 Ø		
	Acute DPCPX ↑		
	Acute fluoxetine Ø		
	Chronic fluoxetine ↓		
	Acute CDP ↓		
	Acute clonazepam ↓		
	Acute diazepam ↓		
	Acute buspirone ↓		
	Acute moclobemide Ø		
	Acute ethanol ↓		
	Acute ketamine ↓		
Entries to dark side	Acute LSD Ø	↑ entries to the dark side indicates ↑ anxiety	[13, 36, 37]
	Restraint stress Ø		
	Acute ibogaine ↓		
Average dark side entry duration	Acute LSD Ø	↑ dark side entry duration indicates ↓ anxiety	[13, 37, 42]
	Acute ibogaine Ø		
	Acute ketamine Ø		
Midline crossings	Chronic fluoxetine Ø	↑ midline crossings indicates ↑ swimming activity	[42, 44]
	Acute CDP Ø		
	Acute clonazepam ↓		
	Acute diazepam Ø		
	Acute buspirone ↓		
	Acute moclobemide Ø		
	Acute ethanol ↑		
	Acute caffeine Ø		
	Acute ketamine ↑		

↑ increased/activated, ↓ reduced/inhibited/impaired, Ø no effect, *CUS* chronic unpredictable stress, *LSD* lysergic acid diethylamide, *CDP* chlordiazepoxide, *ZM241385* an adenosine A_2 antagonist, *DPCPX* an adenosine A_1 antagonist

Table 4 Adult zebrafish anxiety-related behavioral phenotypes: the novel tank test

Phenotype	Treatment + effect	↑ Value indicates	References
Latency to upper half	Chronic fluoxetine ↓	↑anxiety	[24, 41, 45–50]
	Acute alarm pheromone ↑		
	Acute MDMA ↓		
	Acute caffeine Ø		
	Acute ethanol ↓		
	Chronic ethanol Ø		
	Leopard strain ↑		
	Wild-derived Indian strain ↑		
	CUS ↑		
	Acute nicotine ↓		
	Chronic nicotine ↑		
	Acute PCP ↓		
	Acute ketamine ↓		
	Acute ibogaine ↓		
	Acute noribogaine ↓[a]		
	Chronic CDP Ø		
	CDP withdrawal ↑		
	Vmat2 knockdown ↑		
Entries to upper half	Chronic fluoxetine ↑	↓ anxiety	[24, 41, 45–50]
	Acute alarm pheromone ↓		
	Acute MDMA ↓		
	Acute caffeine ↓		
	Acute ethanol ↑		
	Chronic ethanol ↑		
	CUS ↓		
	Wild-derived Indian strain ↓		
	Acute nicotine Ø		
	Chronic nicotine ↓		
	Acute PCP Ø		
	Acute ketamine ↑		
	Acute ibogaine Ø (↑ in first 2 min)		
	Acute MK-801 Ø		
	Chronic CDP Ø		
	CDP withdrawal ↓		
	Vmat2 knockdown ↓		

(continued)

Table 4 (continued)

Phenotype	Treatment + effect	↑ Value indicates	References
Time in upper half	Chronic fluoxetine ↑	↓ anxiety	[24, 41, 45–52]
	Acute alarm pheromone ↓		
	Acute MDMA ↑		
	Acute caffeine ↓		
	Acute ethanol ↑		
	Chronic ethanol ↑		
	Leopard strain ↓		
	Wild-derived Indian strain ↓		
	CUS ↓		
	Acute nicotine ↑		
	Chronic nicotine ↓		
	Acute PCP Ø		
	Acute ketamine ↓		
	Acute ibogaine Ø		
	Acute noribogaine ↑[a]		
	MK-801 ↑		
	Acute buspirone ↑		
	Acute CDP Ø		
	Chronic CDP Ø		
	CDP withdrawal ↓		
	Acute diazepam ↑		
	Vmat2 knockdown ↓		
Erratic movements	Chronic fluoxetine ↓	↑ anxiety	[24, 47]
	Acute alarm pheromone ↑		
	Acute MDMA ↓		
	Acute caffeine Ø		
	Acute ethanol Ø		
	Acute ketamine Ø		
	Acute ibogaine ↑		
	Acute MK-801 Ø		
Freezing bouts	Acute alarm pheromone ↑	↑ anxiety	[24, 45–50, 53]
	Acute MDMA Ø		
	Chronic nicotine Ø		
	Acute PCP ↑		
	Wild-derived Indian strain ↑		
	Acute ketamine Ø		
	Acute ibogaine Ø or ↑		
	Acute MK-801 Ø		
	Chronic CDP Ø		
	CDP withdrawal Ø		
	Vmat2 knockdown Ø		

(continued)

Table 4 (continued)

Phenotype	Treatment + effect	↑ Value indicates	References
Distance traveled (total)	Chronic fluoxetine Ø	↑ hyperactivity or increased exploration	[24, 49]
	Chronic ethanol Ø		
	Acute PCP Ø		
	Acute ketamine Ø		
	Acute ibogaine Ø or ↑		
	Acute MK-801 ↑		
Average velocity	Chronic fluoxetine Ø	Various motor aspects of zebrafish swimming	[24, 49, 51, 52]
	Chronic ethanol Ø		
	Acute PCP Ø		
	Acute ketamine Ø		
	Acute ibogaine Ø		
	MK-801 ↑		
	Acute nicotine ↑		
	Acute buspirone Ø		
	Acute CDP ↓		
	Acute diazepam Ø		
	Vmat2 knockdown ↑		

↑ increased/activated, ↓ reduced/inhibited/impaired, Ø no effect, *MDMA* 3,4-methylenedioxymethamphetamine, *CUS* chronic unpredictable stress, *PCP* phencyclidine, *CDP* chlordiazepoxide, *LSD* lysergic acid diethylamide, *MK-801* dizoclipine, *Vmat2* vesicular monoamine transporter 2
[a]Unpublished data (Maillet, Kalueff, 2015, DemeRx LLC)

Table 5 Larval zebrafish anxiety-related behavioral phenotypes

Phenotype	Treatments + effects	↑ Value indicates	References
Open field test			
Time in periphery	Acute ethanol ↑	↑ time in periphery indicates ↑ anxiety	[54]
	Acute diazepam ↓		
	Acute caffeine ↑		
Distance traveled in periphery	Acute diazepam ↓	↑ distance moved in periphery indicates ↑ anxiety	[55]
	Acute caffeine ↑		
Distance traveled (total)	Acute diazepam Ø	↑ total distance traveled indicates hyperactivity	[56]
	Acute caffeine Ø		
Light dark test			
Latency to dark side	Diazepam ↓	↑ latency to enter the dark side indicates ↑ anxiety	[57]
	Buspirone ↓		
	Ethanol ↓		
	Caffeine ↑		
Time in dark side	Diazepam ↑	↑ time in the dark side indicates ↓ anxiety	[57]
	Buspirone ↑		
	Ethanol ↑		
	Caffeine ↓		
Entries to dark side	Diazepam Ø	↑ entries to the dark side indicates ↓ anxiety	[57]
	Buspirone ↑		
	Ethanol ↑		
	Caffeine ↓		

↑ increased/activated, ↓ reduced/inhibited/impaired, Ø no effect the center in the open field test. Thus, well-defined zones in this test must be consistent and standardized within the laboratory, to ensure valid behavioral phenotypic data. Also importantly, both rodents and zebrafish initially exhibit thigmotaxic anxiety-like behaviors during open field testing, which decrease over time, indicative of intra-session habituation to novelty [63, 64].

Furthermore, although the open field studies are similar in that they each evaluate exploratory behavior when placed into a novel and open environment, differences often exist across laboratories in testing duration, pretest housing conditions, and the size, shape, color and texture of the apparatus (Fig. 1). Zebrafish increase locomotor behavior in a larger open field arena compared to a smaller arena, but, interestingly, the overall temporal activity patterns for their exploratory behaviors remain stable across different arena sizes [34]. Similarly, rodents display differential locomotor behavior depending on the size of the arena, and exhibit a temporal stability in activity

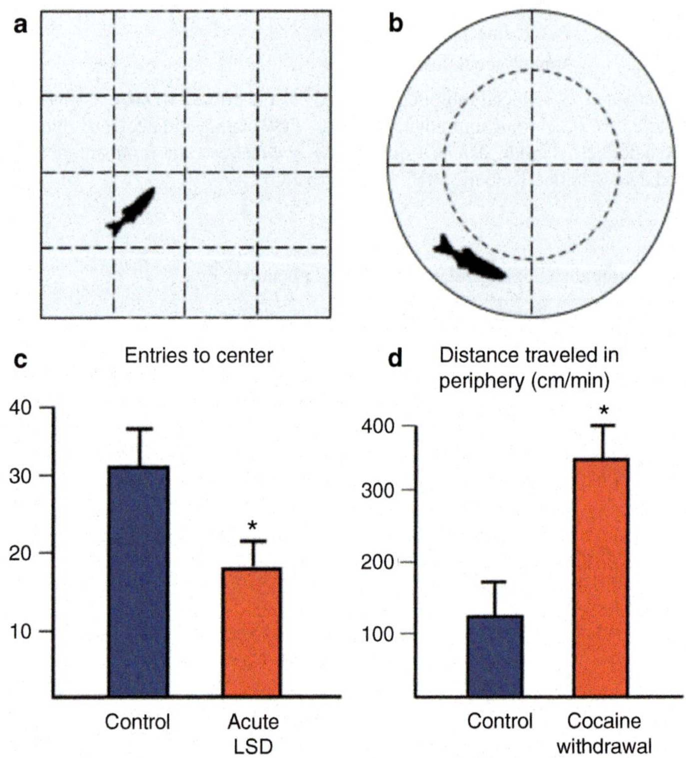

Fig. 1 The zebrafish open-field test (*top view*). This paradigm primarily evaluates horizontal exploratory based behavioral endpoints, such as entries/time spent in the center zones, entries/time spent in the outer zones (i.e., thigmotaxis), total distance traveled, and average velocity. (**a**) A square open field apparatus divided into zones [36]. (**b**) A circular open field apparatus divided into zones [13]. (**c**) Acute lysergic acid diethylamide (LSD) exposure decreased the number of entries to the center of the open field [37]. (**D**) 72-h withdrawal from cocaine (5 intermittent days 1.5 μM) increased distance traveled in the periphery [39]

throughout testing, suggesting that novelty exploration behavior in the open field is well conserved in zebrafish [65, 66]. As already mentioned, like rodent models, zebrafish readily habituate to the open field over time as indicated by a reduction in distance traveled and average velocity by the end of the testing session [36].

Another common phenotype observed during rodent open field testing is the establishment of a homebase, a preferred reference point location commonly seen in rodents [67, 68], which was recently reported in zebrafish [69]. Zebrafish homebase behavior can be measured by dividing the open field arena into quadrants and quantifying average time spent, frequency of visits and distance traveled in each quadrant [69]. This behavior can be sensitive to pharmacological manipulation, since for example, a hallucinogenic drug ibogaine reduces the time spent investigating the entire open field arena before establishing a preferred homebase behavior compared to control fish [13]. Other exploratory behaviors in the open field are sensitive to pharmacological treatment as well, and are summarized in Table 2. Furthermore, open field phenotypes are also sensitive to experimental stressors. For example, an acute stressor such as a 15-min net restraint increases thigmotaxis and average velocity in zebrafish [36].

The light dark test is another paradigm traditionally tested in rodents, and currently extensively applied to zebrafish phenotyping. The light dark test apparatus is typically an aquarium that consists of a light half and a dark half [36, 37, 43]. The test can also take other forms, such as the light dark plus maze with a grey center starting area with two light and two dark arms (Fig. 2) [71]. Rodents are innately aversive to brightly lit environments and exhibit scototaxis (i.e., dark environment preference); a decrease in scototactic behavior indicates anxiolysis [72, 73]. Similarly, adult zebrafish, as well as other fish species (e.g., goldfish, guppies, minnows and tilapia), generally display a robust preference for the dark area of the tank [13, 37, 43, 44]. However, there are early reports of a preference for the white area of the tank in zebrafish [36, 74]. These reported inconsistencies are likely attributable to different housing conditions, lighting, fish sex, age, social status and/or strains, and can be interpreted carefully, keeping in mind a marked and common preference for dark in normal adult zebrafish (and 'reversed' light preference in larval fish [9]). For example, 2 months of rearing in an enriched environment increased time spent in the light environment compared to fish raised in an impoverished environment [75]. Differences in lighting intensity can also alter zebrafish behavior in the light dark test; zebrafish increased scototaxis and spent more time freezing at 500 vs. 250 lux [43]. Further experimentation is necessary to elucidate the factors responsible for differences in baseline scototaxic behavior. The light dark test has been commonly employed with rodent models to evaluate stressor and drug effects on anxiety-related phenotypes [72]. Zebrafish scototaxis is also bidirectionally sensitive to screen such effects (Table 3) [20]. For instance, chronic fluoxetine (an antidepressant with anxiolytic action) and acute benzodiazepine anxiolytics (i.e., chlordiazepoxide, clonazepam and diazepam) all decrease scototaxis [44], while acute caffeine, acute restraint stress and chronic unpredictable stress (CUS) increase scototaxis [36, 41, 44].

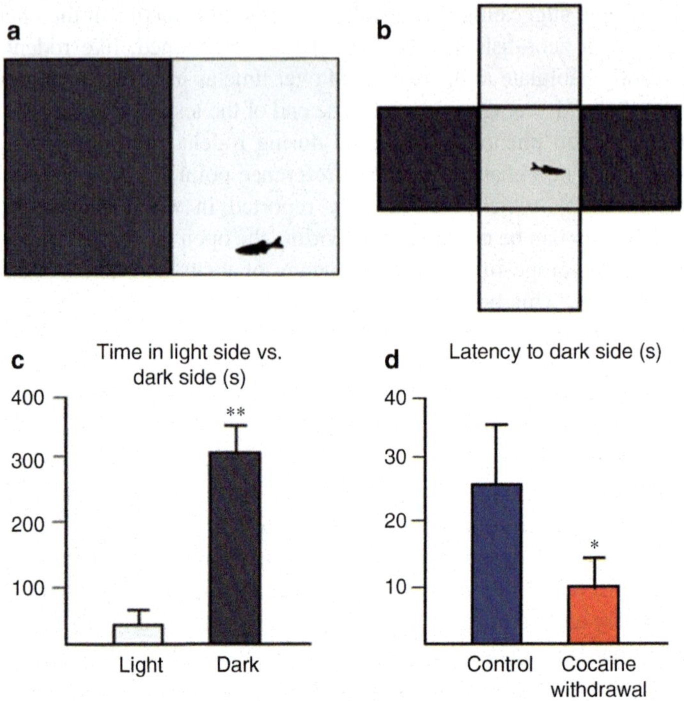

Fig. 2 The zebrafish light dark test. This paradigm primarily evaluates changes in horizontal exploration of light and dark environments, such as the duration of time, number of entries and latency to enter each half. (**a**) A typical light dark test apparatus (*side view*), an aquarium tank with a light colored half and a dark colored half [70]. (**b**) An alternative light dark test apparatus (*top view*) represents a plus maze with a grey center and two light arms and two dark arms [71]. (**c**) Zebrafish commonly display a baseline preference for the darker environment over the light environment (i.e., scototaxis) [70]. (**d**) CUS (chronic unpredictable stress) for 15 days decreased latency to enter the dark side [41]

The novel tank test is a popular novelty-based paradigm that is unique to zebrafish and other aquatic species, and is often used for their behavioral phenotyping and testing drug effects. This test is conceptually similar to the open field test used for rodents, but rather than measuring only horizontal exploration, the novel tank task primarily measures vertical exploration [70]. The novel tank apparatus typically consists of a narrow tank divided horizontally into a top and bottom zone, but may also consist of a three-zone tank (i.e., top, middle and bottom zones; Fig. 3). Upon exposure to a novel tank apparatus, zebrafish initially exhibit a robust anxiety-like response by diving to the bottom of the tank (i.e., geotaxis), also reducing exploration, increasing freezing and erratic movements [21]. Additionally, this paradigm induces stress-related physiological responses, such as elevated cortisol levels, increased breathing and increased heart beat frequency [45]. Habituation to the novel tank occurs over time, as indicated by a decrease in the aforementioned anxiety phenotypes [35]. It is important to note that pre-test housing conditions may

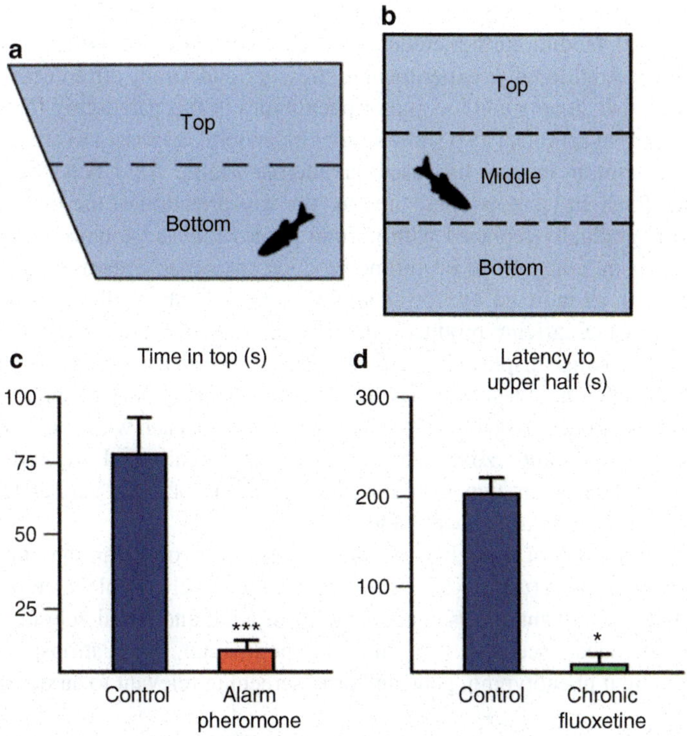

Fig. 3 The zebrafish novel tank test (*side views*). This paradigm primarily evaluates changes in vertical exploration, such as time spent in top and bottom zones, number of entries, latency to enter the top, total distance traveled, and average velocity. Zebrafish initially dive to the bottom of the tank (i.e., geotaxis) and explore upper regions of the tank as habituation occurs (**a**) A typical novel tank test apparatus, consisting of a trapezoidal tank divided into top and bottom zones [70]. (**b**) An alternative novel tank apparatus, divided into top, middle, and bottom zones [71]. Generally, both modifications of this model will be sensitive to zebrafish anxiety-like behaviors, albeit the central zone in model B would mostly reflect a transition aspect between two other zones (top/bottom, the difference between which would be both more relevant behaviorally and most robust phenotypically). (**c**) Anxiogenic effect of acute alarm pheromone on zebrafish behavior in the novel tank test [76]. (**d**) Anxiolytic effect of chronic fluoxetine (100 g/L 2 weeks) in the novel tank test [24]

affect zebrafish behavior in this paradigm. For instance, zebrafish housed in a narrow tank similar to the novel tank apparatus may not display a diving response or changes in swim velocity, but fish housed in a wider tank did, an effect that was likely due to habituation or acclimation to novelty [51].

The novel tank test is an excellent assay for screening anxiotropic (anxiolytic and anxiogenic) agents, as zebrafish anxiety-like behaviors are highly and bidirectionally sensitive to such manipulations (Table 4). For example, chronic fluoxetine reduces, and acute caffeine increases geotaxis [21], similar to these drugs' effects in rodents. The novel tank test can also be used to evaluate anxiety phenotypes evoked experimentally by drug withdrawal (Table 4). Specifically, repeated morphine withdrawal in zebrafish produces a robust anxiogenic profile in the novel tank test in zebrafish [77]. Similar anxiogenic effects of withdrawal are observed in rodent models as well, lending further credence to zebrafish models of withdrawal-evoked anxiety [78]. Additionally, experimental stressors and strain differences produce altered zebrafish anxiety and locomotor phenotypes in this paradigm (Table 4).

Finally, larval zebrafish also show similar behavioral responses to anxiolytic and anxiogenic stimuli in novelty-based paradigms (Table 5), albeit their natural preference for light (scotophobia) 'inverts' the interpretation of the light dark box data, and is gradually replaced with normal photophobia/scototaxis as adults [9]. Acute diazepam, ethanol and buspirone produce anxiolytic responses in the larval light dark test, as indicted by, for example, increased time in the dark side [57]. Conversely, acute caffeine produces an anxiogenic response in the light dark test [57]. Larval zebrafish display characteristic thigmotaxis and avoidance of the center region in the open field test [62]. The larval apparatus may vary in shape, size and color, but typically consists of a 12 or 24-well plate with each well visually divided into an inner and outer zone [55]. Thigmotaxis is enhanced by caffeine and potentiated by diazepam, thus validating the sensitivity of larval zebrafish to study anxiety-like behaviors in the open field.

In summary, each of these behavioral tests do not involve training, are short in experimental duration (usually 5–10 min), and are relatively simple to employ. This, coupled with the advantageous characteristics of adult and larval zebrafish model, provides an ideal scenario for many experimental applications, including high-throughput phenotyping, gene and drug screening relevant to anxiety.

3 Physiological (Endocrine) Response: Cortisol

Robust and quantifiable physiological phenotypes contribute markedly to the utility of zebrafish models for stress and anxiety research. The zebrafish hypothalamus-pituitary-interrenal (HPI) axis is homologous to the human hypothalamus-pituitary-adrenal (HPA) axis, with cortisol being the primary stress hormone in both species (Fig. 4). The evolutionarily conserved stress response between zebrafish and humans establish this aquatic species as a valid model to study cortisol-mediated stress responses [15, 24]. Cortisol can be sampled using different methods, using adult whole-body samples [79], tail vein blood and trunk samples [80], testing water [81], and whole-body larval zebrafish [82]. A temporal-based analysis of whole-body cortisol levels following a net stressor (i.e., acute net handling and air exposure) found increased cortisol at 3 min post-stressor, a linear increase and peak levels at 15 min post-stressor, and return to near control levels 60 min post-stressor [83] (Fig. 5). The analysis of neuroendocrine (i.e., cortisol) responses in zebrafish is a valuable tool complementing behavioral studies. Zebrafish modulate cortisol levels in response to various drug treatments and experimental stressors (Table 6), which often strongly correlate with behavioral responses [21]. For instance, zebrafish treated with chronic fluoxetine decreased whole-body cortisol levels (Fig. 5) and reduced geotaxis in the novel tank test [21]. Conversely, anxiogenic manipulations like morphine withdrawal increased both whole-body cortisol and geotaxis in the novel tank test [77].

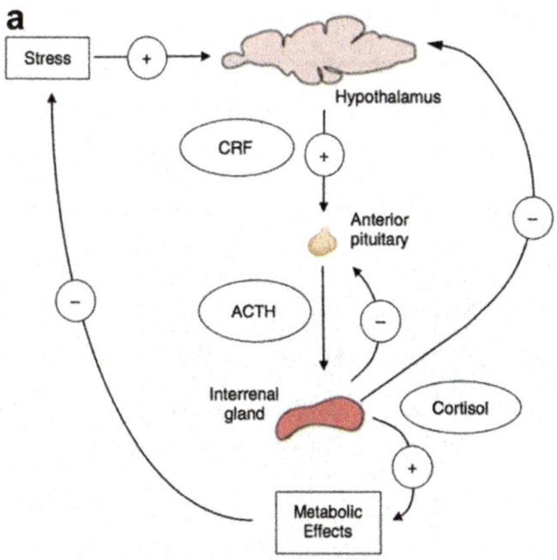

Fig. 4 A striking overall physiological similarity of the endocrine stress axes in zebrafish and humans. "+": activation. "–": inhibition. *CRF* corticotropin-releasing factor, *ACTH* adrenocorticotropic hormone. (**a**) Zebrafish hypothalamus-pituitary-interrenal (HPI) axis. (**b**) Human hypothalamus-pituitary-adrenal (HPA) axis [76]

4 Experimental Stressors: Chronic Unpredictable Stress, Beaker Stress, Predator and Alarm Pheromone Exposure

Zebrafish behavioral and physiological phenotypes are highly sensitive to acute or chronic exposure to a wide range of husbandry, environmental, chemical, mechanical and social stressors (e.g., changes in temperature, pH and lighting, crowding, isolation, restraint, decreasing water level, chasing with net, air exposure, dominant and submissive pairings, predator exposure, and alarm pheromone exposure) [19, 41, 88]. Chronic unpredictable stress (CUS), consisting of a battery of stressors administered over a length of days (see Table 6 for details), increased whole-body cortisol levels as well as anxiety-like behaviors in the novel tank test and light dark test [41]. CUS also down-regulated phosphorylated cAMP response element-binding protein (pCREB), up-regulated corticotropin-releasing factor (CRF) as well as calcineurin mRNA in the zebrafish brain, which are molecular markers that have been observed in human patients with major depressive disorder and rodent models of mood disorders [89–91]. However, brain derived neurotrophic factor (BDNF) is up-regulated in zebrafish following CUS [41], whereas it is commonly down-regulated in rodent models [92, 93]. Notably, BDNF levels are differentially expressed in the rat amygdala and hippocampus, and therefore, it may be useful for future studies to evaluate zebrafish gene expression profiles in a brain region-specific manner [94]. An upregulation of several other molecular markers related to the HPI axis, such as whole brain glucocorticoid receptor (GR), mineralcorticoid receptor (MR), proopiomelanocortin (POMC) hypocretin/orexin, BDNF, as well as c-fos mRNA, has been reported in zebrafish [19]. The immediate early gene *c-fos* acts a reliable biomarker of cellular (e.g., neuronal) activation in various species, including humans [95], rodents [96] and zebrafish [97, 98].

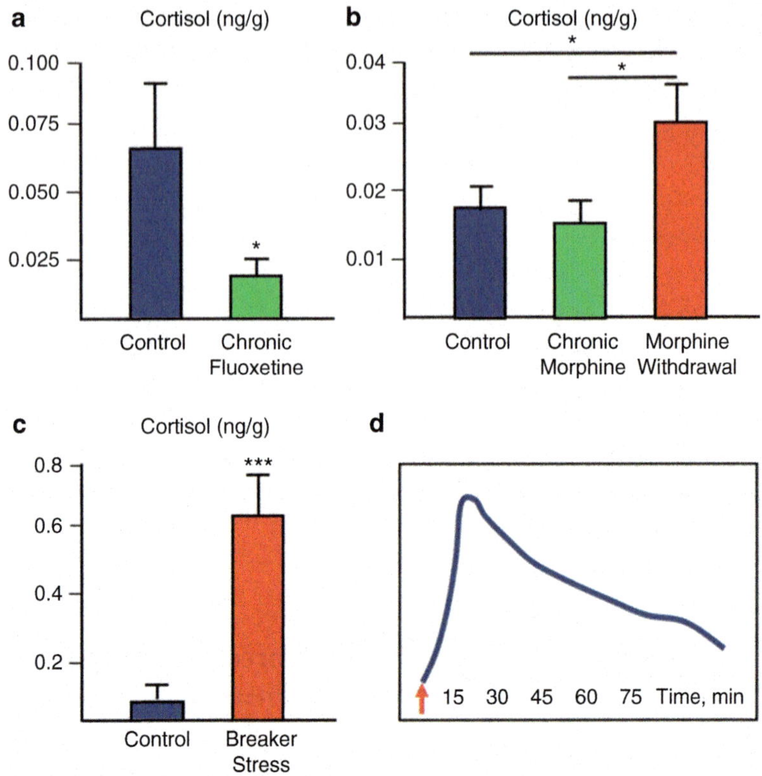

Fig. 5 A typical zebrafish cortisol responses (whole-body cortisol, ng/g fish). (**a**) Exposure to chronic fluoxetine (2 week 100 μg/L). (**b**) Exposure to chronic morphine (2 week 1.5 mg/L) and 24-h morphine withdrawal. (**c**) Exposure to beaker stress paradigm (15 min at 100 mL water in 250 mL beaker). Data are presented as mean ± SEM, $^*p<0.05$, $^{**}p<0.01$, *** $p<0.001$. *U*-test (Modified from [8, 21]). (**d**) A typical time course of zebrafish stress-evoked cortisol response, peaking at 15 min after a stress stimulus (*arrow*) and gradually decreasing over a 1–1.5-h time [83]. Note that this response strikingly resembles the dynamic of human cortisol response to acute stressor

Acute environmental stressors also modulate zebrafish anxiety phenotypes. For example, a recently developed beaker stress model, consisting of confinement for 15 min in 100 mL of water in a 250 mL glass beaker, robustly increases anxiety-like behaviors in the novel tank test and light dark test (own unpublished findings) and whole-body cortisol levels (Fig. 6, Table 6), also see [8]. The robustness of this model is likely due to the combination of confinement in a small environment, a shallow <10 cm water level (stressful for zebrafish), and social isolation from conspecifics. Alarm pheromone exposure also produces behavioral alterations in zebrafish, released by their injured skin cells and detected by the fish olfactory system [23]. Alarm pheromone can be easily extracted from the epidermal cells of euthanized zebrafish and administered to tank water [23]. Acute alarm pheromone

Table 6 Summary of zebrafish cortisol responses to various stimuli

Treatment	Details	Cortisol effect	References
Acute LSD	20-min 250 μg/L	↑ vs. control	[37]
Acute PCP	20-min 3 mg/L	↑ vs. control	[84]
Acute mescaline	20-min 20 mg/L	Ø vs. control	[84]
Acute ibogaine	20-min 10+20 mg/L	Ø vs. control	[13]
Acute ketamine	20-min 20 mg/L+40 mg/L	↓ vs. control	
Chronic nicotine	4 days (2 days 1 mg/L+2 days 2 mg/L)	Ø vs. control	[48]
Chronic fluoxetine	2 week 100 μg/L	↓ vs. control	[21]
Chronic morphine	2 week 1.5 mg/L	Ø vs. control	[21]
Morphine withdrawal	24-h withdrawal from chronic treatment	↑ vs. control ↑ vs. chronic treatment	[21]
Chronic ethanol	1 week 0.3 % v/v	Ø vs. control	[21]
Ethanol withdrawal	24-h withdrawal from chronic treatment	Ø vs. control ↑ vs. chronic treatment	[21]
Chronic diazepam	2 week 72 mg/L	Ø vs. control	[77]
Diazepam withdrawal	72-h withdrawal from chronic treatment	Ø vs. control Ø vs. chronic treatment	[77]
Chronic caffeine	1 week 50 mg/L	Ø vs. control	[77]
Caffeine withdrawal	12-h withdrawal from chronic treatment	Ø vs. control Ø vs. chronic treatment	[77]
Chronic CDP	4-month 100 mg/L	Ø vs. control	[50]
CDP withdrawal	7-day CDP withdrawal from chronic treatment	↑ vs. control ↑ vs. chronic treatment	[50]
Dyadic social stress	Dominant and submissive fish kept in pairs for 5 days	↑ in dominant fish vs. control ↑ in submissive fish vs. control	[80]
Predator exposure (direct contact)	5 min of *Parachromis managuensis* exposure	↑ vs. control	[85]
Predator exposure (visual contact)	60 min of *Parachromis managuensis* exposure	↑ vs. control	[85]
Beaker stressor	15 min in 100 mL within a 250 mL beaker	↑ vs. control	[8]

(continued)

Table 6 (continued)

Treatment	Details	Cortisol effect	References
Acute net handling	3 min net suspension in air + 3 min in tank +3 min suspension in air	↑ vs. control	[83]
	30 s net suspension in air	↑ vs. control	[86]
Acute crowding	40 fish/L for 3 h	↑ vs. control	[87]
Chronic crowding	40 fish/L for 5 days	↑ vs. control	[87]
Acute stress battery	1 day of net chasing + air exposure + water level decrease + crowding (see ref for further details)	↑ vs. control	[19]
Low- grade CUS	12 days of changes in light intensity and spectrum + pH changes + increased water current + crowding + plastic plant introduction + dissolved food extract (see ref for further details)	Ø vs. control	[19]
High-grade CUS	12 days of changes in lighting schedule + net chasing + net restraint + air exposure + crowding + water level decrease + isolation (see ref for further details)	↑ vs. control	[19]

↑ increased/activated, ↓ reduced/inhibited/impaired, Ø no effect, *LSD* lysergic acid diethylamide, *PCP* phencyclidine, *CDP* chlordiazepoxide, *CUS* chronic unpredictable stress

exposure resulted in a robust anxiety-like behavioral response, notably represented through significantly decreased exploration and increased erratic movements and freezing bouts in the novel tank test (Figs. 4 and 6, Table 3) [24]. In contrast, chronic alarm pheromone produces no changes in fish, suggesting that alarm pheromone is only effective acutely, most likely reflecting its natural use as a fast-acting danger signal to nearby shoals [76]. Another study found that acute hypoxanthine 3-N-oxide, a molecule common to the alarm pheromones secreted by several fish species, elicited more erratic movements and jumps as the dose increased [100].

The presence of a predator is another universal stressor for animals [101, 102]. Zebrafish display significant behavioral response to a variety of predator stimuli, such presence and visualization of a sympatric predator, the Indian leaf Fish (*Nandus nandus*, also known as the Gangetic leaf fish). For instance, visual exposure to the Indian leaf Fish resulted in zebrafish geotaxis, unusual tightened grouping (shoaling) of conspecifics and avoidance of the predator by gathering to the opposite corner (Fig. 6) [99]. Experimentally naïve zebrafish respond significantly stronger to a *sympatric predator* from their natural habitat than to an *allopatric predator* (i.e., compressed cichlid, from different non-overlapping natural habitats), suggesting a genetically based predator anxiety [102]. Both acute and chronic exposure to the Indian leaf fish produced similar behavioral responses in the novel tank test [76]. Notably, although the zebrafish displayed a typical response to stress with an

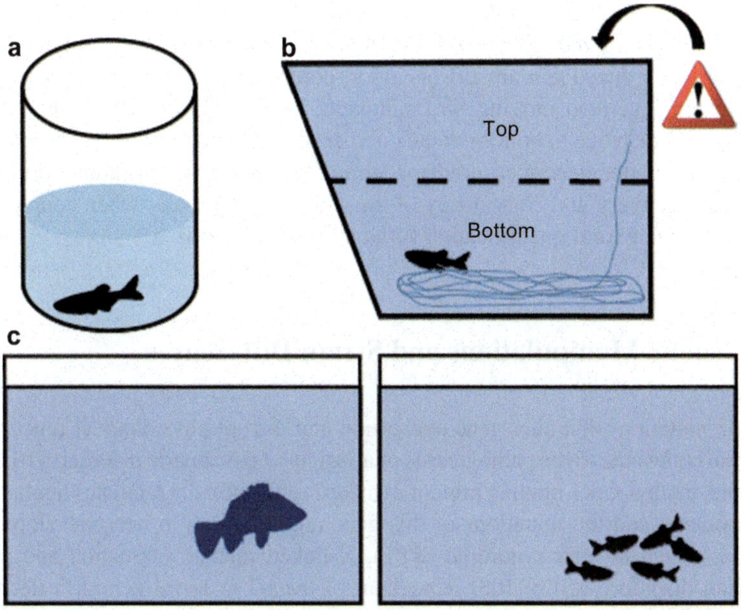

Fig. 6 Experimental stressors commonly used to trigger and assess anxiety-like behaviors in adult zebrafish. (**a**) The beaker stressor test. In this model, zebrafish are removed from their home tank and placed individually in 100 mL of water within a 250 mL beaker for 15-min, resulting in elevated whole-body cortisol levels and anxiety-like behaviors [8]. (**b**) Alarm substance exposure in the novel tank test. Alarm substance is extracted from epidermal cells of euthanized zebrafish and administered to the novel tank test water prior to testing, resulting in increased anxiety-like behaviors [24]. (**c**) Predator exposure paradigm. In this model, a zebrafish tank is placed adjacent to a tank containing a big predator fish (e.g., Indian leaf fish, a natural predator of zebrafish). Visual exposure to a predator fish results in avoidance, tightened shoal cohesion and bottom dwelling [99].

increase of erratic movements, they also displayed shorter latency to enter the upper half and more time spent in the upper half, which are not characteristics associated with stress in the novel tank paradigm. However, as the predator fish spent the majority of the time in the bottom of the tank, it appears that the zebrafish displayed a distinct learned avoidance behavior by moving to the area least likely to be occupied by a predator. In contrast, typical anxiety-like behavior was only significant in the erratic movement endpoint during exposure to an allopatric predator Oscar fish (*Astronotus ocellatus*), indicating weaker responses as compared to Indian leaf fish exposure [76]. This further suggests the importance of a strong genetic 'innate' influence on the zebrafish fear response.

Predator stimuli that are artificial [74], real [8, 99], or computer-generated [103] produce robust and reproducible anxiogenic phenotypes in zebrafish. For example, a recent study found that an animated dot increasing in size presented from above a tank on a computer screen elicited a stronger fear response than other predator-related stimuli (e.g., animated Indian leaf fish, animated needle fish, and a bird silhouette) [104]. The dot stimulus may mimic an approaching fishing bird, another natural predator of zebrafish [105]. Zebrafish treated with acute anxiolytic dose of ethanol show reduced fear/anxiety behaviors compared to control fish in response to a computer generated moving bird silhouette presented from above the tank, as measured by distance to bottom of tank and erratic movements [104]. The approach of using computer generated predator stimuli is particularly attractive due to the automated delivery and consistency of the stimulus, especially when coupled with automated behavioral quantification software [14, 103, 106].

5 Genetic Manipulations and Strain Differences

Genetic mutations that alter gene expression and disrupt physiological functions of the brain contribute to the pathogenesis of a variety of psychiatric disorders [107]. For example, methyl-CpG-binding protein 2 (mecp2) epigentically regulates human brain development, and the mutations of this gene are attributed to neurodevelopmental disorders, such as Rett syndrome (RTT), X-linked mental retardation and autism spectrum disorder (ASD) [108]. Knockout of *mecp2* in larval zebrafish decreased their locomotor activity levels and average velocity in the open field test compared to wild type larvae [109]. Motor impairment is a phenotype commonly observed in mecp2-related disorders in humans, as well as in rodent models [108]. *Mecp2* zebrafish mutants also showed decreased levels of thigmotaxis in the open field [109], an effect that is inconsistent with rodent *mecp2* mutants [110, 111]. This phenotypic difference may be attributable to the larval motor dysfunction, or an avoidance of tactile stimulation from the wall, similar to hyper-responsive ASD patients [112].

One of the main challenges in zebrafish neurophenotyping research is the relatively limited number of outbred or inbred wildtype 'reference' strains, as compared to nearly a hundred of wild-type strains currently available for mice (www.jax.org). As seen in other species, different genetic strains in zebrafish may contribute to varying behavioral phenotypes. For example, in the novel tank test the wild-caught zebrafish from a small river in Bengal (India) exhibited more anxiety-related endpoints (less top transitions, less time in the top, more freezing bouts and increased latency to enter the top), compared to a short-fin (SF) outbred laboratory strain (Table 4) [45]. The leopard zebrafish strain also showed increased anxiety-like behaviors in the novel tank test compared to the wild type SF strain [24]. However, the leopard strain did not show differences in total distance traveled or average velocity, suggesting that differences in anxiety were not due to motor/neurological deficits. In studies using an animated predator stimulus (i.e., Indian leaf fish) the WIK and TU zebrafish strains showed an atypical preference for the side of the tank where the predator stimulus was presented [113], and the AB strain showed avoidance of the stimulus [114]. In a novel tank test study, the WIK zebrafish spend more time in the top of the tank compared to the AB line, suggesting that the WIK strain may be less anxious compared to others [71], most likely reflecting their genetic closeness to the wild zebrafish. The TM1 strain was more likely to approach an artificial painted allopatric predator fish model (cichlid, *Etroplus canarensis*) and took less time to recover after being transferred to a new tank, as measured by latency to feed, when compared to the SH and Nadia strains [115]. Clearly, understanding different behavioral profiles between zebrafish strains is an important method to determine the contribution of genetic background on anxiety and stress related behaviors. Combined with the availability of the growing number of transgenic and mutant zebrafish (some of which show overt differences in anxiety-related behaviors and physiology discussed above), the expansion of this effort and the identification of candidate genes or gene loci will aid in determining genetic susceptibility to stressors in humans.

Additional useful approaches to studying zebrafish anxiety-like traits include quantitative trait loci (QTL)-based analyses and the genetic knockdown of various genes. For example, QTL analysis involves crossing two populations or strains and genotyping the intercross generation, which ultimately reveals the relation between a genomic region and a phenotype [116]. QTL analysis of over 100 mouse behavioral phenotypes in the open field test and light dark test detected 17 QTL accounting for phenotypic variation [117]. In zebrafish, an F2 generation derived from crossing a wild Indian strain with the AB strain revealed QTL mapping of anti-predatory behavior (shoaling) and 'boldness' (approach to a novel object) [118]. The genomic region for anti-predatory behavior was located on chromosome 21, and the region for 'boldness' was located on chromosome 9 and 16 [118]. Zebrafish offer great potential for evaluating behavioral phenotypes at the genetic level using QTL mapping due to low-cost and high fecundity, although few QTL studies with zebrafish have been conducted at this point [116, 118, 119]. Gene knockdown technologies in zebrafish are also valuable systems to elucidate vertebrate gene function that can be achieved using a variety of methods such as zinc-finger nucleases [120], transcription activator-like-effector nuclease (TALEN) [27, 121] and clustered regularly-interspaced short palindromic repeats (CRISPR) [121]. For example, CRISPR knockdown of vesicular monoamine transporter 2 (Vmat2) in zebrafish results in decreased levels of dopamine, serotonin, norepinephrine and their metabolites [46]. A similar reduction in monoamines is seen in Vmat2 heterozygous mice [122]. *Vmat2* mutant zebrafish also increased geotaxis in the novel tank test, and female mutants are more anxious than males. Interestingly, chronic pharmacological blockage of Vmat2 in zebrafish by reserpine treatment cased general hypoactivity in zebrafish [123] and elevated cortisol levels, generally consisted with increased affective (albeit not purely or necessarily anxiogenic) tone.

6 Conclusion

Despite anxiety-related disorders being one of the most widespread neuropsychiatric conditions, their pathological mechanisms are poorly understood, and their treatments remain essentially the same for the last 50 years [124, 125]. Discovering the underlying mechanisms of psychopathology is fundamental to treatment, reversal, or prevention of complex brain disorders, including stress/anxiety-related illnesses [126, 127]. A substantial challenge faced by phenotype-based screening is expensive and inefficient mammalian models that require large quantities of compounds and time during experimentation [124, 125]. With a clear benefit of genetic and physiological similarity, the use of zebrafish as an alternative model mitigates these limitations [128–130]. Recent circuitry-based studies in zebrafish continue to unravel complex neural regulation of anxiety-related states in this species [131, 132]. Together with a robust sensitivity to drugs and acute/chronic stressors, novelty-based paradigms, endocrine correlates, and an ease of genetic manipulation makes high-throughput phenotyping and pharmacological screens in zebrafish a promising possibility in translational neuroscience of anxiety "from tank to bedside" [8, 133, 134].

References

1. Hettema JM, Prescott CA, Myers JM, Neale MC, Kendler KS. The structure of genetic and environmental risk factors for anxiety disorders in men and women. Arch Gen Psychiatry. 2005;62:182–9.
2. Baxter A, Scott K, Vos T, Whiteford H. Global prevalence of anxiety disorders: a systematic review and meta-regression. Psychol Med. 2013;43:897–910.
3. American Psychiatric Association. Diagnostic and statistical manual of mental disorders. Washington, DC: American Psychiatric Association; 2013.
4. Lang PJ, Davis M, Öhman A. Fear and anxiety: animal models and human cognitive psychophysiology. J Affect Disord. 2000;61:137–59.
5. Pellow S, Chopin P, File SE, Briley M. Validation of open: closed arm entries in an elevated plus-maze as a measure of anxiety in the rat. J Neurosci Methods. 1985;14:149–67.
6. Donner NC, Johnson PL, Fitz SD, Kellen KE, Shekhar A, Lowry CA. Elevated tph2 mRNA expression in a rat model of chronic anxiety. Depress Anxiety. 2012;29:307–19.
7. Nieto-Gonzalez JL, Holm MM, Vardya I, Christensen T, Wiborg O, Jensen K. Presynaptic plasticity as a hallmark of rat stress susceptibility and antidepressant response. PLoS One. 2015;10:e0119993.
8. Kalueff AV, Echevarria DJ, Stewart AM. Gaining translational momentum: more zebrafish models for neuroscience research. Prog Neuropsychopharmacol Biol Psychiatry. 2014; 55:1–6.
9. Kalueff AV, Gebhardt M, Stewart AM, Cachat JM, Brimmer M, Chawla JS, et al. Towards a comprehensive catalog of zebrafish behavior 1.0 and beyond. Zebrafish. 2013;10:70–86.
10. Bruni G, Lakhani P, Kokel D. Discovering novel neuroactive drugs through high-throughput behavior-based chemical screening in the zebrafish. Front Pharmacol. 2014;5:153.
11. Gerlai R. High-throughput behavioral screens: the first step towards finding genes involved in vertebrate brain function using zebrafish. Molecules. 2010;15:2609–22.
12. Cachat J, Stewart A, Utterback E, Hart P, Gaikwad S, Wong K, et al. Three-dimensional neurophenotyping of adult zebrafish behavior. PLoS One. 2011;6:e17597.
13. Cachat J, Kyzar EJ, Collins C, Gaikwad S, Green J, Roth A, et al. Unique and potent effects of acute ibogaine on zebrafish: the developing utility of novel aquatic models for hallucinogenic drug research. Behav Brain Res. 2013;236:258–69.
14. Stewart AM, Grieco F, Tegelenbosch RA, Kyzar EJ, Nguyen M, Kaluyeva A, et al. A novel 3D method of locomotor analysis in adult zebrafish: implications for automated detection of CNS drug-evoked phenotypes. J Neurosci Methods. 2015;255:66–74.
15. Alsop D, Vijayan MM. Development of the corticosteroid stress axis and receptor expression in zebrafish. Am J Physiol Regul Integr Comp Physiol. 2008;294:R711–9.

16. Panula P, Sallinen V, Sundvik M, Kolehmainen J, Torkko V, Tiittula A, et al. Modulatory neurotransmitter systems and behavior: towards zebrafish models of neurodegenerative diseases. Zebrafish. 2006;3:235–47.
17. Panula P, Chen Y-C, Priyadarshini M, Kudo H, Semenova S, Sundvik M, et al. The comparative neuroanatomy and neurochemistry of zebrafish CNS systems of relevance to human neuropsychiatric diseases. Neurobiol Dis. 2010;40:46–57.
18. Howe K, Clark MD, Torroja CF, Torrance J, Berthelot C, Muffato M, et al. The zebrafish reference genome sequence and its relationship to the human genome. Nature. 2013;496:498–503.
19. Pavlidis M, Theodoridi A, Tsalafouta A. Neuroendocrine regulation of the stress response in adult zebrafish, *Danio rerio*. Prog Neuropsychopharmacol Biol Psychiatry. 2015;60: 121–31.
20. Maximino C, da Silva AWB, Araújo J, Lima MG, Miranda V, Puty B, et al. Fingerprinting of psychoactive drugs in zebrafish anxiety-like behaviors. PLoS One. 2014;9:e103943.
21. Cachat J, Stewart A, Grossman L, Gaikwad S, Kadri F, Chung KM, et al. Measuring behavioral and endocrine responses to novelty stress in adult zebrafish. Nat Protoc. 2010;5: 1786–99.
22. Gerlai R. Antipredatory behavior of zebrafish: adaptive function and a tool for translational research. Evol Psychol. 2013;11:591–605.
23. Speedie N, Gerlai R. Alarm substance induced behavioral responses in zebrafish (*Danio rerio*). Behav Brain Res. 2008;188:168–77.
24. Egan RJ, Bergner CL, Hart PC, Cachat JM, Canavello PR, Elegante MF, et al. Understanding behavioral and physiological phenotypes of stress and anxiety in zebrafish. Behav Brain Res. 2009;205:38–44.
25. Hwang WY, Fu Y, Reyon D, Maeder ML, Tsai SQ, Sander JD, et al. Efficient genome editing in zebrafish using a CRISPR-Cas system. Nat Biotechnol. 2013;31:227–9.
26. Shah AN, Davey CF, Whitebirch AC, Miller AC, Moens CB. Rapid reverse genetic screening using CRISPR in zebrafish. Nat Methods. 2015;12:535–40.
27. Bedell VM, Wang Y, Campbell JM, Poshusta TL, Starker CG, Krug II RG, et al. In vivo genome editing using a high-efficiency TALEN system. Nature. 2012;491:114–8.
28. Kurt M, Arik AC, Celik S. The effects of sertraline and fluoxetine on anxiety in the elevated plus-maze test. J Basic Clin Physiol Pharmacol. 2000;11:173–80.
29. Belzung C. Measuring rodent exploratory behavior. Tech Behav Neural Sci. 1999;13: 738–52.
30. Sousa N, Almeida O, Wotjak C. A hitchhiker's guide to behavioral analysis in laboratory rodents. Genes Brain Behav. 2006;5:5–24.
31. File SE. Factors controlling measures of anxiety and responses to novelty in the mouse. Behav Brain Res. 2001;125:151–7.
32. Treit D, Fundytus M. Thigmotaxis as a test for anxiolytic activity in rats. Pharmacol Biochem Behav. 1988;31:959–62.
33. Kallai J, Makany T, Csatho A, Karadi K, Horvath D, Kovacs-Labadi B, et al. Cognitive and affective aspects of thigmotaxis strategy in humans. Behav Neurosci. 2007;121:21.
34. Stewart AM, Gaikwad S, Kyzar E, Kalueff AV. Understanding spatio-temporal strategies of adult zebrafish exploration in the open field test. Brain Res. 2012;1451:44–52.
35. Wong K, Elegante M, Bartels B, Elkhayat S, Tien D, Roy S, et al. Analyzing habituation responses to novelty in zebrafish (*Danio rerio*). Behav Brain Res. 2010;208:450–7.
36. Champagne DL, Hoefnagels CC, de Kloet RE, Richardson MK. Translating rodent behavioral repertoire to zebrafish (*Danio rerio*): relevance for stress research. Behav Brain Res. 2010;214:332–42.
37. Grossman L, Utterback E, Stewart A, Gaikwad S, Chung KM, Suciu C, et al. Characterization of behavioral and endocrine effects of LSD on zebrafish. Behav Brain Res. 2010;214:277–84.
38. Peitsaro N, Sundvik M, Anichtchik OV, Kaslin J, Panula P. Identification of zebrafish histamine H 1, H 2 and H 3 receptors and effects of histaminergic ligands on behavior. Biochem Pharmacol. 2007;73:1205–14.

39. López Patiño MA, Yu L, Yamamoto BK, Zhdanova IV. Gender differences in zebrafish responses to cocaine withdrawal. Physiol Behav. 2008;95:36–47.
40. López-Patiño MA, Yu L, Cabral H, Zhdanova IV. Anxiogenic effects of cocaine withdrawal in zebrafish. Physiol Behav. 2008;93:160–71.
41. Chakravarty S, Reddy BR, Sudhakar SR, Saxena S, Das T, Meghah V, et al. Chronic unpredictable stress (CUS)-induced anxiety and related mood disorders in a zebrafish model: altered brain proteome profile implicates mitochondrial dysfunction. PLoS One. 2013;8: e63302.
42. De Campos EG, Bruni AT, De Martinis BS. Ketamine induces anxiolytic effects in adult zebrafish: a multivariate statistics approach. Behav Brain Res. 2015;292:537–46.
43. Stewart A, Maximino C, de Brito TM, Herculano AM, Gouveia Jr A, Morato S, et al. Neurophenotyping of adult zebrafish using the light/dark box paradigm. In: Zebrafish neurobehavioral protocols. New York: Springer; 2011. p. 157–67.
44. Maximino C, da Silva AWB, Gouveia A, Herculano AM. Pharmacological analysis of zebrafish (*Danio rerio*) scototaxis. Prog Neuropsychopharmacol Biol Psychiatry. 2011;35: 624–31.
45. Kalueff AV, Echevarria DJ, Homechaudhuri S, Stewart AM, Collier AD, Kaluyeva AA, et al. Zebrafish neurobehavioral phenomics for aquatic neuropharmacology and toxicology research. Aquat Toxicol. 2016;170:297–309.
46. Wang Y, Li S, Liu W, Wang F, Hu L-F, Zhong Z-M, et al. Vesicular monoamine transporter 2 (Vmat2) knockdown elicits anxiety-like behavior in zebrafish. Biochem Biophys Res Commun. 2016;470:792–7.
47. Stewart A, Riehl R, Wong K, Green J, Cosgrove J, Vollmer K, et al. Behavioral effects of MDMA ('ecstasy') on adult zebrafish. Behav Pharmacol. 2011;22:275–80.
48. Stewart AM, Grossman L, Collier AD, Echevarria DJ, Kalueff AV. Anxiogenic-like effects of chronic nicotine exposure in zebrafish. Pharmacol Biochem Behav. 2015;139:112–20.
49. Neelkantan N, Mikhaylova A, Stewart AM, Arnold R, Gjeloshi V, Kondaveeti D, et al. Perspectives on zebrafish models of hallucinogenic drugs and related psychotropic compounds. ACS Chem Neurosci. 2013;4:1137–50.
50. Stewart A, Wong K, Cachat J, Gaikwad S, Kyzar E, Wu N, et al. Zebrafish models to study drug abuse-related phenotypes. Rev Neurosci. 2011;22:95–105.
51. Bencan Z, Sledge D, Levin ED. Buspirone, chlordiazepoxide and diazepam effects in a zebrafish model of anxiety. Pharmacol Biochem Behav. 2009;94:75–80.
52. Bencan Z, Levin ED. The role of alpha7 and alpha4beta2 nicotinic receptors in the nicotine-induced anxiolytic effect in zebrafish. Physiol Behav. 2008;95:408–12.
53. Yamamoto K, Ruuskanen JO, Wullimann MF, Vernier P. Two tyrosine hydroxylase genes in vertebrates new dopaminergic territories revealed in the zebrafish brain. Mol Cell Neurosci. 2010;43:394–402.
54. Lockwood B, Bjerke S, Kobayashi K, Guo S. Acute effects of alcohol on larval zebrafish: a genetic system for large-scale screening. Pharmacol Biochem Behav. 2004;77:647–54.
55. Schnörr SJ, Steenbergen PJ, Richardson MK, Champagne DL. Assessment of thigmotaxis in larval zebrafish. In: Zebrafish protocols for neurobehavioral research. New York: Springer; 2012. p. 37–51.
56. Schnörr SJ, Steenbergen PJ, Richardson MK, Champagne DL. Measuring thigmotaxis in larval zebrafish. Behav Brain Res. 2012;228:367–74.
57. Steenbergen PJ, Richardson MK, Champagne DL. Patterns of avoidance behaviours in the light/dark preference test in young juvenile zebrafish: a pharmacological study. Behav Brain Res. 2011;222:15–25.
58. Borsini F, Podhorna J, Marazziti D. Do animal models of anxiety predict anxiolytic-like effects of antidepressants? Psychopharmacology (Berl). 2002;163:121–41.
59. Katz RJ, Roth KA, Carroll BJ. Acute and chronic stress effects on open field activity in the rat: implications for a model of depression. Neurosci Biobehav Rev. 1981;5:247–51.
60. Prut L, Belzung C. The open field as a paradigm to measure the effects of drugs on anxiety-like behaviors: a review. Eur J Pharmacol. 2003;463:3–33.

61. Hall CS. Emotional behavior in the rat. I. Defecation and urination as measures of individual differences in emotionality. J Comp Psychol. 1934;18:385.
62. Ahmad F, Richardson MK. Exploratory behaviour in the open field test adapted for larval zebrafish: impact of environmental complexity. Behav Processes. 2013;92:88–98.
63. Lamprea M, Cardenas F, Setem J, Morato S. Thigmotactic responses in an open-field. Braz J Med Biol Res. 2008;41:135–40.
64. Lynn DA, Brown GR. The ontogeny of exploratory behavior in male and female adolescent rats (*Rattus norvegicus*). Dev Psychobiol. 2009;51:513.
65. Kalueff AV, Keisala T, Minasyan A, Kuuslahti M, Tuohimaa P. Temporal stability of novelty exploration in mice exposed to different open field tests. Behav Processes. 2006;72:104–12.
66. Eilam D, Dank M, Maurer R. Voles scale locomotion to the size of the open-field by adjusting the distance between stops: a possible link to path integration. Behav Brain Res. 2003;141: 73–81.
67. Eilam D, Golani I. Home base behavior of rats (*Rattus norvegicus*) exploring a novel environment. Behav Brain Res. 1989;34:199–211.
68. Horev G, Benjamini Y, Sakov A, Golani I. Estimating wall guidance and attraction in mouse free locomotor behavior. Genes Brain Behav. 2007;6:30–41.
69. Stewart A, Cachat J, Wong K, Gaikwad S, Gilder T, DiLeo J, et al. Homebase behavior of zebrafish in novelty-based paradigms. Behav Processes. 2010;85:198–203.
70. Stewart A, Kadri F, DiLeo J, Min Chung K, Cachat J, Goodspeed J, et al. The developing utility of zebrafish in modeling neurobehavioral disorders. Int J Comp Psychol. 2010;23: 104–21.
71. Sackerman J, Donegan JJ, Cunningham CS, Nguyen NN, Lawless K, Long A, et al. Zebrafish behavior in novel environments: effects of acute exposure to anxiolytic compounds and choice of *Danio rerio* Line. Int J Comp Psychol. 2010;23:43–61.
72. Bourin M, Hascoët M. The mouse light/dark box test. Eur J Pharmacol. 2003;463:55–65.
73. Hascoët M, Bourin M, Dhonnchadha BÁN. The mouse ligth-dark paradigm: a review. Prog Neuropsychopharmacol Biol Psychiatry. 2001;25:141–66.
74. Gerlai R, Lahav M, Guo S, Rosenthal A. Drinks like a fish: zebra fish (*Danio rerio*) as a behavior genetic model to study alcohol effects. Pharmacol Biochem Behav. 2000;67:773–82.
75. Maximino C, de Brito TM, de Mattos Dias CAG, Gouveia A, Morato S. Scototaxis as anxiety-like behavior in fish. Nat Protoc. 2010;5:209–16.
76. Cachat JM, Canavello PR, Elegante MF, Bartels BK, Elkhayat SI, Hart PC, et al. Modeling stress and anxiety in zebrafish. In: Kalueff AV, Cachat JM, editors. Zebrafish models in neurobehavioral research. New York: Humana Press; 2011. p. 73–88.
77. Cachat J, Canavello P, Elegante M, Bartels B, Hart P, Bergner C, et al. Modeling withdrawal syndrome in zebrafish. Behav Brain Res. 2010;208:371–6.
78. Zelena D, Barna I, Mlynarik M, Gupta O, Jezova D, Makara GB. Stress symptoms induced by repeated morphine withdrawal in comparison to other chronic stress models in mice. Neuroendocrinology. 2005;81:205–15.
79. Canavello PR, Cachat JM, Beeson EC, Laffoon AL, Grimes C, Haymore WA, et al. Measuring endocrine (cortisol) responses of zebrafish to stress. In: Zebrafish neurobehavioral protocols. New York: Springer; 2011. p. 135–42.
80. Pavlidis M, Sundvik M, Chen Y-C, Panula P. Adaptive changes in zebrafish brain in dominant–subordinate behavioral context. Behav Brain Res. 2011;225:529–37.
81. Félix AS, Faustino AI, Cabral EM, Oliveira RF. Noninvasive measurement of steroid hormones in zebrafish holding-water. Zebrafish. 2013;10:110–5.
82. Yeh C-M, Glöck M, Ryu S. An optimized whole-body cortisol quantification method for assessing stress levels in larval zebrafish. PLoS One. 2013;8:e79406.
83. Ramsay JM, Feist GW, Varga ZM, Westerfield M, Kent ML, Schreck CB. Whole-body cortisol response of zebrafish to acute net handling stress. Aquaculture. 2009;297:157–62.
84. Kyzar EJ, Collins C, Gaikwad S, Green J, Roth A, Monnig L, et al. Effects of hallucinogenic agents mescaline and phencyclidine on zebrafish behavior and physiology. Prog Neuropsychopharmacol Biol Psychiatry. 2012;37:194–202.

85. Barcellos LJG, Ritter F, Kreutz LC, Quevedo RM, da Silva LB, Bedin AC, et al. Whole-body cortisol increases after direct and visual contact with a predator in zebrafish, *Danio rerio*. Aquaculture. 2007;272:774–8.
86. Tran S, Chatterjee D, Gerlai R. Acute net stressor increases whole-body cortisol levels without altering whole-brain monoamines in zebrafish. Behav Neurosci. 2014;128:621.
87. Ramsay JM, Feist GW, Varga ZM, Westerfield M, Kent ML, Schreck CB. Whole-body cortisol is an indicator of crowding stress in adult zebrafish, *Danio rerio*. Aquaculture. 2006;258:565–74.
88. Clark KJ, Boczek NJ, Ekker SC. Stressing zebrafish for behavioral genetics. Rev Neurosci. 2011;22:49–62.
89. Krishnan V, Nestler EJ. Linking molecules to mood: new insight into the biology of depression. Am J Psychiatry. 2010;167:1305–20.
90. Krishnan V, Nestler EJ. Animal models of depression: molecular perspectives. In: Molecular and functional models in neuropsychiatry. Berlin: Springer; 2011. p. 121–47.
91. Sulser F. The role of CREB and other transcription factors in the pharmacotherapy and etiology of depression. Ann Med. 2002;34:348–56.
92. Duman RS, Monteggia LM. A neurotrophic model for stress-related mood disorders. Biol Psychiatry. 2006;59:1116–27.
93. Krishnan V, Nestler EJ. The molecular neurobiology of depression. Nature. 2008;455: 894–902.
94. Lakshminarasimhan H, Chattarji S. Stress leads to contrasting effects on the levels of brain derived neurotrophic factor in the hippocampus and amygdala. PLoS One. 2012;7:e30481.
95. Zhang P, Hirsch EC, Damier P, Duyckaerts C, Javoy-Agid F. c-fos protein-like immunoreactivity: distribution in the human brain and over-expression in the hippocampus of patients with Alzheimer's disease. Neuroscience. 1992;46:9–21.
96. Moreno JL, Holloway T, Albizu L, Sealfon SC, González-Maeso J. Metabotropic glutamate mGlu2 receptor is necessary for the pharmacological and behavioral effects induced by hallucinogenic 5-HT2A receptor agonists. Neurosci Lett. 2011;493:76–9.
97. Lau BYB, Mathur P, Gould GG, Guo S. Identification of a brain center whose activity discriminates a choice behavior in zebrafish. Proc Natl Acad Sci U S A. 2011;108:2581–6.
98. Chatterjee D, Tran S, Shams S, Gerlai R. A simple method for immunohistochemical staining of zebrafish brain sections for c-fos protein expression. Zebrafish. 2015;12:414–20.
99. Stewart AM, Braubach O, Spitsbergen J, Gerlai R, Kalueff AV. Zebrafish models for translational neuroscience research: from tank to bedside. Trends Neurosci. 2014;37:264–78.
100. Parra KV, Adrian JC, Gerlai R. The synthetic substance hypoxanthine 3-N-oxide elicits alarm reactions in zebrafish (*Danio rerio*). Behav Brain Res. 2009;205:336–41.
101. Kats LB, Dill LM. The scent of death: chemosensory assessment of predation risk by prey animals. Ecoscience. 1998;5:361–94.
102. Bass SL, Gerlai R. Zebrafish (*Danio rerio*) responds differentially to stimulus fish: the effects of sympatric and allopatric predators and harmless fish. Behav Brain Res. 2008;186: 107–17.
103. Gerlai R, Fernandes Y, Pereira T. Zebrafish (*Danio rerio*) responds to the animated image of a predator: towards the development of an automated aversive task. Behav Brain Res. 2009;201:318–24.
104. Luca RM, Gerlai R. In search of optimal fear inducing stimuli: differential behavioral responses to computer animated images in zebrafish. Behav Brain Res. 2012;226:66–76.
105. Spence R, Fatema M, Reichard M, Huq K, Wahab M, Ahmed Z, et al. The distribution and habitat preferences of the zebrafish in Bangladesh. J Fish Biol. 2006;69:1435–48.
106. Luca RM, Gerlai R. Animated bird silhouette above the tank: acute alcohol diminishes fear responses in zebrafish. Behav Brain Res. 2012;229:194–201.
107. Mitchell KJ. The genetics of neurodevelopmental disease. Curr Opin Neurobiol. 2011;21:197–203.
108. Gonzales ML, LaSalle JM. The role of MeCP2 in brain development and neurodevelopmental disorders. Curr Psychiatry Rep. 2010;12:127–34.

109. Pietri T, Roman A-C, Guyon N, Romano SA, Washbourne P, Moens CB, et al. The first mecp2-null zebrafish model shows altered motor behaviors. Front Neural Circuits. 2013;7:118.
110. Shahbazian MD, Young JI, Yuva-Paylor LA, Spencer CM, Antalffy BA, Noebels JL, et al. Mice with truncated MeCP2 recapitulate many Rett syndrome features and display hyperacetylation of histone H3. Neuron. 2002;35:243–54.
111. Stearns N, Schaevitz L, Bowling H, Nag N, Berger U, Berger-Sweeney J. Behavioral and anatomical abnormalities in Mecp2 mutant mice: a model for Rett syndrome. Neuroscience. 2007;146:907–21.
112. Belmonte MK, Cook E, Anderson GM, Rubenstein JL, Greenough WT, Beckel-Mitchener A, et al. Autism as a disorder of neural information processing: directions for research and targets for therapy. Mol Psychiatry. 2004;9:646–63.
113. Pannia E, Tran S, Rampersad M, Gerlai R. Acute ethanol exposure induces behavioural differences in two zebrafish (*Danio rerio*) strains: a time course analysis. Behav Brain Res. 2014;259:174–85.
114. Ahmed O, Seguin D, Gerlai R. An automated predator avoidance task in zebrafish. Behav Brain Res. 2011;216:166–71.
115. Moretz JA, Martins EP, Robison BD. Behavioral syndromes and the evolution of correlated behavior in zebrafish. Behav Ecol. 2007;18:556–62.
116. Wright D. QTL mapping using behavioral traits in the adult zebrafish. In: Zebrafish protocols for neurobehavioral research. New York: Springer; 2012. p. 301–12.
117. Henderson ND, Turri MG, DeFries JC, Flint J. QTL analysis of multiple behavioral measures of anxiety in mice. Behav Genet. 2004;34:267–93.
118. Wright D, Nakamichi R, Krause J, Butlin RK. QTL analysis of behavioral and morphological differentiation between wild and laboratory zebrafish (*Danio rerio*). Behav Genet. 2006;36:271–84.
119. Wright D, Butlin RK, Carlborg Ö. Epistatic regulation of behavioural and morphological traits in the zebrafish (*Danio rerio*). Behav Genet. 2006;36:914–22.
120. Ekker SC. Zinc finger-based knockout punches for zebrafish genes. Zebrafish. 2008;5:121–3.
121. Auer TO, Del Bene F. CRISPR/Cas9 and TALEN-mediated knock-in approaches in zebrafish. Methods. 2014;69(2):142–50.
122. Fon EA, Pothos EN, Sun B-C, Killeen N, Sulzer D, Edwards RH. Vesicular transport regulates monoamine storage and release but is not essential for amphetamine action. Neuron. 1997;19:1271–83.
123. Kyzar E, Stewart AM, Landsman S, Collins C, Gebhardt M, Robinson K, et al. Behavioral effects of bidirectional modulators of brain monoamines reserpine and d-amphetamine in zebrafish. Brain Res. 2013;1527:108–16.
124. Cryan JF, Holmes A. The ascent of mouse: advances in modelling human depression and anxiety. Nat Rev Drug Discov. 2005;4:775–90.
125. Griebel G, Holmes A. 50 years of hurdles and hope in anxiolytic drug discovery. Nat Rev Drug Discov. 2013;12:667–87.
126. Duman RS, Heninger GR, Nestler EJ. Molecular psychiatry adaptations of receptor-coupled signal transduction pathways underlying stress-and drug-induced neural plasticity. J Nerv Ment Dis. 1994;182:692–700.
127. Nestler EJ. The origins of molecular psychiatry. J Mol Psychiatry. 2013;1:1–2.
128. McCammon JM, Sive H. Addressing the genetics of human mental health disorders in model organisms. Annu Rev Genomics Hum Genet. 2015;16:173–97.
129. McCammon JM, Sive H. Challenges in understanding psychiatric disorders and developing therapeutics: a role for zebrafish. Dis Model Mech. 2015;8:647–56.
130. Stewart AM, Ullmann JFP, Norton WHJ, Parker MO, Brennan CH, Gerlai R, et al. Molecular psychiatry of zebrafish. Mol Psychiatry. 2015;20:2–17.
131. Okamoto H, Agetsuma M, Aizawa H. Genetic dissection of the zebrafish habenula, a possible switching board for selection of behavioral strategy to cope with fear and anxiety. Dev Neurobiol. 2012;72:386–94.

132. Lee A, Mathuru AS, Teh C, Kibat C, Korzh V, Penney TB, et al. The habenula prevents helpless behavior in larval zebrafish. Curr Biol. 2010;20:2211–6.
133. Stewart AM, Gerlai R, Kalueff AV. Developing highER-throughput zebrafish screens for in-vivo CNS drug discovery. Front Behav Neurosci. 2015;9:14.
134. Valentim AM, Félix LM, Carvalho L, Diniz E, Antunes LM. A new anaesthetic protocol for adult zebrafish (*Danio rerio*): propofol combined with lidocaine. PLoS One. 2016;11:e0147747.

Assessing Cognitive Phenotypes in Zebrafish

David J. Echevarria, Adam D. Collier, and Elizabeth A. Lamb

Abstract Zebrafish are becoming increasingly utilized in behavioral studies as a model of human learning and memory. The existing studies indicate that this species is capable of a variety of cognitive processes, demonstrating its value as a model organism. Popular zebrafish behavioral paradigms to assess various aspects of learning and memory include the startle response, novel tank, learned spatial alternation, three-chamber tank, T-maze and plus maze. Each paradigm is sensitive to pharmacological, genetic and/or experimental manipulations, and within each test specific methodologies and apparatuses have been developed to accommodate the abilities and limitations of this aquatic model, as reviewed here.

Keywords Zebrafish • Habituation • Learning • Memory • Startle response • Novel tank • Spatial alternation • Three-chamber tank • T-maze • Plus maze

1 Introduction

Although understanding the concept of "memory" is a foundational component of studying animal models of learning [1], defining an internal cognitive construct is not an easy task. Early definitions originated from philosophical discussions within the emerging field of psychology, and evolved as more established experimental techniques were developed to test and measure these definitions. The rise of behaviorism and cognitive psychology coincided with advancements in technology and the advent of multidisciplinary perspectives like neuroscience and biological psychiatry. Today, one widely accepted definition of memory is "the retention of experience-dependent internal representations over time" [2]. Memory may then be subdivided into short-term memory, and long-term memory; the latter is further divided into explicit (declarative) memory and implicit (non-declarative) memory; see Table 1 for details [10, 11].

D.J. Echevarria (✉) • A.D. Collier • E.A. Lamb
Department of Psychology, University of Southern Mississippi, 118 College Drive, Hattiesburg, MS 39406, USA
e-mail: david.echevarria@usm.edu

A.V. Kalueff (ed.), *The rights and wrongs of zebrafish: Behavioral phenotyping of zebrafish*, DOI 10.1007/978-3-319-33774-6_4

Table 1 Memory subdivisions and relevant cognitive tests

Type of memory	Definition	Human test	Rodent test	Zebrafish test	Zebrafish ref
Short-term memory	Maintaining relevant information for a short period of time	Change detection, delayed recognition	Mazes, delayed spatial alternation, fear conditioning, place conditioning, operant conditioning	Mazes, delayed spatial alternation, fear conditioning, place conditioning, operant conditioning	[3–7]
Long-term memory: explicit	Conscious recollection of facts, knowledge and events	Explicit retrieval, free recall, digit span	NA	NA	NA
Long-term memory: implicit	Unconscious acquisition, storage, retention, retrieval and extinction of memories	Priming, emotional conditioning	Mazes, delayed spatial alternation, fear conditioning, place conditioning, operant conditioning	Mazes, delayed spatial alternation, fear conditioning, place conditioning, operant conditioning	[3–7]
Implicit memory: non-associative	Reflexive response memory	Habituation, startle response	Habituation, startle response	Habituation, startle response	[8, 9]

Short-term memory may be defined as the maintenance of relevant information for a short period of time that is susceptible to decay without rehearsal or sustained attention [12]. Explicit memories are those that most fit with a lay definition of memory, such as conscious recollection of specific facts, past events, and semantic knowledge [13]. Conversely, implicit memory is related to the *unconscious* acquisition, storage, retention, retrieval, and extinction of memories [13]. Implicit memories occur rather automatically and manifest through the performance of a behavior after information has been encoded via experience [12]. Implicit memory and the corresponding behaviors are the target of many of the cognitive tasks that are used to assess learning and memory in animal models [14]. These cognitive tasks often involve training animal subjects to perform discrimination tasks or navigate spatial mazes, with the animal associating a particular cue with the appropriate response behavior (Table 1). This related external behavior is measured to indicate that the animal has learned the requirements of the task and that implicit memory formation has occurred.

Non-associative learning is another sub-set of implicit memory and centers around reflexive responses, representing a basic form of learning (Table 1). Of particular interest to simple learning research is habituation, which may be assessed to characterize cognitive phenotypes [15]. Habituation was succinctly first defined as a "response decrement as a result of repeated stimulation" [16]; that is to say, habituation is the reduction of a reflexive response that is initially automatic after the presentation of a stimulus. Successive stimulus presentation results in dampened and potentially extinguished behavioral responses, as well as a decrease in synaptic transmission [17, 18]. However, habituation must be differentiated from mere sensory adaptation to the stimulus (in which the stimulus can no longer be appropriately perceived) or fatigue (when the animal is physically incapable of performing the response behavior). Importantly, habituation is a form of behavioral adaptation in which the animal learns not to respond to the recurring stimulus. Therefore, a number of behavioral phenotypes may be assessed to conclude that true habituation has occurred, such as changes in the rate of habituation corresponding to changes in the inter-stimulus intervals and spontaneous recovery of the response behavior after cessation of regular stimulus presentation. Additionally, dishabituation may be evaluated, in which after habituating to one stimulus the response behavior reappears at full strength upon presentation of a new stimulus [18, 19].

In line with the ultimate goal of understanding learning, memory and the neural mechanisms underlying these processes, modern research has shifted towards utilizing animal models. The neurotransmitters glutamate, GABA, acetylcholine, norepinephrine, dopamine, enkephalin, vasopressin, oxytocin, and galanin play significant roles in the formation of memories [20]. Administration of drugs that target these neurotransmitters is one of the primary experimental techniques to assess the biochemical aspects of learning and memory in model species; see Table 2 for details [31, 32]. Rodent models have been the gold standard in cognitive research for decades; a host of behavioral behavioral paradigms have been established and validated to test learning and memory processes in mice and rats. A recent growing trend in the field of neurobehavioral research has been to translate these cognitive tasks to those that can be performed by zebrafish (*Danio rerio*).

Table 2 Pharmacology of zebrafish cognitive phenotypes

Drug	Mechanism	Major effect	Ref
Startle response test			
Rolipram	PDE4 inhibitor	↑ASR	[8]
		↓Habituation	
Donepezil	Acetylcholinesterase inhibitor	↑ASR	[8]
		↓Habituation	
Memantine	NMDA receptor antagonist	↑ASR	[8]
		↓Habituation	
Mecamylamine	Nicotinic antagonist	ØASR	[8]
		ØHabituation	
Mecamylamine + donepezil	Nicotinic antagonist + acetylcho linesterase inhibitor	Blocked donepezil effects	[8]
Atropine	Muscarinic antagonist	↑ASR	[8]
		↓Habituation	
Atropine + donepezil	Muscarinic antagonist + acetylch olinesterase inhibitor	Did not block donepezil effects	[8]
Developmental chlorpyrifos	Acetylcholinesterase inhibitor	↑ASR	[21]
		↓Habituation	
Developmental pilocarpine	Muscarinic agonist	↑ASR	[21]
		↓Habituation	
Developmental nicotine	Nicotinic agonist	↑ASR	[21]
		↓Habituation	
Novel tank test			
Acute ethanol	Multiple	↓Anxiety	[15]
		ØHabituation	
Chronic ethanol	Multiple	↓Anxiety	[15]
		↑Habituation	
Acute morphine	Opioid antagonist	↓Anxiety	[15]
		ØHabituation	
Chronic fluoxetine	Selective serotonin reuptake inhibitor	↓Anxiety	[15]
		↑Habituation	
Acute caffeine	Adenosine antagonist	↑Anxiety	[15]
		↓Habituation	
Acute PTZ	GABA antagonist	↑Anxiety	[15]
		↓Habituation	
Alarm pheromone	?	↑Anxiety	[15]
		↓Habituation	
Acute nicotine + mecamylamine	Nicotinic agonist + nicotinic antagonist	↓Anxiety	[22]
		↑Habituation	
Acute nicotine	Nicotinic agonist	Blocked nicotine effect	[22]
Chronic nicotine	Nicotinic agonist	↑Anxiety	[23]
		↓Habituation	

(continued)

Table 2 (continued)

Drug	Mechanism	Major effect	Ref
Acute piracetam (100 and 400 mg/L)	?	ØAnxiety	[24]
		ØHabituation	
Acute piracetam (700 mg/L)	?	↑Anxiety	[24]
		↓Habituation	
Chronic piracetam (200 mg/L)	?	ØAnxiety	[24]
		ØHabituation	
Learned spatial alternation			
Developmental ethanol	Multiple	↓Performance	[25]
Three-chamber choice			
Acute nicotine	Nicotinic agonist	↑Performance	[26]
		↑DOPAC	
Acute mecamylamine	Nicotinic agonist	Ø DOPAC	[26]
Acute nicotine + mecamylamine	Nicotinic agonist + nicotinic antagonist	Blocked DOPAC increase	[26]
Acute nicotine (50, 100, 150, 200, 400 and 800 mg/L)	Nicotinic agonist	↑Performance at 50, 100 mg/L	[27]
		↓Performance at 150, 200, 400, 800 mg/L	
T-maze			
MK-801	NMDA antagonist	↓Performance	[28]
		Ø	
Chronic ethanol	Multiple	↓Performance	[29]
LSD	5HT2A agonist	↓Performance	[30]
Plus maze			
Alarm pheromone	?	↓Performance	[5]
Chronic piracetam	?	↑Performance	[24]

Ø no effect, ↑ increased effect, ↓ decreased effect, ? not well understood

Zebrafish possess an ideal balance of simplicity and complexity in both behavior and physiology, and albeit the zebrafish nervous system is simpler (e.g., than in rodents), this aquatic species is still able to modulate complex behaviors related to, for instance, learning [33, 34], addiction [35, 36], aggression [37, 38] and social behavior [39, 40]. There are many advantages of utilizing zebrafish as a model for learning and memory. For example, the entire zebrafish genome has been sequenced, which provides a wealth of opportunity for testing the effects of genetic mutations on learning capabilities [7]. Also, zebrafish have a rapid development and may be subjected to behavioral tests at both larval and adult stages, with larvae being capable of exhibiting forms of both associative and non-associative learning [8, 41]. This is particularly useful for high throughput screening, in which drug effects can be tested on large numbers of larvae simultaneously. Drug administration is also convenient with zebrafish, as most compounds can be added to the ambient water and quickly absorbed [42]. Alternatively, drugs may be administered via injection (e.g., intraperitoneal (IP) injection or intramuscular (IM) injection) which allows for precision in dosing and is useful for compounds that are not water soluble or are less economical to use in the larger doses required for water solutions [43, 44].

In light of the benefits of this aquatic model, scientists have spent recent years adapting preexisting rodent experimental designs to suit zebrafish morphology, selected natural tendencies, and physical capabilities [45]. Rodent tasks can logically be transferred to zebrafish, as the exploratory behaviors that serve as the basis for many of the cognitive tasks have been evolutionarily conserved between zebrafish and rodents [46, 47] Moreover, even when seemingly major components of tasks must be altered to facilitate use on an aquatic species, the conclusions that can be drawn regarding learning should be valid as long as the test maintains focus on the same cognitive processes as the original version [45]. Zebrafish cognitive phenotypes are sensitive to pharmacological manipulation to screen for cognitive enhancers and investigate the neural basis of learning and memory, and in many cases, this may be completed in a high-throughput manner (Table 2). The zebrafish adaptations of these tests may even prove more informative than the original rodent versions in some instances, as a function of the added third dimension of movement within aquatic environments [48, 49]. Assessing zebrafish cognitive phenotypes in conjunction with rodent studies is important, particularly in regards to confirming that behaviors and mechanisms are conserved between these two vertebrate species, which adds strength to their role as a model for human conditions.

Although the field of zebrafish behavioral and cognitive research is rapidly growing, it is still a relatively new area. Thus, although zebrafish have still not yet been tested as thoroughly as classic vertebrate (i.e., rodents) models, the continuation of behavioral research on this species may yield significant insight into the fundamental neural mechanisms of learning and memory. This chapter introduces several behavioral paradigms to assess zebrafish learning and memory. While not exhaustive, the selected models represent the state of neurobehavioral research in this up-and-coming animal model.

2 Behavioral Paradigms

2.1 Startle Response Test

As habituation studies are dependent on analyzing changes in reflexive responses, tests have been developed focusing on the zebrafish startle reflex. This response is evolutionarily conserved, as it is protective against predatory threats and its automatic nature makes it a suitable behavior for studying habituation, a simple form of non-associative learning [50, 51]. This measure has been thoroughly studied in rodent models and found to be a sensitive indicator of drug effects [52]. For example, NMDA receptor antagonists, glycine receptor ligands and the benozdiazepines diazepam and flurazepam attenuate startle response behavior in rats [50, 53]. This relatively simple behavior manifests early in zebrafish development (i.e., 4–5 dpf), allowing startle responses to be studied in both larval and adult animals [9, 54]. For example, the evaluation of 96 larval zebrafish may be completed simultaneously in a well plate by exposing fish to a sound stimulus and measuring their acoustic startle response (ASR). ASR may be quantified as changes in distance travelled following a sound presentation, with an increase in distance traveled corresponding to an increased ASR. Larval zebrafish exhibit increased ASR following the first stimulus presentation, and show attenuated ASR over repeated stimulus presentations, indicating a habituation response [8]. Following the establishment of habituation, a single presentation of a secondary startling stimulus (e.g., a light pulse) restores the initial ASR response, suggestive of dishabituation [8].

Pharmacological manipulation aids in the elucidation of pathways underlying zebrafish startle responses and habituation phenotypes. This also helps to identify potential cognitive enhancing drugs, for example, rolipram and memantine (see Table 2 for details) increase ASR and decrease habituation in larval zebrafish, which may be interpreted as enhanced cognition via increased alertness [8]. Donepezil, an acetylcholinesterase inhibitor is used as a therapy for human Alzheimer's patients and has been found to improve cognitive ability [55, 56]. Similar to rolipram and memantine, donepezil increases ASR and decreases habituation in larval zebrafish [8]. The effects of donepezil are blocked when combined with mecamylamine (nicotinic antagonist) but not when combined with atropine (muscarinic antagonist), suggesting that the potentiated ASR effects from donepezil are mediated by nicotinic receptors [8]. This finding is further supported by the cognitive enhancing effects of acute nicotine in adult zebrafish within a spatial position discrimination learning task [57]

A related response to the ASR is the reflexive startle following a mechanical tap stimulus. Automated systems provide equal tap stimulus intensity to multiple animals simultaneously (Fig. 1). An automated assessment of the reflexive startle response has been employed to evaluate persisting neurobehavioral and neurochemical changes following exposure to the environmentally salient organophosphate pesticide chlorpyrifos (CPF) during early zebrafish development [21]. CPF exposure during early development in rodent models causes a variety of lasting neurotoxic effects during adulthood [58, 59]. Larval zebrafish exposed to CPF 0–5 dpf do not habituate to the tapping stimulus and display an increased startle response when tested as adults. This behavioral change has been interpreted as hyperactivity rather than cognitive enhancement, which is supported by impaired performance in a spatial discrimination task during adulthood following developmental CPF exposure [21, 60]. Larval animals tested at 6 dpf show reduced serotonin and dopamine, and adult zebrafish developmentally exposed to CPF show reduced dopamine levels [21]. Developmental nicotine and pilocarpine (muscarinic agonist) exposure increase the startle response and decrease habituation in zebrafish tested in adulthood, albeit not as potently as CPF, suggesting that the nicotinic and muscarinic systems may synergistically modulate the persisting effects of CPF or other transmitter systems may be involved [21]. Evaluating changes in startle response behavior and non-associative learning (i.e., habituation) in zebrafish, especially larval zebrafish, is a powerful system for screening the cognitive enhancing/impairing and neurotoxic effects of drugs in a high-throughput manner. However, it would be best to employ multiple behavioral paradigms to fully understand how drugs modulate cognition.

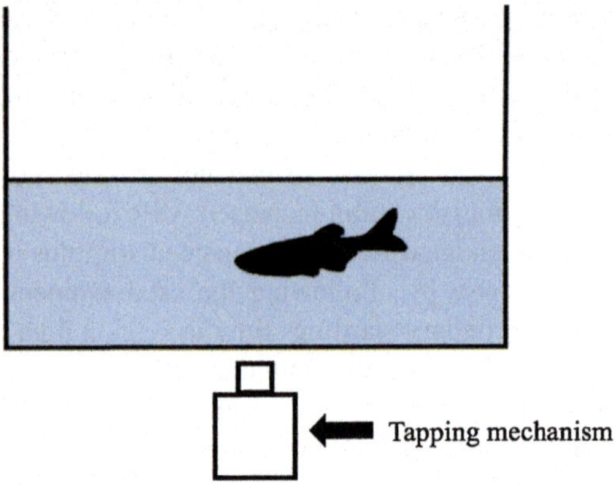

Fig. 1 Startle response test. *Side view*. An automated system delivers standardized tap stimuli to elicit the reflexive startle response to tapping, as measured by the fish swim velocity for 5 s past stimulus. This task is used to analyze habituation as indicated by change in the startle response over repeated presentations of the stimulus (from Eddins et al. [21])

2.2 *Novel Tank Test*

As an alternative to the startle response, the propensity of zebrafish to occupy the bottom of a novel tank (geotaxis) is another target for habituation studies. In their natural habitat, the upper water column likely makes zebrafish vulnerable to aerial predators [61]. As a result, this potential predation threat motivates zebrafish to seek safety in deeper water within a novel environment. This change in vertical exploration can be easily measured in a laboratory setting as a test of habituation by introducing a fish into a novel tank. The novel tank test apparatus often consists of a 1.5-L trapezoidal tank delineated horizontally into two equal sections that mainly limits fish to vertical exploration (Fig. 2). The novel tank test has been likened to the open field test [51] and elevated plus maze [22] used in rodent behavioral research. The similarity comes from the assessment of habituation in terms of the attenuation of animals' natural anxiety to open and novel spaces, as shown via increased exploration over time. Behavioral parameters of anxiety and habituation assessed in the novel tank test include, for example, latency to enter the top, time spent in and number of entries into the top section, as well as additional indicators of anxiety (e.g., erratic movements and duration and number of freezing events) [62]

The novel tank test assesses both intra- and inter-session habituation phenotypes in zebrafish. Intra-session (i.e., within trial) habituation in this task indicates the function of working spatial memory, while inter-session (i.e., between trials) habituation is suggestive of long-term memory capabilities. Adult zebrafish display robust intra-session habituation in both 6- and 30-min trials, as well as inter-session habituation to 6-min trials over a period of 7-days [15]. Anxiogenic

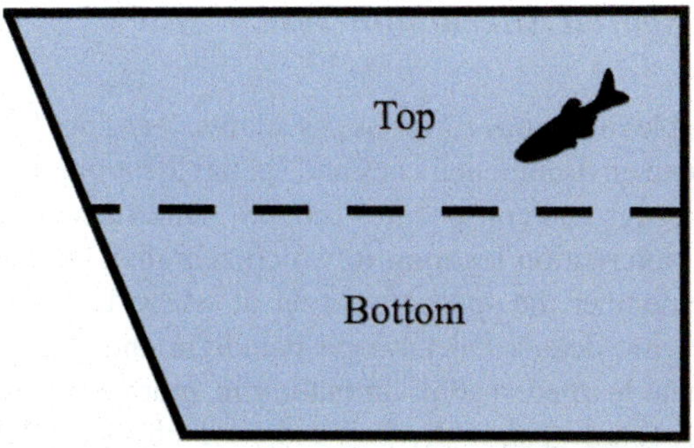

Fig. 2 The novel tank test. *Side view*. A trapezoidal 1.5-gallon housing tank is divided into two horizontal halves. This test may be used to assess intra- and inter-session habituation via changes in anxiety-like and exploratory behavioral parameters throughout the test session, such as the per-minute distribution of the time spent in the upper half, transitions to upper half, swim velocity and distance traveled. These behaviors may be pharmacologically manipulated by treating animals with drug prior to testing (from Cachat et al. [62])

treatments (e.g., caffeine, pentylenetetrazole (PTX) and alarm pheromone) attenuate intra-session habituation. Select anxiolytic treatments (e.g., chronic ethanol and chronic fluoxetine) increase habituation, albeit others (e.g., acute ethanol and acute morphine) have no effect on habituation [15]. Piracetam, a cognitive enhancer, reduces anxiety and improves learning and memory in both rodents and humans [63–65]. Lower doses (i.e., 100 and 400 mg/L) of acute piracetam in adult zebrafish have no effect on anxiety-like behaviors or intra-session habituation and a higher dose (i.e., 700 mg/L) impairs intra-session habituation and increases geotaxis but not other anxiety-like behaviors [24]. Chronic treatment of 200 mg/L piracetam has no effect on intra-session or inter-session habituation or anxiety-like behaviors in the novel tank test. However, in an alternative test of anxiety-like behaviors, the light–dark test, chronic piracetam does reduce time spent in the dark half of the tank, indicative of anxiolysis. Furthermore, chronic piracetam improves zebrafish learning and memory in a cued learning plus-maze test (see further). Acute nicotine decreases bottom dwelling in the novel tank, indicating decreased anxiety and habituation to the novel environment over time [22]. Chronic nicotine however increases bottom dwelling and decreases habituation in the novel tank test, an effect opposite to acute treatment [23]. Co-administration of mecamylamine (nicotine antagonist) with acute nicotine 5-min before behavioral testing attenuates habituation and top dwelling, but not when co-administered 20-min before testing [22]. These findings are useful examples of the importance of recognizing that drug effects on cognitive phenotpyes (e.g., habituation) may differ based on a number of factors, including mechanism of action, dose, behavioral paradigm, length of exposure time, and latency between exposure and behavioral testing.

2.3 Learned Spatial Alternation Test

Zebrafish are capable of various forms as associative learning. For instance, zebrafish learn to associate environmental cues and spatial locations with rewarding stimuli, such as food, drugs and conspecifics [66–68]. One such domain of associative learning is spatial alternation learning in which zebrafish are tested, for example, their ability to remember the spatial location of where they last received a food reward. The following details a behavioral paradigm and experimental procedure designed to evaluate learned spatial alternation in zebrafish [69]. A 2-gallon tank was fitted with a central opaque divider, under which the zebrafish had space to swim to access the sides of the tank, with one side marked with a red card to aid in spatial orientation (Fig. 3). For a series of 28 trials, food was presented at alternating sides of the tank 5 s after the divider was tapped as a cue. The time after the cue, but prior to food administration, was considered the choice phase. The location of the fish at the time of food administration was recorded to serve as a measure of choice accuracy. The fish were considered to have successfully learned the task when performing at above 75 % correct, which they were able to do within the first 14 trials. Additionally, zebrafish were capable of remembering how to perform the task after a 10-day delay. When the fish were tested without the food reward, there was evidence of extinction, as choice accuracy decreased. Both juvenile fish aged 6–8 weeks old and adults (>8 weeks) proved successful at this learning task, albeit juveniles 3–4 weeks old did not perform better than chance, suggesting that the latter group may have been too young to learn and remember the task. In a separate study zebrafish were applied as a model of fetal alcohol syndrome and treated with ethanol (10 and 30 mM) for 24-h postfertilization [25]. Animals were then raised to

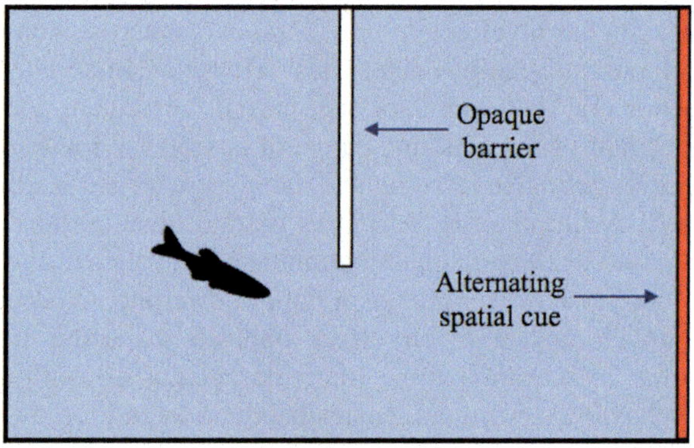

Fig. 3 Learned spatial alternation test. *Side view*. A 2-gallon tank is divided by an opaque barrier beneath which the fish can swim to access either side of the apparatus. A visual cue card (e.g., a red card) is presented on alternating sides of the tank along with a food reward, serving as a spatial orientation device. Food is presented 5 s after the central barrier is tapped as a cue. Using this apparatus spatial learning is measured by the location of the fish at the time of food administration (from Williams et al. [69])

adulthood using standard husbandry procedures and tested in a learned spatial alternation task as outlined above. Zebrafish developmentally exposed to 10 mM ethanol performed worse than controls and animals exposed to 30 mM ethanol showed the greatest deficits in this task, indicating that zebrafish cognitive phenotypes are sensitive to developmental ethanol exposure.

Other stimuli that are rewarding to zebrafish may be applied in learned spatial alteration paradigms. As a shoaling species, zebrafish prefer to spend time in proximity to conspecifics, and as such, the sight of conspecifics may be employed as a rewarding stimulus [66, 70, 71]. Zebrafish learn to anticipate the presentation of computer generated images of conspecifics on alternating sides of a shuttle box, as indicated by movement from one side of the tank to the other during the time between stimulus presentations [72]. This particular paradigm is important for behavioral zebrafish studies empowering high throughput screening since multiple set-ups can be run simultaneously and the task can be entirely automated. The stimulus presentation is computer-controlled and all behaviors are recorded and can be scored by video tracking software, eliminating the need for human involvement and the potential introduction of any biases or human errors. Learned spatial alternation paradigms may be strengthened however by employing an elongated tank would increase the effort required for zebrafish to travel to the correct side, indicating that its presence on that side was an intentional choice and not simply due to the natural meandering of the fish around the tank.

2.4 Three-Chamber Tank Test

The three-chamber task often uses a tank with two clear Plexiglas walls inserted to divide it into three equal chambers (Fig. 4). The central portion serves as the start area and the divider walls contain doors that can slide up to provide access to the side choice chambers. This apparatus is a valuable tool in zebrafish behavioral research as it can be used in a wide variety of contexts to test different components of learning and memory. For example, an early study investigated both spatial and non-spatial visual discrimination learning using the three-chamber task apparatus [73]. Zebrafish were trained to go to either the left or right side to avoid a threatening stimulus (i.e., a moving fishnet) that was inserted into the central chamber, and the choice accuracy and latency were measured. The back wall of the apparatus was marked with a dark colored panel to be used as a directional reference point. If the fish entered the correct chamber, the net was removed; however, if an incorrect choice was made, the net disturbance was added to the choice chamber for 60 s as punishment. The zebrafish were shown to be successful at this learning task and they improved with repeated testing, as indicated by the number of correct choices made. Zebrafish also accomplished reversal learning when the side that was considered correct was changed. This supports the conclusion that zebrafish are capable of spatial learning and provides support for their cognitive flexibility. A second task in this study assessed non-spatial learning and used the three-chamber tank to test performance in a visual color

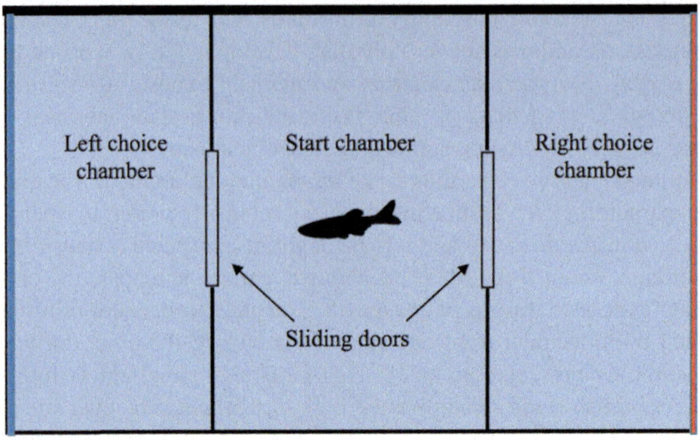

Fig. 4 Three-chamber tank test. *Top view*. Clear Plexiglas walls equally divide a tank into a start area and two side choice chambers lined with removable color inserts (e.g., *red* and *blue*) on each far wall. Gates in the dividing walls are lifted to allow the fish access to the choice chambers. Entry into the incorrect chamber results in a confinement punishment by sliding the wall to decrease the area of the choice chamber (from Levin and Chen [27])

discrimination test. The protocol was similar to the previous experiment, except that instead of going to a particular side, fish were trained to go into the chamber lined with their assigned color (e.g., red or blue). Again, the zebrafish exhibited learning and could relearn the task with newly switched color parameters. The authors note this research could be expanded by including drug treatments or gene manipulations to tease apart the possible explanations for success (or failure) in this task. For example, if altered fish are able to escape from the net stimulus and enter a chamber, yet cannot correctly complete the reversal task, it can be clearly concluded that the deficit in performance is due to a cognitive impairment and not, for example, a failure in motivation or motor function.

In another variation of the three-chamber tank, a round-bottom tank evaluated the role of nicotinergic and dopaminergic mechanisms on zebrafish performance [26]. The fish first established a preferred chamber (based on three consecutive entries into the same side), and subsequent entries into this chamber were then deemed an 'incorrect' choice and zebrafish were punished with confinement in a 1 cm swimming space. Seven subsequent trials were run to examine if the fish would learn to choose the correct (un-preferred) side. Acute nicotine improved the learning rate in this task, accompanied by increased brain levels of the primary dopamine metabolite, DOPAC, which serves as an indirect indicator of dopamine activity, the levels of which correlated with zebrafish choice accuracy. The nicotinic antagonist mecamylamine administered in conjunction with nicotine prevented the nicotine-induced DOPAC increase, but not when administered alone, allowing the overall conclusion that dopamine systems are involved in the neural processes underlying zebrafish learning in this type of task. These results confirm previous work with rodent models and the involvement of dopamine systems in learning and memory [74–77]. Therefore, further efforts by the pharmaceutical and medical communities can be devoted to developing novel nicotinic agonists to treat cognitive impairments associated with Alzheimer's disease, ADHD, and schizophrenia [78–80]. Based on the data gleaned from this set of experiments, it seems that this species and behavioral paradigm would be appropriately suited to preliminary in vivo tests of such novel potential pharmaceutical candidates.

The three-chamber tank apparatus has been used in a subset of the research to test zebrafish performance in a spatial alternation task [27, 57]. In this task, the correct choice chamber alternates on successive trials, as previously discussed. [27] tested the effects of nicotine on zebrafish performance in this task. Using this task, nicotine was administered in six different doses prior to behavioral testing to determine a dose-effect function [27]. Control fish were successful at remembering which chamber had previously been rewarded and showed a significant linear trend of improvement. The drug test results indicated that the lower doses of nicotine (i.e., 50 and 100 mg/L) improved memory function and high doses (i.e., 150, 200, 400 and 800 mg/L) diminished learning ability. This biphasic nicotine effect in increasing memory performance at lower doses and impairing performance at higher doses has also been shown in rodents, primates, and humans [81, 82]; once again this supports the validity of zebrafish as a model organism for studying the molecular mechanisms underlying nicotinic effects on learning and memory.

2.5 T-Maze Test

The T-maze is another testing apparatus that may be applied in a variety of ways, as has been shown in the rodent literature. Historically, a primary utility of this apparatus is in testing visual discrimination learning in rats in a simple maze format, especially in evaluating which extra-maze or kinesthetic cues affect the animal's ability to learn the task [83]. Additionally, it has been used to study what, if any, cognitive mapping underlies the observable maze learning behavior in rodents [84]. The T-maze can be employed to study various components of associative learning, including acquisition, extinction and reversal. The zebrafish T-maze apparatus is often a clear Plexiglas tank with an elongated stem with two shorter arms extending perpendicularly from one end that may be covered with colored or patterned sleeves. The aquatic apparatus also contains equally spaced slots throughout the maze to allow gates or walls to be inserted to limit zebrafish motility, most often in the formation of acclimation start boxes, or blocking off of particular areas (Fig. 5) [85]. The T-maze was first applied as a test of learning and memory in zebrafish to serve as a control for other variables of the study, for example, to show that changes in conditioned place preference were the result of drug or genetic manipulations and not due to detriments in the subjects' memory [6, 28, 86, 87]. A deep area with substrate and plants can be placed at the end of one of the arms of the T-maze to evaluate how well experimentally manipulated animals can learn and remember the location

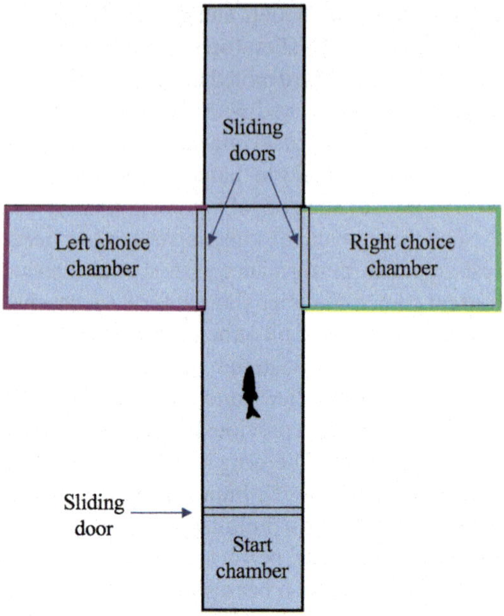

Fig. 5 T-Maze test. *Top view*. A Plexiglas offset cross maze is configured into a T-maze with a drop-in wall for use in a visual discrimination task (e.g., color discrimination). After being released from the start chamber at the end of the long stem, fish swim into either of the colored arm choice chambers and upon entry into the correct chamber arm fish are provided a food reward. Incorrect choices are followed by a correction trial (from Colwill et al. [4])

of this desirable enriched environment; control fish exhibit a preference for this environment and learn its spatial location [28, 86]. Pharmacological treatments may reduce the motivation of zebrafish to enter an enriched arm of the T-maze rather than impairing cognitive function, requiring a careful interpretation of results [28].

The T-maze was first applied to study visual discrimination learning in zebrafish by evaluating how well subjects learn to go into the arm of the maze with a particular color or pattern sleeve (e.g., purple/green, red/blue, horizontal/vertical stripes) that has been paired with delivery of a food reward [4]. The fish underwent a series of acclimation pre-training trials prior to the actual task to eliminate any confounding effects of stress or anxiety associated with isolation in a novel environment. For the actual discrimination task, the fish began each trial acclimating in the start box at the end of the long stem of the maze. After a gate was raised, a single fish was free to swim into the maze and choose between the two arms. Once the fish swam into one of the arms, it was confined to that area with a lowered gate for 30 s before being netted and returned to the start box for subsequent trials. If the choice was correct, the fish was rewarded with food and if it was incorrect, a correction trial would follow, with the previously chosen side blocked off. Once the fish had learned to choose their assigned color, a series of extinction sessions were conducted to see if the preference for their particular color would diminish when no rewards given. Finally, following extinction, trials were run in which the alternate color would be rewarded, testing discrimination reversal. For each of these trials, the latency to enter one of the choice arms and whether it was correct was recorded. In each of the three versions of the task, the zebrafish were shown to learn to discriminate between the two choices and select the correct arm, and over the course of the trials the latency to choose decreased. Additionally, the conclusion that the fish learned to choose a certain color/pattern based on being given a food reward, was strengthened by the observed extinction of success with retiring food reward administration. The fish were also capable of learning to reverse their preference.

Manipulation of zebrafish T-maze cognitive phenotypes can help reveal the pathways underlying spatial learning. For example, chronic ethanol impairs zebrafish T-maze performance, but has no effect in an active-avoidance learning task, suggesting that alcohol specifically impairs spatial learning, but not CS-US associative learning [29]. Lyrsergic acid diethylamide (LSD) decreases the ability of zebrafish to associate an arm of the T-maze with a rewarding conspecific exposure, but has no effect in a social preference test, suggesting that LSD may modulate spatial discrimination learning but not social preference [30]

2.6 Plus Maze Test

The plus maze is similar to the T-maze, only differing in shape as it consists of four equal length arms extending from a central juncture point that may be placed on a rotating platform (Fig. 6), representing a zebrafish equivalent of the radial arm maze used to test spatial and non-spatial associative learning in rodents [7]. In a test of non-spatial learning, zebrafish performed a cued memory task [7]. Zebrafish learned to associate an internal maze cue (a red card) with a food reward and exhibited a preference for entering the arm marked with the red card, regardless of position. This is essentially a more complex version of the aforementioned learned spatial

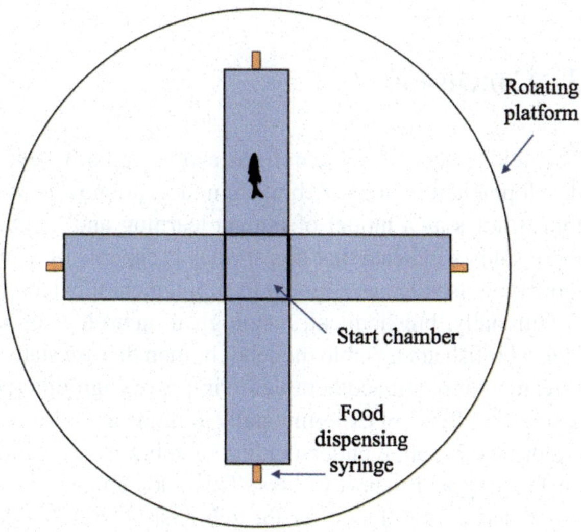

Fig. 6 Plus maze test. *Top view.* Fish typically begin trials in the central start box and then explore the maze after the walls have been lifted via a pulley system. Positioning of the food dispensing syringes at the end of each arm allow for control of which choices result in reward. Fish can be tested on their ability to use intra- or extra-maze cues to determine the location of rewards (from Sison and Gerlai [7])

alteration task [69] Spatial memory was also evaluated in this study using the plus maze, similar to the classic Morris Water Maze task for rodents [88]. This task also required the fish to learn to associate a cue with the location in which they received a reward, but this time the cues were external cues from the room, such as the position of lights or laboratory equipment; zebrafish are also successful at this task [7]. The task assessed which arm was selected and how much time passed before entering a choice chamber.

The effects of acute stress stimuli (i.e., alarm pheromone and predator fish) on learning were assessed using the same protocol [5]. Both stressors impaired zebrafish performance in the spatial and non-spatial tasks. This study analyzed a number of measurements, which allowed them to rule out alternative explanations for the findings (e.g., decreased foraging motivation or increased escape-seeking). These extra analyses are important to ensure that the apparent results are not just due to alternate factors, for example, decreased locomotor activity.

Chronic piracetam significantly increased performance in the plus maze learning task, an important finding as it mirrors piracetam's effects previously established in rats and humans, further validating the use of the zebrafish model in screening for cognitive enhancers [24]. Another variation of this task replaced food reward with the reward of proximity to conspecifics [66]. Small tanks were placed next to an opening in each arm of the maze, only one of which contained zebrafish. The results show that the fish learned to associate the red card with the conspecific reward and continued to spend more time near the cue even when the conspecifics were no longer present. Note that for replacing the reward type, there may be multiple reasons why food rewards may not meet the needs of a proper incentive for an associative learning task and is thus why using conspecifics may act as a viable alternative [66]. For example, the amount of food reward consumed by zebrafish cannot be controlled and variation in satiation from previous feedings may be present and vary on the size and sex of subjects.

3 Conclusion

The field of zebrafish research has evolved greatly beyond its origins in genetics and developmental biology. Zebrafish models are now being utilized in cognitive behavioral studies as a model of human learning and memory. The collective results of these studies indicate that this species is capable of learning and memory processes, indicating its value as a model to supplement the data from rodent studies (Table 1). Additionally, biochemical, histological, neurological, and anatomical data suggest that zebrafish are a viable model of human disease states and a solid candidate for the screening of pharmacotherapies to improve cognition [24, 35]. In order to effectively assess the effects of experimentally administered drugs, it is imperative that we have a thorough baseline understanding of zebrafish cognitive behavior and performance on learning and memory tasks. This adds emphasis to the fact that not only must these tests be developed for the zebrafish model, but they must also be empirically assessed and found to be both reliable and valid. It is critical that direct and parsimonious behavioral paradigms are developed and validated to investigate and characterize cognitive behavioral phenomena in zebrafish [27, 66, 73]. The behavioral paradigms and models discussed here show that significant steps have been taken to fulfill that very need. As shown in Tables 1 and 2, many types of learning can be studied and manipulated in the zebrafish; ranging from quick and efficient studies of habituation (amenable to high-throughput drug screens) to more elaborate tests of associative learning (that require more time for training, but can yield insight into the processes of acquisition, retention, recall, and more detailed aspects of memory).

While the presented studies indicate that these tasks are appropriate for use in zebrafish, each paradigm has the potential to provide more data on specifics of the learning process if methods and apparatuses are amended, new measures are introduced, and the general development of these tasks continues [89]. With the recent sequencing of the zebrafish genome, along with the ease and availability of genetic mutant models, the opportunity exists to expand upon the knowledge of the genetic origins and the influencing factors of genes on neurobehavioral components, most notably, learning and memory. The potential for continued expansion and rapid growth of the field of neurobehavioral research with zebrafish is immense. This is particularly evident when considering the fact that investigators using this species are privy to the benefit of following the path previously forged by the classic neurobehavioral studies of rodent models, which experienced rapid growth upon the advent of abundant genetic mutants. In addition to genetic modifications, pharmacological administration provides a wealth of opportunities to test the functioning of a non-normal system (Table 2). The value in this comes from the ability to utilize drug exposure as both a method of producing a model of a particular brain disease, as well as a method of testing the unknown behavioral effects of a new compound. Testing in zebrafish various compounds with well-established effects on learning and memory from previous rodent studies is also valuable (Table 2). By utilizing the zebrafish model for studying the role of neurotransmitters in learning behavior, it is likely that the underlying molecular mechanisms may more quickly be revealed by studying them in a more simplistic system. Also, due to the decreased time necessary to run these behavioral experiments in zebrafish, more drugs may be tested in a shorter amount of time, thus hastening the progress towards a fuller understanding of the role of neurotransmitters in learning and memory.

This brief review of the zebrafish cognitive behavioral literature sought not only to present the current state of the available neurobehavioral paradigms, but also to highlight the potential development of novel zebrafish models of learning and memory through use of these methodologies. As many of these zebrafish behavioral paradigms have multiple iterations, with one apparatus or task being able to be used to study multiple behaviors with can be analyzed in a number of ways, this further increases the value of each paradigm. The latest progression in the field of zebrafish research has been a shift in emphasis to study the behavior of these animals, an important transition which will allow for a more informative data set regarding learning and memory, through the study of mutant or wild type strain differences; the effects of genetic manipulations, in vivo drug screens, or other treatments.

References

1. Spear NE, Miller JS, Jagielo JA. Animal memory and learning. Annu Rev Psychol. 1990;41(1): 169–211.
2. Dudai Y. The neurobiology of memory: concepts, findings, trends. New York: Oxford University Press; 1989.
3. Bilotta J, Risner ML, Davis EC, Haggabloom SJ. Assessing appetitive choice discrimination learning in zebrafish. Zebrafish. 2005;2(4):259–68.
4. Colwill RM, Raymond MP, Ferreira L, Escudero H. Visual discrimination learning in zebrafish (*Danio rerio*). Behav Processes. 2005;70(1):19–31. doi:10.1016/j.beproc.2005.03.001.
5. Gaikwad S, Stewart A, Hart P, Wong K, Piet V, Cachat J, Kalueff AV. Acute stress disrupts performance of zebrafish in the cued and spatial memory tests: the utility of fish models to study stress-memory interplay. Behav Processes. 2011;87(2):224–30. doi:10.1016/j.beproc.2011.04.004.
6. Ninkovic J, Bally-Cuif L. The zebrafish as a model system for assessing the reinforcing properties of drugs of abuse. Methods. 2006;39(3):262–74. doi:10.1016/j.ymeth.2005.12.007.
7. Sison M, Gerlai R. Associative learning in zebrafish (*Danio rerio*) in the plus maze. Behav Brain Res. 2010;207(1):99–104. doi:10.1016/j.bbr.2009.09.043.
8. Best JD, Berghmans S, Hunt JJ, Clarke SC, Fleming A, Goldsmith P, Roach AG. Non-associative learning in larval zebrafish. Neuropsychopharmacology. 2008;33(5):1206–15. doi:10.1038/sj.npp.1301489.
9. Chanin S, Fryar C, Varga D, Raymond J, Kyzar E, Enriquez J, et al. Assessing startle responses and their habituation in adult zebrafish. In: Zebrafish protocols for neurobehavioral research. New York: Springer; 2012. p. 287–300.
10. Atkinson RC, Shiffrin RM. Human memory: a proposed system and its control processes. In: Spence KW, Spence JT, editors. Psychology of learning and motivation, vol. 2. New York, NY: Academic; 1968. p. 89–195.
11. Karami NR. Explicit and implicit memory. Advances in cognitive science. 2002;3(4):57–65.
12. Koob GF, Le Moal M, Thompson RF. Encyclopedia of behavioral neuroscience, three-volume set, 1–3: online version. Oxford: Newnes; 2010.
13. Banks WP. Encyclopedia of consciousness, vol. 1. San Diego: Academic Press; 2009.
14. Squire LR. Memory systems of the brain: a brief history and current perspective. Neurobiol Learn Mem. 2004;82(3):171–7.
15. Wong K, Elegante M, Bartels B, Elkhayat S, Tien D, Roy S, et al. Analyzing habituation responses to novelty in zebrafish (*Danio rerio*). Behav Brain Res. 2010;208(2):450–7. doi:10.1016/j.bbr.2009.12.023.
16. Harris JD. Habituatory response decrement in the intact organism. Psychol Bull. 1943;40: 385–422.
17. Castellucci V, Pinsker H, Kupfermann I, Kandel ER. Neuronal mechanisms of habituation and dishabituation of the gill-withdrawal reflex in aplysia. Science. 1970;167:1745–8.
18. Thompson RF, Spencer WA. Habituation: a model phenomenon for the study of neuronal substrates of behavior. Psychol Rev. 1966;73(1):16–43.
19. Rankin CH, Abrams T, Barry RJ, Bhatnagar S, Clayton DF, Colombo J, et al. Habituation revisited: an updated and revised description of the behavioral characteristics of habituation. Neurobiol Learn Mem. 2009;92(2):135–8. doi:10.1016/j.nlm.2008.09.012.
20. Crawley JN. What's wrong with my mouse? Behavioral phenotyping of transgenic and knockout mice. New York: Wiley-Liss; 2000.
21. Eddins D, Cerutti D, Williams P, Linney E, Levin ED. Zebrafish provide a sensitive model of persisting neurobehavioral effects of developmental chlorpyrifos exposure: comparison with nicotine and pilocarpine effects and relationship to dopamine deficits. Neurotoxicol Teratol. 2010;32(1):99–108. doi:10.1016/j.ntt.2009.02.005.
22. Levin ED, Bencan Z, Cerutti DT. Anxiolytic effects of nicotine in zebrafish. Physiol Behav. 2007;90(1):54–8. doi:10.1016/j.physbeh.2006.08.026.

23. Stewart AM, Grossman L, Collier AD, Echevarria DJ, Kalueff AV. Anxiogenic-like effects of chronic nicotine exposure in zebrafish. Pharmacol Biochem Behav. 2015;139(Pt B):112–20.
24. Grossman L, Stewart A, Gaikwad S, Utterback E, Wu N, Dileo J, et al. Effects of piracetam on behavior and memory in adult zebrafish. Brain Res Bull. 2011;85(1–2):58–63. doi:10.1016/j.brainresbull.2011.02.008.
25. Carvan 3rd MJ, Loucks E, Weber DN, Williams FE. Ethanol effects on the developing zebrafish: neurobehavior and skeletal morphogenesis. Neurotoxicol Teratol. 2004;26(6):757–68. doi:10.1016/j.ntt.2004.06.016.
26. Eddins D, Petro A, Williams P, Cerutti DT, Levin ED. Nicotine effects on learning in zebrafish: the role of dopaminergic systems. Psychopharmacology (Berl). 2009;202(1–3):103–9. doi:10.1007/s00213-008-1287-4.
27. Levin ED, Chen E. Nicotinic involvement in memory function in zebrafish. Neurotoxicol Teratol. 2004;26(6):731–5. doi:10.1016/j.ntt.2004.06.010.
28. Swain HA, Sigstad C, Scalzo FM. Effects of dizocilpine (MK-801) on circling behavior, swimming activity, and place preference in zebrafish (*Danio rerio*). Neurotoxicol Teratol. 2004;26(6):725–9. doi:10.1016/j.ntt.2004.06.009.
29. Yang S, Kim W, Choi BH, Koh HY, Lee CJ. Alcohol impairs learning of T-maze task but not active avoidance task in zebrafish. Kor J Biol Sci. 2003;7(4):303–7.
30. Grossman L, Utterback E, Stewart A, Gaikwad S, Chung KM, Suciu C, et al. Characterization of behavioral and endocrine effects of LSD on zebrafish. Behav Brain Res. 2010;214(2):277–84.
31. Doguc DK, Delibas N, Vural H, Altuntas I, Sutcu R, Sonmez Y. Effects of chronic scopolamine administration on spatial working memory and hippocampal receptors related to learning. Behav Pharmacol. 2012;23(8):762–70.
32. Frussa-Filho R, de Lima Patti C, Fukushiro DF, Ribeiro LTC, Kameda SR, de Cassia Carvalho R. The plus-maze discriminative avoidance task: an ethical rodent model for concomitant evaluation of learning, memory, anxiety, motor activity and their interactions. Rodent model as tools in ethical biomedical research. Cham: Springer; 2016. p. 327–44.
33. Bailey JM, Oliveri AN, Levin ED. Pharmacological analyses of learning and memory in zebrafish (*Danio rerio*). Pharmacol Biochem Behav. 2015;139(Pt B):103–11.
34. Blaser R, Vira D. Experiments on learning in zebrafish (*Danio rerio*): a promising model of neurocognitive function. Neurosci Biobehav Rev. 2014;42:224–31.
35. Kalueff AV, Stewart AM, Gerlai R. Zebrafish as an emerging model for studying complex brain disorders. Trends Pharmacol Sci. 2014;35(2):63–75.
36. Tran S, Chatterjee D, Gerlai R. An integrative analysis of ethanol tolerance and withdrawal in zebrafish (*Danio rerio*). Behav Brain Res. 2015;276:161–70.
37. Toms CN, Echevarria DJ. Back to basics: searching for a comprehensive framework for exploring individual differences in zebrafish (*Danio rerio*) behavior. Zebrafish. 2014;11(4):325–40.
38. Way GP, Ruhl N, Snekser JL, Kiesel AL, McRobert SP. A comparison of methodologies to test aggression in zebrafish. Zebrafish. 2015;12(2):144–51.
39. Parker MO, Annan LV, Kanellopoulos AH, Brock AJ, Combe FJ, Baiamonte M, et al. The utility of zebrafish to study the mechanisms by which ethanol affects social behavior and anxiety during early brain development. Prog Neuropsychopharmacol Biol Psychiatry. 2014;55:94–100.
40. Qin M, Wong A, Seguin D, Gerlai R. Induction of social behavior in zebrafish: live versus computer animated fish as stimuli. Zebrafish. 2014;11(3):185–97.
41. Roberts AC, Bill BR, Glanzman DL. Learning and memory in zebrafish larvae. Front Neural Circuits. 2013;7:126.
42. Echevarria DJ, Hammack CM, Pratt DW, Hosemann JD. A novel behavioral test battery to assess global drug effects using the zebrafish. Int J Comp Psychol. 2008;21:19–34.
43. Braida D, Limonta V, Pegorini S, Zani A, Guerini-Rocco C, Gori E, Sala M. Hallucinatory and rewarding effect of salvinorin A in zebrafish: k-opioid and CB1-cannabinoid receptor involvement. Psychopharmacology (Berl). 2007;190(4):441–8.

44. Kinkel MD, Eames SC, Philipson LH, Prince VE. Intraperitoneal injection into adult zebrafish. J Vis Exp. 2010;(42). doi:10.3791/2126.
45. D'Mello GD, Steckler T. Animal models in cognitive behavioural pharmacology: an overview. Cogn Brain Res. 1996;3:345–52.
46. Stewart A, Cachat J, Wong K, Gaikwad S, Gilder T, DiLeo J, et al. Homebase behavior of zebrafish in novelty-based paradigms. Behav Processes. 2010;85:198–203. doi:10.1016/j.beproc.2010.07.009.
47. Stewart AM, Gaikwad S, Kyzar E, Kalueff AV. Understanding spatio-temporal strategies of adult zebrafish exploration in the open field test. Brain Res. 2012;1451:44–52. doi:10.1016/j.brainres.2012.02.064.
48. Cachat J, Stewart A, Utterback E, Hart P, Gaikwad S, Wong K, et al. Three-dimensional neurophenotyping of adult zebrafish behavior. PLoS One. 2011;6(3):e17597.
49. Stewart AM, Grieco F, Tegelenbosch RA, Kyzar EJ, Nguyen M, Kaluyeva A, et al. A novel 3D method of locomotor analysis in adult zebrafish: implications for automated detection of CNS drug-evoked phenotypes. J Neurosci Methods. 2015;255:66–74.
50. Anthony EW, Nevins ME. Anxiolytic-like effects of N-methyl-D-aspartate-associated glycine receptor ligands in the rat potentiated startle test. Eur J Pharmacol. 1993;250(2):317–24.
51. Stewart A, Kadri F, DiLeo J, Min Chung K, Cachat J, Goodspeed J, et al. The developing utility of zebrafish in modeling neurobehavioral disorders. Int J Comp Psychol. 2010;23(1): 104–21.
52. Grillon C, Baas J. A review of the modulation of the startle reflex by affective states and its application in psychiatry. Clin Neurophysiol. 2003;114(9):1557–79.
53. Davis M. Diazepam and flurazepam: effects on conditioned fear as measured with the potentiated startle paradigm. Psychopharmacology (Berl). 1979;62(1):1–7.
54. Kimmel CB, Patterson J, Kimmel RO. The development and behavioral characteristics of the startle response in the zebra fish. Dev Psychobiol. 1974;7(1):47–60.
55. Peskind ER, Potkin SG, Pomara N, Ott BR, Graham SM, Olin JT, et al. Memantine treatment in mild to moderate Alzheimer disease: a 24-week randomized, controlled trial. Am J Geriatr Psychiatry. 2006;14(8):704–15.
56. Takeda A, Loveman E, Clegg A, Kirby J, Picot J, Payne E, Green C. A systematic review of the clinical effectiveness of donepezil, rivastigmine and galantamine on cognition, quality of life and adverse events in Alzheimer's disease. Int J Geriatr Psychiatry. 2006;21(1):17–28.
57. Levin ED, Limpuangthip J, Rachakonda T, Peterson M. Timing of nicotine effects on learning in zebrafish. Psychopharmacology (Berl). 2006;184(3–4):547–52. doi:10.1007/s00213-005-0162-9.
58. Aldridge JE, Seidler FJ, Slotkin TA. Developmental exposure to chlorpyrifos elicits sex-selective alterations of serotonergic synaptic function in adulthood: critical periods and regional selectivity for effects on the serotonin transporter, receptor subtypes, and cell signaling. Environ Health Perspect. 2004;112(2):148.
59. Icenogle LM, Christopher NC, Blackwelder WP, Caldwell DP, Qiao D, Seidler FJ, et al. Behavioral alterations in adolescent and adult rats caused by a brief subtoxic exposure to chlorpyrifos during neurulation. Neurotoxicol Teratol. 2004;26(1):95–101.
60. Levin ED, Chrysanthis E, Yacisin K, Linney E. Chlorpyrifos exposure of developing zebrafish: effects on survival and long-term effects on response latency and spatial discrimination. Neurotoxicol Teratol. 2003;25(1):51–7. doi:10.1016/s0892-0362(02)00322-7.
61. Luca RM, Gerlai R. In search of optimal fear inducing stimuli: differential behavioral responses to computer animated images in zebrafish. Behav Brain Res. 2012;226(1):66–76. doi:10.1016/j.bbr.2011.09.001.
62. Cachat J, Stewart A, Grossman L, Gaikwad S, Kadri F, Chung KM, et al. Measuring behavioral and endocrine responses to novelty stress in adult zebrafish. Nat Protoc. 2010;5(11): 1786–99.
63. Malykh AG, Sadaie MR. Piracetam and piracetam-like drugs. Drugs. 2010;70(3):287–312.
64. Salimov R, Salimova N, Shvets L, Shvets N. Effect of chronic piracetam on age-related changes of cross-maze exploration in mice. Pharmacol Biochem Behav. 1995;52(3): 637–40.

65. Waegemans T, Wilsher CR, Danniau A, Ferris SH, Kurz A, Winblad B. Clinical efficacy of piracetam in cognitive impairment: a meta-analysis. Dement Geriatr Cogn Disord. 2002;13(4):217–24.
66. Al-Imari L, Gerlai R. Sight of conspecifics as reward in associative learning in zebrafish (*Danio rerio*). Behav Brain Res. 2008;189(1):216–9.
67. Lau B, Bretaud S, Huang Y, Lin E, Guo S. Dissociation of food and opiate preference by a genetic mutation in zebrafish. Genes Brain Behav. 2006;5(7):497–505.
68. Mathur P, Lau B, Guo S. Conditioned place preference behavior in zebrafish. Nat Protoc. 2011;6(3):338–45.
69. Williams FE, White D, Messer WSJ. A simple spatial alternation task for assessing memory function in zebrafish. Behav Processes. 2002;58:125–32.
70. Engeszer RE, Ryan MJ, Parichy DM. Learned social preference in zebrafish. Curr Biol. 2004;14(10):881–4.
71. Saverino C, Gerlai R. The social zebrafish: behavioral responses to conspecific, heterospecific, and computer animated fish. Behav Brain Res. 2008;191(1):77–87.
72. Pather S, Gerlai R. Shuttle box learning in zebrafish (*Danio rerio*). Behav Brain Res. 2009;196(2):323–7. doi:10.1016/j.bbr.2008.09.013.
73. Arthur D, Levin ED. Spatial and non-spatial visual discrimination learning in zebrafish (*Danio rerio*). Anim Cogn. 2001;4(2):125–31.
74. Beninger RJ. The role of dopamine in locomotor activity and learning. Brain Res Rev. 1983;6(2):173–96.
75. El-Ghundi M, Fletcher PJ, Drago J, Sibley DR, O'Dowd BF, George SR. Spatial learning deficit in dopamine D 1 receptor knockout mice. Eur J Pharmacol. 1999;383(2):95–106.
76. El-Ghundi M, O'Dowd BF, George SR. Insights into the role of dopamine receptor systems in learning and memory. Rev Neurosci. 2007;18(1):37–66.
77. Wise RA. Dopamine, learning and motivation. Nat Rev Neurosci. 2004;5(6):483–94.
78. Freedman R, Olincy A, Buchanan RW, Harris JG, Gold JM, Johnson L, et al. Initial phase 2 trial of a nicotinic agonist in schizophrenia. Am J Psychiatry. 2008;165(8):1040–7.
79. Potter A, Corwin J, Lang J, Piasecki M, Lenox R, Newhouse PA. Acute effects of the selective cholinergic channel activator (nicotinic agonist) ABT-418 in Alzheimer's disease. Psychopharmacology (Berl). 1999;142(4):334–42.
80. Wilens TE, Verlinden MH, Adler LA, Wozniak PJ, West SA. ABT-089, a neuronal nicotinic receptor partial agonist, for the treatment of attention-deficit/hyperactivity disorder in adults: results of a pilot study. Biol Psychiatry. 2006;59(11):1065–70.
81. Levin ED, Simon BB. Nicotinic acetylcholine involvement in cognitive function in animals. Psychopharmacology (Berl). 1998;138(3–4):217–30.
82. Ortega LA, Tracy BA, Gould TJ, Parikh V. Effects of chronic low-and high-dose nicotine on cognitive flexibility in C57BL/6J mice. Behav Brain Res. 2013;238:134–45.
83. Restle F. Discrimination of cues in mazes: a resolution of the "place-vs.-response" question. Psychol Rev. 1957;64(4):217–28.
84. Tolman EC. Cognitive maps in rats and men. Psychol Rev. 1948;55(4):189–208.
85. Gould GG. Modified associative learning T-maze test for zebrafish (*Danio rerio*) and other small teleost fish. Neuromethods. 2011;51:61–73.
86. Darland T, Dowling JE. Behavioral screening for cocaine sensitivity in mutagenized zebrafish. Proc Natl Acad Sci U S A. 2001;98(20):11691–6. doi:10.1073/pnas.191380698.
87. Ninkovic J, Folchert A, Makhankov YV, Neuhauss SC, Sillaber I, Straehle U, Bally-Cuif L. Genetic identification of AChE as a positive modulator of addiction to the psychostimulant D-amphetamine in zebrafish. J Neurobiol. 2006;66(5):463–75. doi:10.1002/neu.20231.
88. Morris R. Developments of a water-maze procedure for studying spatial learning in the rat. J Neurosci Methods. 1984;11(1):47–60.
89. Kalueff AV, Echevarria DJ, Stewart AM. Gaining translational momentum: more zebrafish models for neuroscience research. Prog Neuro-Psychopharmacol Biol Psychiatry. 2014;55:1–6.

Social Phenotypes in Zebrafish

Ana Rita Nunes, Nathan Ruhl, Svante Winberg, and Rui F. Oliveira

Abstract Zebrafish are an established model organism in developmental and behavioral neuroscience, also recently emerging as an excellent model to study social behavior. Zebrafish are highly social, forming groups (shoals) with structured social relationships, dominance hierarchies and overt territoriality. Moreover, social behavior in zebrafish exhibits considerable plasticity both within- (i.e., as a context-dependent behavior) and between individuals (e.g., sex-differences, personality and coping styles) of the same strain, as well as between strains. This richness and plasticity of social behavior, together with the genetic tools available to visualize and manipulate neural circuits in zebrafish places it in the forefront of studying the neurobiological mechanisms underlying complex social behavior. Here, we review the cognitive abilities involved in social behavior, as well as the different functional classes of social behavior present in zebrafish and their variation. We also highlight recent ground-breaking methodological developments in the field, including automated image-based tracking and classification of behavior coupled with video-animated social stimuli, which collectively foster the development of future high-throughput screens of zebrafish social phenotypes.

Keywords Social behavior • Social cognition • Shoaling • Aggression • Mating

A.R. Nunes
Instituto Gulbenkian de Ciência, Oeiras, Portugal

Champalimaud Neuroscience Program, Lisbon, Portugal

N. Ruhl
Rowan University, Glassboro, NJ, USA

S. Winberg
Uppsala University, Uppsala, Sweden

R.F. Oliveira (✉)
Instituto Gulbenkian de Ciência, Oeiras, Portugal

Champalimaud Neuroscience Program, Lisbon, Portugal

ISPA – Instituto Universitário, Lisbon, Portugal
e-mail: ruiol@ispa.pt

A.V. Kalueff (ed.), *The rights and wrongs of zebrafish: Behavioral phenotyping of zebrafish*, DOI 10.1007/978-3-319-33774-6_5

1 Introduction

Social behavior—any behavior expressed toward another animal—is typically directed towards conspecifics, but in some species may involve heterospecifics [1]. In the latter case, the vasotocinergic system known to regulate social interactions among conspecifics seems to have been evolutionary co-opted to modulate behavior directed towards heterospecific clients [2]. Similarly, aggression towards conspecifics and heterospecific intruders also seems to share common underlying physiological mechanisms [3]. This emphasizes the importance of considering behavioral expression toward both hetero- and conspecifics in order to fully understand the mechanisms of a behavioral trait. While there have been a few studies that assess zebrafish social behavior (especially shoaling) relative to heterospecifics [4–6], most studies of zebrafish have focused on social behavior relative to conspecifics, which will be discussed here.

The zebrafish is a highly social species for which several aspects of social behavior have already been described. Most of this knowledge comes from studies in laboratory conditions and following paradigms adapted from other species (mainly rodents) despite major differences in the biology of zebrafish relative to those other species. Therefore, field observations of zebrafish behavior and ecology are critical to enable laboratory observations in a naturalistic framework.

With a geographic distribution in India, Pakistan, Nepal, Bangladesh and Myanmar, zebrafish occur in streams and pools with macrophytes and a substrate composed of mud, sand or gravel. The spatial complexity of the natural environment most likely affects social interactions in the wild [7, 8]. Environmental enrichment has an impact on zebrafish behavior and on cognitive abilities in a laboratory setting [9–11]. Thus, the manner in which zebrafish are housed may have an impact on the behaviors observed in captive individuals. In the wild, zebrafish form mixed-sex groups with similar numbers of females and males ranging from two to several hundred fish per group [12].

Shoaling serves several adaptive functions, including protection from predators, as well as increasing foraging efficiency and mating success [5, 13]. Zebrafish natural predators include piscivorous fish such as the Indian Leaf Fish (*Nandus nandus*), snakeheads (*Channa spp*,), freshwater garfish (*Xennentodon spp.*), catfish (e.g., *Mystus bleekeri*) and knifefish,(e.g., *Notopterus notopterus*), as well as avian predators that live in the floodplains, such as the Indian pond heron (*Ardeola grayii*) and the kingfisher (*Alcedo atthis*) (reviewed in [7]). In the laboratory, the presence of the Indian Leaf Fish, or animated images of this predator, are effective in inducing fear-like responses in zebrafish [14, 15]. Animated images of other sympatric predators can also induce fear-like responses, but with different amplitudes, suggesting that zebrafish may recognize threats and adjust defensive behaviors to the type of threat [16]. Zebrafish appear to recognize novel predators as a threat as well, since zebrafish whole-body cortisol increases after visual contact with a neotropical cichlid [17].

As previously noted by Spence et al. [12], zebrafish in both the laboratory (domesticated strain) and field-based mesocosm (wild strain) prefer to spawn in shallow gravel vegetated areas that offer eggs protection from predators while also allowing for circulation of water for egg oxygenation [18]. Most spawning occurs at dawn [18], although it may take place at other times. In nature, there is a peak of spawning during periods of heavy rains [12]. Females produce clutches of several hundred eggs released in a single spawn [19], but spawning behavior depends on a variety of factors, including shoal density [20] and sex ratio [21, 22]. Zebrafish lack parental care, so they are unable to visually imprint on their parents or to use them as models for appropriate social behavior, which allows for the study of intrinsic and extrinsic factors that shape social behaviors (reviewed in [12]).

The rest of this chapter presents a detailed discussion of the cognitive ability of zebrafish relative to social behaviors, major functional behaviors of zebrafish (aggregation, aggression, and reproduction), factors influencing variation in behavioral expression, and future considerations.

2 Social Cognition

Here, we will describe the basic cognitive processes underlying social skills in zebrafish, including the ability to collect information from others (social attention); recognize conspecifics (social recognition); and learn from and about others (social learning) [23].

2.1 *Social Attention*

Social attention is an important behavioral mechanism to collect information from others. Once obtained, this information becomes available for learning (e.g., social recognition, social eavesdropping, and social learning) and decision-making processes.

Social attention can be measured by quantifying the engagement of a bystander toward a social interaction between two conspecifics [24]. For example, zebrafish males of the AB strain are more attentive towards social interactions than non-interacting conspecifics when placed in a small test tank facing demonstrators (either a pair of conspecific males fighting, two males non-interacting, or, as a control, an empty tank; Fig. 1; [24]). Attentiveness of the bystander fish is inferred from its position in the arena (spatial distribution and time spent in the vicinity of the stimulus), and from its body orientation and directional focus towards the stimuli fish, while the social interaction among demonstrators takes place. The bystander´s behavior is video-recorded and these videos are subsequently analyzed using a custom-made video tracking system (e.g., see http://github.com/joseaccruz/fish-tracker) that allows the tracking of three points in the fish (head, centroid, tail). Thus,

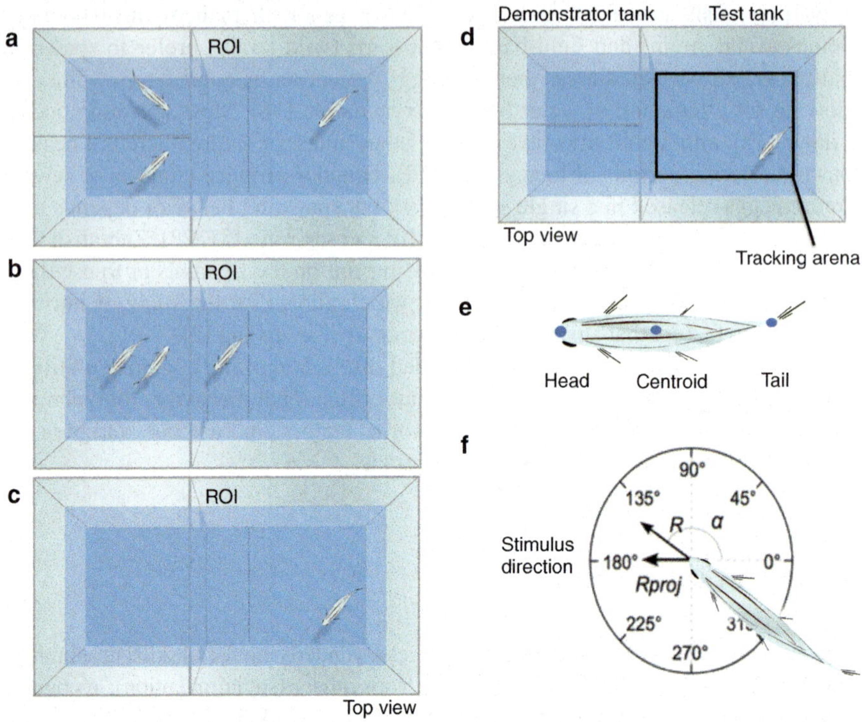

Fig. 1 Social attention in Zebrafish. Social attention in zebrafish is measured by the bystander position in the arena (spatial distribution and time spent in the vicinity of the stimulus (the region of interest, ROI)), and from its body orientation and directional focus towards the stimuli fish. The bystander fish is placed in a test tank, facing the stimuli, either (**a**) non-interacting conspecifics (two males isolated from each other), (**b**) a pair of conspecifics fighting or (**c**) a control tank with no conspecifics (empty tank). The bystander's behavior is video-recorded for 30 min. (**d**) Top view of the tracking arena, defined post-test for offline tracking of the recorded videos. (**e**) For the focal fish tracking, three points (*blue dots*) are used for coordinates extraction (head, centroid and tail). (**f**) Schematic of the focal fish mean orientations measured by its centroid-to-head axis angle α. R represents the mean resultant vector's length and Rproj its projection onto the stimulus direction; adapted from Abril-de-Abreu et al. [24]

for each video frame a vector that conveys information both on the position and orientation of the bystander is obtained [24]. Commonly used attentional measures include the time spent close to the demonstrator fish [i.e., in an arbitrarily defined region of interest (ROI) in the tank], and directional focus defined as the projection of the fish´s resultant vector from the directional vectors length R onto the demonstrator tank´s direction (180°) (see [24] for more details). Directional focus towards demonstrators ranges from −1 to 1, with positive values indicating directionality towards the stimulus direction, negative values indicating directionality away from it, and null values indicating no directional focus. Overall, zebrafish are attentive to both live conspecifics and video playbacks of conspecific stimuli, and are more attentive towards a video of fighting than non-interacting conspecifics [24]. Therefore, videos that manipulate social stimuli can be used to dissect specific aspects of social attention. For instance, replacing fighting fish with fighting dots has shown that attention is not influenced by activity, but rather by form features of the fish during the pre-resolution phase of the fight and by biological movement features of the dominant fish chasing the subordinate during the post-resolution phase [24].

2.2 *Social Recognition*

Zebrafish can discriminate between con- and heterospecifics of similar body size and shape in the choice preference test [25]. Briefly, when a group of zebrafish is placed in a tank together with another group of either conspecifics or heterospecifics (open field test), zebrafish tend to stay closer to each other than to heterospecific individuals [5].

Zebrafish can also discriminate between different classes of individual conspecifics. In rodents, social recognition is typically investigated using a three-chambered social recognition paradigm to assess preference between a familiar and a novel conspecific, in which focal individuals prefer to associate with a novel conspecific [26]. An adapted version of this paradigm [27] demonstrated that zebrafish not only recognize conspecifics, but that discriminatory preferences for novel or familiar fish varies between strains (e.g., AB and Gold strains preferred a novel fish while WIK and Petco strains did not discriminate). A version of the social recognition test in zebrafish currently used in the Oliveira lab is illustrated in Fig. 2. The three-chamber social recognition paradigm generally allows for the focal zebrafish to utilize both olfactory and visual cues. However, zebrafish also recognize conspecifics using only visual or olfactory cues. In the choice preference test, males are able to distinguish between unmodified and modified images of females, suggesting that there are specific visual features that zebrafish use to recognize conspecifics [5]. In an odor flume test, juveniles showed preference towards the area that contained odor cues from kin vs. non-kin fish [28]. Olfactory kin recognition is based on imprinting that occurs during a critical period in development at day 6 post-fertilization [29]. Visual imprinting occurs at day 4 post-fertilization in zebrafish [30], and therefore may also influence findings obtained in social recognition studies.

2.3 *Social Learning*

Social learning is influenced by another conspecific, either by observing it or by interacting with it [31], and can be classified into two major types: (1) learning about others and (2) learning from others. Social learning of both types has been demonstrated in zebrafish.

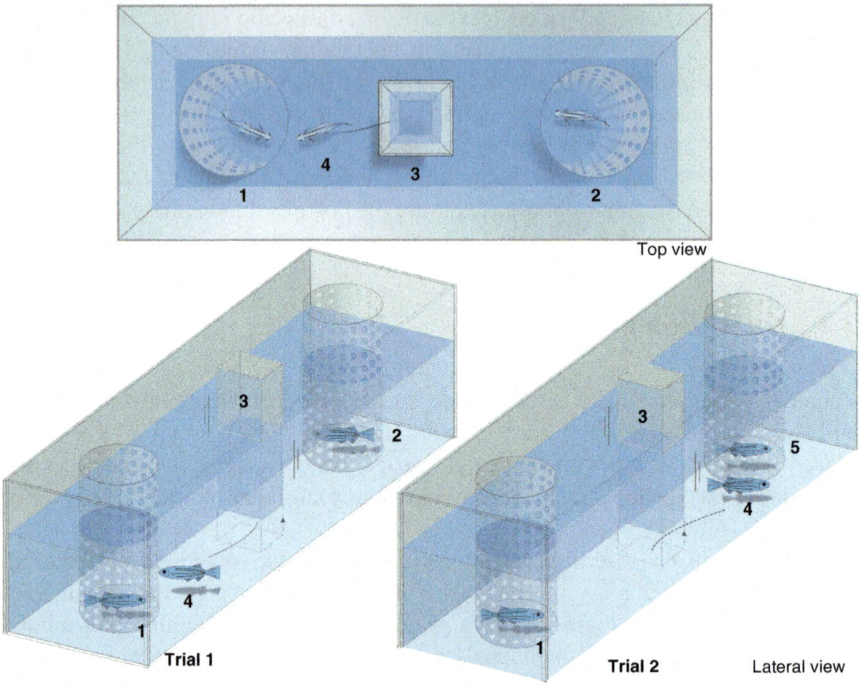

Fig. 2 Social recognition in zebrafish. Zebrafish discriminate between a novel and familiar conspecific in the presence of visual and olfactory cues. In the present paradigm, a conspecific-containing cylindrical tube (1 and 2) is placed in each of the arena end-compartments. A start-box (3) placed in the center of the arena contains the focal zebrafish (4). These cylindrical tubes are either transparent to test only for visual cues, perforated and opaque to test for olfactory cues or perforated and transparent to test for both olfactory and visual cues. The test is divided in two trials. After an acclimatization period, the start box (3) is removed and the focal fish is allowed to explore the cylindrical tubes containing the conspecific fishes (1 and 2) for 20 min, while the fish behavior is video-recorded. The focal fish is then collected and returned to its home tank. On the second trial, one of the conspecifics from the previous trial is replaced by a novel conspecific fish (5). This time, the zebrafish is allowed to discriminate between a familiar conspecific (1, fish presented in trial 1) and a novel fish (5). The time spent in close proximity to each stimulus is taken as a measurement of discrimination

2.3.1 Learning About Others

Things that can be learned about another conspecific include information about competitive ability and mate quality. For instance, it has been recently shown that bystanders of agonistic interactions extract information from the observed interaction that they will use subsequently when they are exposed to the individuals they had observed [32]. This phenomenon has been described in other species (e.g., Siamese fighting fish, *Betta splendens*; [33]) and has been termed *social eavesdropping* [34]. In zebrafish, the use of eavesdropped information depends on the social status of the bystander male. After observing a fight, dominant, but not subordinate

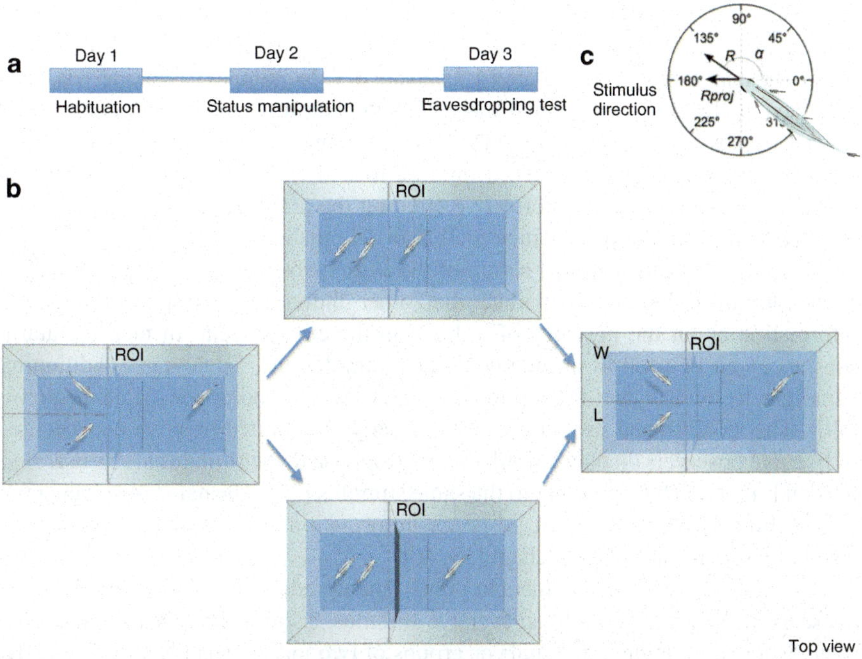

Fig. 3 Social eavesdropping in zebrafish. (**a**) After being subject to arena habituation (Day 1) and to a fight where dominance status was established (Day 2) the bystander, either a winner or a loser from the previous fight, is placed in a test arena facing a demonstrator tank that contains a pair of demonstrators separated by a partition (**a**, Day 3 and **b**). When the partition is removed, the demonstrators start to fight and the bystander observes the fight for 30 min. After the fight, the demonstrators' dominance status is identified (winner versus loser) and they are separated again by a partition. Zebrafish eavesdropping is measured using attentional measurements of directionality (**c**) and proximity (ROI) towards the winner and the loser (see social attention, section above), to assess if the bystander can discriminate between them; adapted from Abril-de-Abreu et al. [32]

bystander male zebrafish, become more attentive to the losers than winners of the observed fights (Fig. 3). Thus, bystander males not only pay attention to social interactions, but also collect information to subsequently use when interacting with the observed individuals.

2.3.2 Learning from Others

Perhaps, the most important thing that can be learned from a conspecific is information about a threat. When zebrafish are attacked or startled by the presence of a predator, they frequently display a fright response, consisting of a burst of rapid erratic swimming followed by freezing. Injured zebrafish release a chemical substance from their skin, known as alarm substance ("Schreckstoff" is the original

German term proposed by Karl von Frisch who first described it in minnows; [35]), that elicits a fright response in other individuals [36]. The chemical nature of the alarm substance has been a matter of debate. Hypoxanthine-3 N-oxide has been proposed as the active compound that elicits the alarm response [37]. However, it has not been reliably detected in the skin where many other possible active components are present. More recently, a mixture that includes the glycosaminoglycan (GAG) chondroitin, was also effective eliciting the alarm response, and has been proposed to be the alarm substance present in zebrafish skin [38].

Irrespective of the chemical nature of the alarm substance, it can be regarded as a cue produced by one individual that other individuals subsequently use as information about the presence of a threat in the environment. In fact, the alarm substance can be used as an unconditioned stimulus in a Pavlovian conditioning paradigm to elicit a fear-conditioned response (i.e., conditioned fright response [39]). This conditioned response can be visually transmitted to naïve conspecifics [40]; naïve observers display a similar fright response to the stimulus after observing the conditioned fish responding to the same stimulus [39]. The naïve observers may then communicate this fright response information to additional individuals, forming a social transmission chain [39, 40].

Zebrafish can learn from others not only that a predator is present, but also how to escape it. When exposed to an aversive stimulus, a shoal of fish learns an escape route faster than single individuals or groups of two individuals [41]. In this study, two types of behaviors were measured in a classic conditioning paradigm: (1) avoidance (fish swims to a safe compartment before the unconditioned stimulus appears) and (2) escape (fish swims to a safe compartment after the unconditioned stimulus appears). The latency of response after the onset of the conditioned stimulus was also measured. The results obtained in this study indicate that learning efficiency is higher in shoals than in singles or in pairs of fish, suggesting an effect of transfer of information between individuals in the learning process. Thus, living in larger shoals may provide protection from threats in a social information context, in addition to other predator-protection benefits such as dilution of risk (see further).

Zebrafish can also learn a specific escape route from others, to flee from an artificial predator [42]. In this paradigm, a group of naïve zebrafish (observers) was exposed to a moving net, which simulates chasing by a predator. The 'observer' fish were placed together with 'demonstrator' that had been trained in advance to use one out of two possible escape routes through the net (Fig. 4a). Measurements of escape latency and escape route choice (calculated as a score, subtracting the observers using one route versus the other, divided by all observers escaping) showed that the observers learned the escape route used by the demonstrators and escaped faster than fish of a control group. This effect persists after demonstrators are removed but, although reduced escape latencies remained stable along the transmission chain, fidelity to the original demonstrated route collapsed [42].

Finally, zebrafish can utilize social information about current threat levels to alter behaviors associated with boldness and risk-taking. A recent study [43] found that wild zebrafish become emboldened after interacting with domesticated zebrafish, but domesticated bold fish did not change their behavior after interacting with

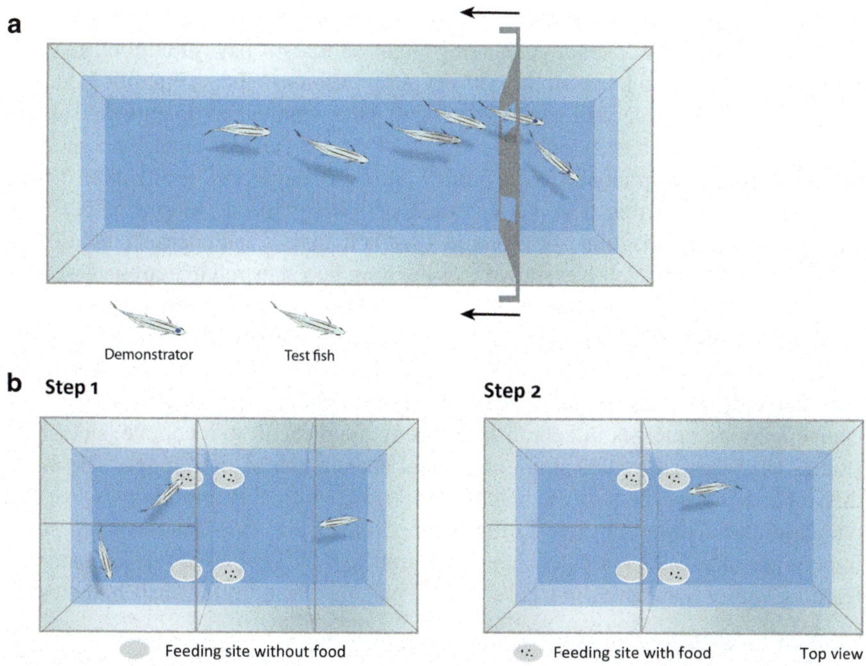

Fig. 4 Examples of social learning tests. (**a**) Zebrafish can learn from others conspecifics to escape, or the location of an escape route. In this paradigm the moving trawl mimics a predator. The observers (test fish) are placed together with the demonstrators that were previously trained to take a certain route. The observers learn to take the correct route by observing the demonstrators, an effect that is maintained when the demonstrators are removed. Figure adapted from Lindeyer and Reader [42]. (**b**) Zebrafish learns from others the location of the food. The focal fish observe two feeding sites: one with conspecific feeding and the other with conspecifics not feeding. When the demonstrator fish are removed, focal fish choose to eat in the feeding site close to the feeding site with a demonstrator conspecific feeding

domesticated shy fish. These results suggest that zebrafish can change their own risk-taking behavior based on information obtained from conspecifics, but also that shy behavior has a cost (e.g., reduced exploration of the environment) that is not worth paying unless there are social cues from conspecifics reinforcing it. In summary, zebrafish can learn from others about the levels of threat in the environment, how to escape threats, and how to adjust its behavior accordingly.

In addition to threats, zebrafish can also learn from others about the location of food. Zala and Määttänen [44] trained demonstrator zebrafish to associate a color (red) with food, allowed an untrained observer fish to interact (shoal and feed) with the demonstrators, and then tested whether a latent preference for red existed in the observer fish relative to a control group. As the observers preferred the red stimulus, zebrafish can be conditioned to a feeding site used by conspecifics. In another learning paradigm, the presence of conspecifics feeding was used successfully as a conditioned stimulus, again indicating that zebrafish pay attention and use information about the location of food via observation of others (Fig. 4b) [45].

Despite the occurrence of social learning in adult zebrafish as documented above, no studies have so far assessed the ontogeny of their social learning. Since shoaling behavior increases gradually in zebrafish from 10 days post fertilization to the juvenile stage [46], this may be a critical period for the development of social learning. A characterization of the ontogeny of both classic and operant conditioning in the same AB zebrafish has shown that learning becomes reliable around week 3, and reaches adult performance levels at week 6 [47]. Another study has shown that visual access to a group of conspecifics could be used successfully as an unconditioned rewarding stimulus in a classic conditioning paradigm in 6- to 8-day-old larval zebrafish [48]. The discrepancies between these two studies in the age at which associative learning emerges in zebrafish ontogeny are difficult to reconcile, since both studies used the same (AB) strain and visual cues in fish of similar age. The major difference between the two was that an aversive conditioning stimulus was used in one case and a rewarding one in the other. Collectively, these studies suggest the importance of social learning changes over time, and the window for development of learning ability (i.e., between weeks 1 and 3 post fertilization) for further studies.

3 Functional Social Behaviors

Here, we will describe zebrafish social behavior following an ethological functional classification into affiliative, aggressive and mating behavior. Since zebrafish do not exhibit parental care, reproductive behavior is restricted to mating.

3.1 Affiliative Behaviors

Affiliative behaviors comprise all behaviors that promote group cohesion. Thus, physical proximity between individuals not competing for a resource is typically taken as an indicator of affiliative behavior. Zebrafish individuals swim closely together in mixed-sex shoals [12, 25, 49]. The accepted theory for the evolution of shoals is related to their anti-predatory function. On one hand, shoals with multiple individuals may confuse predators, unable to focus on a single target fish. On the other hand, larger shoals may detect predators sooner and evade an approaching predator more efficiently. Shoaling may also have adaptive advantages due to dilution of risk (i.e., safety in numbers). Apart from their role on anti-predator defenses, shoals may also facilitate foraging efficiency and mating success [5, 13].

In the laboratory, affiliative behavior has been quantified in two different ways: (1) by the preference of individual zebrafish to associate with conspecifics in the social preference test; or (2) by measuring the behavior of freely moving individuals in a group and their tendency to spontaneously aggregate (e.g., in the shoaling test). While the former test does not provide an insight into collective behavior, it measures how rewarding the social stimuli are to individual fish at the time, and enables the use of imaging tools to explore the mechanisms underlying social preferences.

3.1.1 Affiliation in Social Preference Tests

The social preference test (shoal preference test) has been widely used to assess zebrafish sociability. It measures the motivation of the zebrafish to approach conspecifics (either real ones or their computer-animated images) in the two-choice preference test paradigm, where the time spent close to the stimulus fish is recorded and taken as a measure of preference. A detailed step-by-step protocol for this test is provided in [50] and summarized in Fig. 5. Zebrafish will readily shoal using this methodology and may even thrash along the glass divider between the stimulus shoal and the central arena [5, 51]. While this methodology has the drawback of utilizing only visual cues, recent evidence suggests that visual cues are the most important type of cue in the two-choice preference paradigm [52].

Social preference is a very robust behavior with high intra-individual repeatability [53], and hence it has been widely used to test the effect of different factors on zebrafish sociality. Increased concentrations of oxytocin-related peptides (i.e., oxytocin, vasopressin, isotocin, vasotocin) induce an increase in social preference, whereas their antagonists produced a dose-dependent inhibition of social preference [51]. Administration of a dopamine D1R antagonist (SCH23390) significantly decreases social preference [54]; as does ethanol, in a dose-dependent manner from 0 to 1 % [15, 55, 56]). Caffeine (0–50 mg/L) does not affect social preference [57], and ketamine (20–40 mg/L) reduces the number of entries in the social zone (Fig. 5 (2a)) but does not influence the social preference [58].

The social preference test can also be used to study which visual cues affect the preference of zebrafish to approach conspecifics. Social preference may be influenced by stimulus shoal body-size/age, shoal-size, gender, and strain. For example, zebrafish express social preference towards larger and similar sized individuals, but not towards shoals made up of smaller individuals then themselves [59]. During development, a 3-week larvae, at the stage where sociality seems to be already developed [25, 60], does not exhibit strong social preference for smaller/younger fish [60]. Zebrafish shoal with individuals and large groups, but the association with a single fish is of lower magnitude [59]. For fish preferring to shoal with a single individual over a larger group, the social preference is likely mediated by factors associated with social recognition, aggression or mating. Female zebrafish prefer to shoal with larger shoals (all else equal), but males may not show a strong group-size preference [61]; these sex-linked shoaling preferences are likely influenced by operational sex-ratio [21, 22].

When given a choice, zebrafish generally prefer to associate with shoals of the same phenotype (but see [6]). In the two-choice preference test, zebrafish can discriminate between a shoal of its own vs. a different strain (i.e., nacre vs.

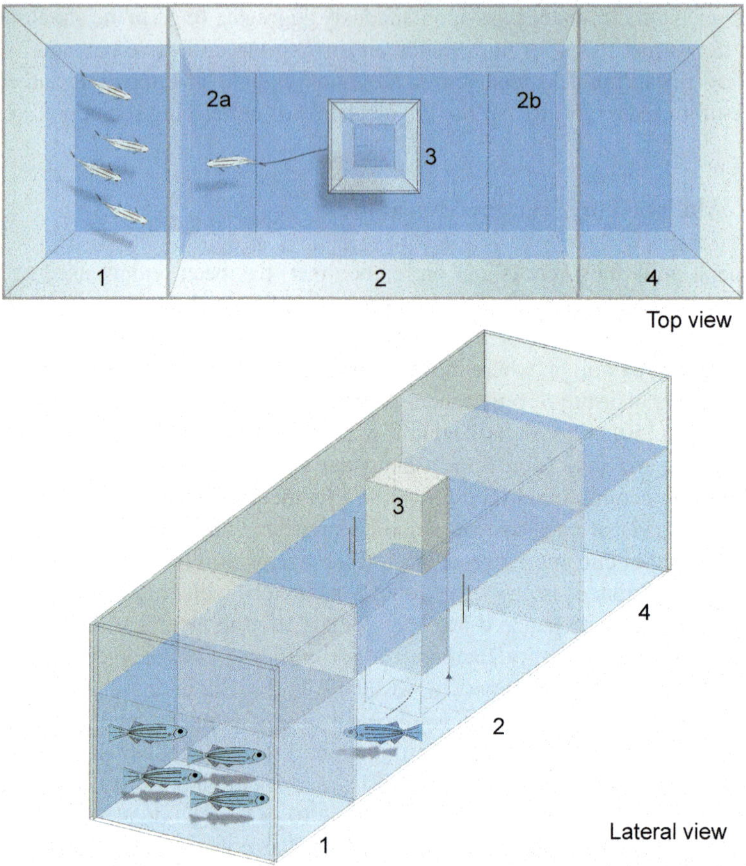

Fig. 5 Social preference test. The test arena is composed of two side compartments (1 and 4) and a central compartment (2) separated by clear glass. One side compartment contains a conspecific shoal (1) and the other one (4) is empty or contains an alternative stimulus. To avoid spatial bias, stimuli are randomized to compartments (1) and (4). To avoid the confounding effect of olfactory cues, the glass partitions are sealed to the walls of the tank. After an acclimatization period, the focal fish is released from a start box (3) placed in the middle of the central compartment, allowed to explore the arena, and its behavior is video-recorded for subsequent analysis. Social preference measures include the amount of time spent by the focal fish near the shoal (2a, as a proximity measurement), cumulative time in the non-social zone (2b), sociality score (cumulative time spent near the shoal divided by the sum of cumulative time in the shoal and non-shoal zone), latency to approach the shoal and frequency of entries in the shoal zone

AB shoal) using only visual cues [25]. This preference may depend on the early social environment experienced by the individual. For instance, wild-type AB fish raised together with nacre fish, later prefer to associate with a nacre shoal, whereas nacre individuals raised with wild type AB fish prefer to associate with the latter' shoals as adults [62]. Thus, assortative shoaling (i.e., aggregation of individuals with the same genotype/phenotype) based on visual imprinting seems to be present in zebrafish, and its underlying evolutionary mechanisms may be linked to gender differences in visual perception and imprinting [4].

Given the preference of zebrafish to associate with conspecifics, the hypothesis that the presence of others can act as a reward by itself has been raised. This was tested using a shoal of conspecifics as a social reward (i.e., unconditioned stimulus) in a classic conditioning paradigm [63]. Using a plus maze, a red card (conditioned stimulus) was either paired with a shoal (paired group) or presented in a separate arm to where the red card was presented (unpaired group). Following the training phase, fish were presented with the red card alone (probe trial) and the percentage of time spent close to the stimulus was taken as a measure of associative learning. Zebrafish in the paired group spent significantly more time near the red card than the fish from the unpaired group, suggesting that it acquired the association between the cue card and the stimulus fish, and that the presence of a shoal was rewarding.

The rewarding effect of the sight of a conspecific is further supported by increased brain dopamine levels in response to conspecific images [64]. Conspecifics have also been used as social reward in a learning paradigm using 6–8 days post-fertilization larvae [48], suggesting that conspecifics represent a social reward from very early on during ontogeny. However, other studies indicate that the expression of a social preference in the shoal preference paradigm only appears 3 weeks post-fertilization [25, 60]. These contrasting results may be due to differences in the geometry of the testing arenas, since Engeszer et al. [25] and Dreosti et al. [60] used test arenas with larger open areas, whereas [48] used channel-shaped arenas which help orient the movement of the fish along an axis. This latter design may have helped fish to move along an approach-avoidance axis, and hence facilitated the detection of a social preference. Thus, further studies are needed to clarify this issue.

The preference to associate with a shoal is also present, and with a similar magnitude, when computer-animated images of a shoal, or a video-recording of live fish are used instead of live fish [52]. This suggests that induction of robust shoaling responses does not necessarily require live stimulus fish or even 3D movement of the presented stimuli. In contrast, zebrafish show higher preference to associate with a shoal of live than robotic fish resembling conspecifics in dimension, morphology and swimming patterns [57, 65]. Thus, video playbacks, rather than robotics, seem to be a promising approach to best explore social preference in zebrafish.

3.1.2 Affiliation in Freely Moving Groups

Fish aggregations can be classified either as shoals, where conspecifics are closer to each other than what would be expected in case of a stochastic distribution, or as schools, if individuals within the aggregation align their direction of movement and move synchronously, resulting in a polarized group [66]. Although zebrafish groups can exhibit some degree of polarization and coordinated movement, most of the time their behavior corresponds to that of a shoal [12].

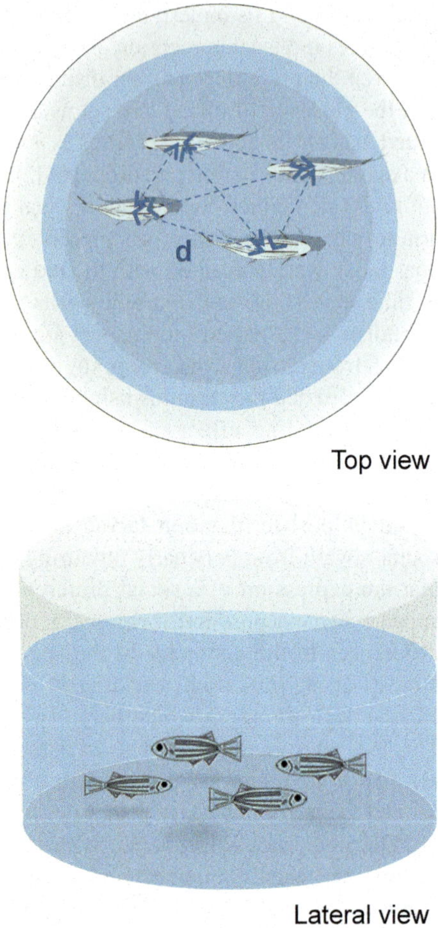

Fig. 6 Shoaling behavior test. The test consists in placing a group of conspecific fish into a novel tank and quantifying social cohesion in the group of fish, which is often measured by the average mean distance (d) among members (represented by *arrows*)

Shoaling behavior can be measured in the lab using the shoaling test that consists in placing a group of fish into a novel tank and quantifying their spatial behavior and movement patterns (Fig. 6). After an acclimatization period, which can go from 3 min (already sufficient for the anxious fish to reestablish their natural shoaling behaviors) up to 10 min, the shoal is video-recorded for behavioral analysis [67]. Social cohesion is usually measured by the average mean distance among members, with four body lengths commonly used as a criterion for shoal membership in cyprinid species [66]. The main behavioral measures in the shoaling test include average inter-fish distance (distance between the centroid of each member of the shoal); average nearest neighbor distance (distance of the centroid of each fish to the closest neighboring fish); average farthest neighbor (distance from the centroid of each fish to the farthest neighboring fish); top dwelling (percent of fish in the upper half of the tank); average distance between the group centroid and the center of the tank (which is an indicator of the inverse of thigmotaxis); variance of inter-fish distance (an index reflecting how homogeneously the fish are distributed within the shoal); shoal area (the size of the shoal: width and lengh); excursions from shoal (number of excursions of individual fish away from the shoal, i.e., above four body lengths); and duration of excursions (duration of excursions of individual fish away from the shoal) [68]. Collecting and analyzing these behavioral measures in a shoal of behaving fish is challenging since all individuals are simultaneously interacting with each other. Even with video-tracking software, this is not an easy task given that most software packages have problems reliably tracking every fish due to the targeted fish overlapping each other as they move (but see [69]). To address this limitation, some researchers have used alternative approaches, such as using a grid to divide the experimental tank and assessing the number of fish in each cell of the grid [70, 71]. Others have quantified shoaling by measuring the distances between each fish and its nearest neighbor in a series of screen shots taken from the video images (e.g., [68]). Fish that are apart from each other up to four average fish lengths are considered to be part of the same shoal [68]. However, this strategy does not distinguish the fish within the group, and individual tagging is needed for individual identification [5]. Overall, these methods to quantify shoaling are labor intensive and they do not adequately quantify the dynamics of the shoal. Recent advances in the video-tracking of multiple individuals leaves open the possibility that these challenges can be overcome and that automated analysis of freely shoaling and schooling zebrafish can be reliably achieved [69, 72]. The shoaling test can also be used as a measure of stress since group cohesion reflects zebrafish stress/anxiety; stressed fish tend to swim closer together, with smaller inter-fish distance, than non- stressed fish [58].

Similar to the shoal preference tests, shoaling has also been assessed in an ontogenetic context in freely interacting shoals. For example, newly hatched larvae do not shoal, but group cohesion increases from 10 days post fertilization to the juvenile stage, with adults expressing robust shoaling [46]. These developmental changes in shoaling behavior parallel changes in dopamine and serotonin levels, implicating these neurochemical systems in the regulation of shoaling [73].

3.2 *Aggression*

Aggression is an adaptive trait in many organisms, including humans [74]. It plays an important role in the social life history of a zebrafish population, influencing shoal-cohesion, mating and predator evasion [75]. Zebrafish behaviors associated with aggression are frequent and easy to elicit experimentally, making zebrafish a strong candidate model-organism for unlocking the neural and genetic basis for aggression in humans [75, 76]. While numerous studies have established a genetic basis for aggression in zebrafish [77, 78], behavioral plasticity in trait expression has been observed in the lab [79] and in natural populations [9], suggesting that epigenetic factors are also important. Likewise, environmental pollutants [80, 81] and habitat characteristics [9] have also been linked to changes in the expression of aggressive behaviors.

3.2.1 Aggressive Behaviors

Functionally, aggressive behaviors may be directed at shoal-mates (con- or heterospecific) in order to establish social dominance [82], ensure access to a mating opportunity [83, 84], or control a contested resource such as a territory [20] or food [85]. Behaviors associated with aggression include bites, lateral displays, charges or strikes, and darts or retreats [86, 87]. Bites consist of closing the mouth against either another individual or against a simulated individual. A lateral display consists of an approach to the stimulus followed by a turn to the left or right with fins erect. Charges are similar to lateral displays in that they involve an approach to the stimulus, but charges occur very quickly and do not involve the lateral display of erect fins. Darts are the same as charges, but instead of a rapid approach to the stimulus, a darting fish moves in some other direction. These behaviors have been well studied in artificial stimulus-based laboratory studies (see further), but when direct fish-to-fish interactions take place, additional aggressive behaviors have been identified [86]. These additional behaviors include circling (a form of lateral display that may last an extended period and during which the fish rise in the water column), freezing (remaining immobile with retracted fins), chasing (an extended charge), and fleeing (an extended dart in which the fleeing individual moves away from a pursuing individual for an extended period of time).

In addition to these behaviors, another important behavior is the amount of time spent directly interacting with a stimulus. Time spent interacting is somewhat artificial because of the nature of evoking aggressive behaviors in the lab, but nonetheless, significantly correlates with the frequency of bites and displays [87]. Some studies report an aggressive behavior termed "thrashing" [88], but this behavior is poorly defined and appears to be a combination of the bite and charge behaviors.

Behaviors associated with aggression can be elicited via a number of different techniques which vary in (1) the frequency of behavioral expression, (2) the amount of variation expressed between individuals, and (3) the types of behaviors that can be assessed. Which method is most appropriate is a study-specific question. Researchers need to balance the goals of their experiment with the practical concerns of conducting their study and the behaviors they are interested in assessing. Here, we compare different methods of studying aggression and attempt to provide practical advice on assay selection.

3.2.2 Mirror Elicited Aggression

One of the most commonly used methods to elicit aggressive behaviors in fish is with a mirror image stimulus (MIS; Fig. 7a), which was first used in zebrafish [89] to rapidly assess aggressive behavior using video-tracking techniques. Gerlai et al. [89] used an inclined mirror design in which the focal fish could only see its reflection in certain areas of the experimental arena (thereby limiting the area of the fish tank where aggressive behaviors can be expressed). MIS can also be implemented with the mirror placed flat against the wall of an experimental arena, although in this

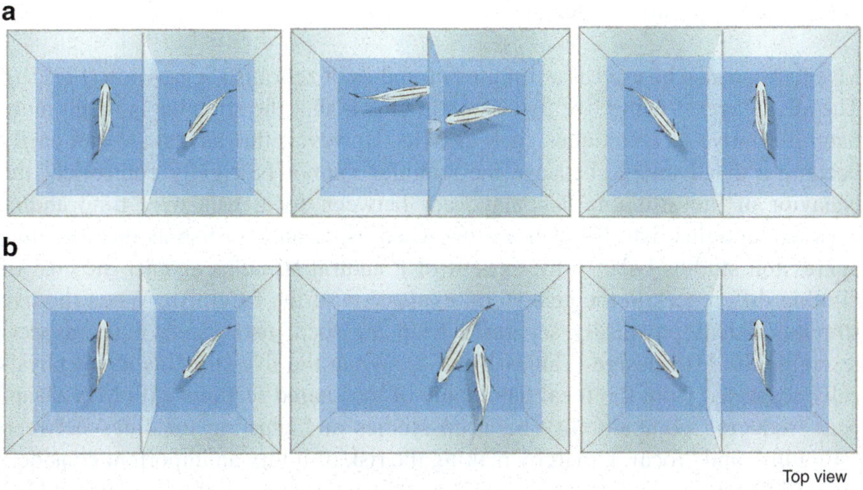

Fig. 7 Aggressive behaviors in zebrafish. (**a**) For mirror-elicited fights (MIS, mirror image stimulus) the arena is divided by a partition containing one mirror in each side, and a second partition (outer partition) is placed in front of each mirror to cover it. To elicit the mirror fight, the most outer partitions are removed and the fish are allowed to interact with their own image in the mirror. Then, partitions are placed again to cover the mirror. (**b**) For a real opponent fight two male conspecifics are placed in an arena and separated by a removable opaque partition. When the opaque partition is removed, the fish are allowed to interact. After post-resolution phase, winner and loser fish are identified and separated again by a removable opaque partition

Table 1 Qualitative assessment of methods to measure aggression

Method	Risk of injury	Technical challenges	Cost	Number of behaviors	Frequency of behavior	Variation in expression
MIS	Low	Low	Low	Low	High	High
Video	Low	High	High	Low	Low	Low
Model	Low	Medium	Low-Medium	Low	Low	Low
Live separated	Low	Low	Medium	Low	High	Medium
Live in contact	High	Medium	Medium	High	Not assessed	Not assessed

Frequency of behavior and variation in expression after Way et al. [90]

design the focal fish can always see the stimulus. The availability of a refuge (where the focal fish can cease interacting with the mirror stimulus) may influence behavioral expression [80]. The MIS design is popular because it combines a low risk of injury to the focal fish (allowing the individual fish to more easily be tested in multiple assays), simplicity, low cost, and high interactivity between the focal fish and the stimulus (Table 1). The main limitations of MIS studies are that the full repertoire of aggressive behaviors are not assessed and the stimulus (a perfect reflection of the actions of the focal fish) is unlikely to be naturally encountered.

3.2.3 Real Opponent Aggression

Live fish can also be used as a stimulus in studies of zebrafish aggression (Fig. 7b). The advantage of using live fish as a stimulus is that the stimulus is much more natural relative to techniques such as MIS. However, this advantage can easily become a disadvantage because the stimulus cannot be rigidly controlled; the behavior of the stimulus fish may vary between focal fish, over time and/or between stimulus fish. In other words, when using another fish as the stimulus, aggression studies balance the benefit of a natural stimulus against the cost of diluting differences in the behavioral expression of the focal fish. When the live stimulus fish is physically separated from the focal fish tested, many aspects resemble the MIS design (Tables 1 and 2). When the live stimulus is not physically separated from the focal fish, many of the shared features with MIS disappear (reducing comparability between studies and blurring the lines between "stimulus" and "focal"), thereby making the risk of injury an important consideration (Table 1). This risk is also an ethical concern, and a factor in experimental design, since an injured or stressed fish may behave differently in subsequent behavioral assays. Despite increased injury risk, studying aggression using two directly interacting individuals—a dyad [86]—has the distinct benefit of assessing the full repertoire of aggressive behavior (Table 2).

3.2.4 Other Methodological Approaches to Study Aggression in Zebrafish

Methods that use video or models as a stimulus for the focal fish have an advantage over MIS studies in that the stimulus can be customized to the needs of a particular study. For instance, models and videos can be altered to move in different ways, and the physical make-up of the stimulus (e.g., body size, fin size, color) can be varied as well. However, studies using these techniques may face a weak response from the focal fish to the stimulus [87] (Table 2) and can be technologically, fiscally and methodologically more complicated than MIS studies (Table 1).

Table 2 Aggressive behaviors that can be measured using different experimental designs

Method	Bites	Displays	Time	Charges	Darts	Circling	Freezing	Chasing	Fleeing
MIS	**Yes**	**Yes**	**Yes**	Yes	Yes	*No*	*Yes*	*No*	*No*
Video	Yes	Yes	**Yes**	Yes	Yes	*No*	*Yes*	*No*	*No*
Model	Yes	Yes	**Yes**	Yes	Yes	*No*	*Yes*	*No*	*No*
Live separated	**Yes**	**Yes**	**Yes**	Yes	Yes	*No*	*Yes*	*No*	*No*
Live in contact	*Yes*	*Yes*	*Yes*	*Yes*	*Yes*	*Yes*	*Yes*	*Yes*	*Yes*

Bold text indicates measures of aggression that are highly reliable and repeatable [90]. The efficacy and repeatability of behaviors has not been assessed for the italicized contexts

3.2.5 Social Dominance

Dominance is directly related to aggression of individuals [86] and can be predicted by their boldness [82]. It is unclear why behaviors associated with boldness would be important in a dyadic contest to establish dominance, suggesting that the power of boldness traits to predict social status may be linked to an underlying positive relationship between boldness and aggression, hence representing a behavioral syndrome (i.e., a co-expression of behavioral traits). Numerous studies have linked aggression and boldness behaviors into an aggression-boldness behavioral syndrome [91–93], while others have identified situations with little connection between the two [90, 94]. More than anything, this disagreement in the literature indicates the importance of environmental variation and natural selection in determining individual personality [91], only recently reviewed for zebrafish in [95].

If an individual expresses the right combination of aggressive and bold behaviors in complement with epigenetic factors, that individual will be dominant (Fig. 8). Epigenetic factors that may contribute to dominance include body size, age and experience [32, 79, 84]. Subordinate zebrafish experience greater stress than dominant individuals, but the magnitude of their stress is sex-linked. Males experience greater stress levels (as measured by the concentration of 5-HT metabolites) than females [96]. Dominance is also important in fish mating contests, where aggressive interactions are associated with an increase in fertilization success, but not the number of eggs released by females [21, 22]. Because both aggression and boldness are heritable zebrafish traits that predict dominance, dominance may also be considered a heritable trait. However, its heritability is restricted to the variation in aggression and boldness that can be attributed to genetic inheritance, as opposed to variation in environmental or contextual influences (e.g., social history and age [79]).

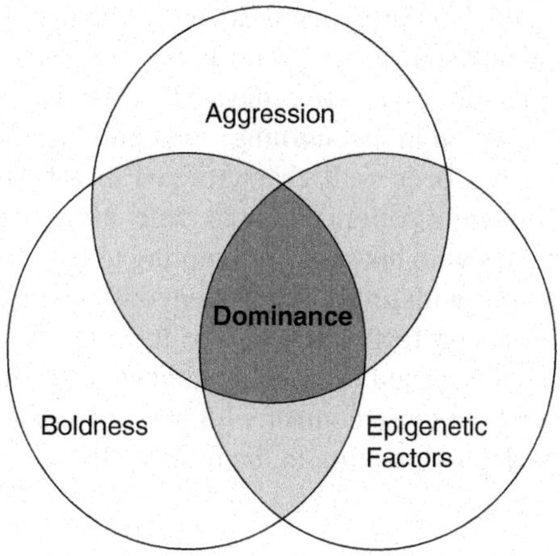

Fig. 8 Conceptual diagram summarizing the relationship between aggression, boldness, epigenetic factors and social dominance

3.2.6 Environmental Modulation of Aggression

Expression of aggression is highly plastic and varies considerably depending on the context in which this behavior is measured [9]. While aggression-boldness behavioral syndromes exist in wild populations, a positive relationship between both behavioral axes may be rare and/or linked to habitat or community characteristics [94].

When wild fish are first introduced to a laboratory setting, aggressive interactions may increase [94], making it difficult to study "natural" aggressive behaviors under controlled circumstances, and emphasizing the behavioral plasticity of aggression [9]. Once acclimated to the laboratory environment, aggression (and other measures of stress) may not be influenced by variation in lab conditions, such as light intensity, stocking density, mechanical sounds/vibration and diet, but the influence of these stressors likely varies between individuals [97].

While stocking density may not influence the number of aggressive interactions all else equal, this factor may interact with sex ratio resulting in increased aggression, especially in the early morning, i.e., when mating occurs [20, 22]. Although spatial complexity in holding tanks generally decreases aggression [10, 85, 98], this may not be true for wild fish newly introduced to the lab [9]. Perhaps more than any other consideration in the lab, how and when food is supplied can influence zebrafish aggression. For example, unfed males are more aggressive than unfed females, with little differences in aggression when all fish are well fed [99]. Furthermore, laboratory strains of zebrafish are often kept separately from one another, but when mixed together, may alter the expression of aggressive behaviors that persist for weeks [100]. Since water temperature may modulate zebrafish aggression [101], modeling a natural temperature cycle is also important.

3.3 Mating Behavior

In the laboratory, domesticated zebrafish strains breed all year round, with a single female spawning several hundred eggs at a time, with an average inter-spawning intervals of about 2 days [12]. Mating starts immediately after illumination is turned on in the morning, and continues for about an hour [102]. Courtship behavior has been well characterized in zebrafish, with the initial phase including the following patterns: (1) the male chases and touches the ventral part of the female flanks with his snout, attempting to lead the female to a spawning site; (2) the male swims with fins raised in front of the female or circling her; and (3) the male swims back and forth between the female and the spawning site. Once the female has been attracted to the spawning site, the male aligns his body with the female and, while in close contact with her, quivers at high frequency and low amplitude triggering oviposition in the female [102].

3.3.1 Mate Choice

Zebrafish display mate preferences, which can be studied in a laboratory context using the two-choice preference test (similar to the social recognition test) with only visual [103] or with both visual and chemical cues [104]. The time that the female spends in close proximity to a male with a particular trait value (e.g., body or fin size, courtship behavior, ornament size, coloration) reflects female mate choice (Fig. 9). Female zebrafish are also responsive to video playbacks, which enables efficient testing of her preference for specific male features [105].

Egg production can also reflect female mate choice [19, 106]. In this paradigm, a female and two males (of different sizes or of different dominance status) are isolated overnight by a net partition that allows for visual and olfactory communication. The next morning, the female is exposed for 10 min to each male, and at the end the number of eggs released with each of them is taken as a measurement of mate preference [19, 106]. Likewise, the female latency to mate can be used to infer her partner preference, with the shorter latency reflecting the stronger preference for the mate [103].

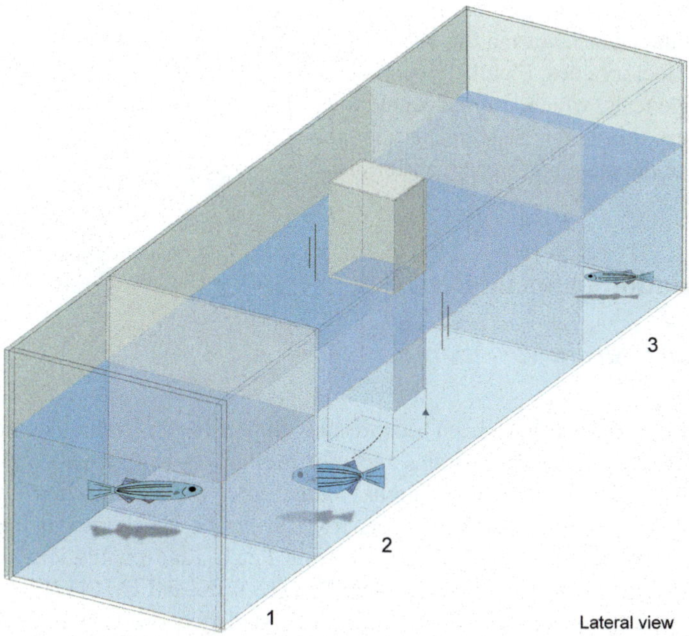

Fig. 9 Matting behaviors in zebrafish. Example of the mate choice test: a female (2) (often distinguished from the males by a protruding belly and silvery-blue color, while the males have a pinkish-yellow cast and a slender body shape) is allowed to discriminate, in the two-choice paradigm, between two male conspecifics differing in relevant trait values (e.g., male body size: big male fish (1) and small male fish (3)). The female male choice is inferred by the time that the female spends in close association with a particular male

Mounting evidence shows that female zebrafish prefer to associate with [107], and allocate more eggs to, larger mates [106]. However, there was no significant difference in preference between males with larger or smaller fins [108]. In line with this, wild-type females do not differentiate between wild-type and mutant short-fin or long-fin males [104]. Similarly, short-fin females do not show a preference based on fin length, whereas long-fin females prefer long-fin males over wild-type males [104]. Interestingly, female zebrafish do not choose mates based on their social status, since the number of spawned eggs does not change when females are allowed to mate with males differing in dominance rank [20].

Important methodological considerations also exist for using the two-choice preference test to assess mate choice preferences. First, the time of the day may be critical since spawning occurs naturally only during the first hour of light. Thus, the mate choice tests should be run during this period [103]. In contrast, association with males later during the day may reflect affiliative (e.g., shoaling) motivation rather than a preference for a mating partner. Second, although most mate choice test in zebrafish have so far used only visual cues from the males (but see [104]), both olfactory and visual cues are used in sex recognition [109]. It is also known that male pheromones induce ovulation in females [110]. Moreover, olfactory cues are involved in kin recognition, and may also play a role in inbreeding avoidance by reducing the chances of mating with a close kin [28]. Thus, mate choice may also rely on olfactory cues. Finally, most mate choice studies reported so far have only used single tests, which are inadequate to establish a stable preference for specific traits. Thus, repeated measurements should be performed in order to assess the consistency of the mate preference in zebrafish.

4 Variation in Social Behavior

4.1 Inter-Strain Variation

Many different strains have been used to study the behavior of zebrafish. For example, behavioral studies use strains recently brought from the wild (i.e., Nadia, Gaighatta and Bangladesh populations, named after the places where they were collected); the commercial populations from Scientific Hatcheries and pet-shops (most commonly the wild type short fin from Petco), and populations with long laboratory breeding histories, such as AB, TU, TL, WIK, and TM1 [12]. The laboratory strains are genetically well characterized, and display overt differences in microsatellite variations, single nucleotide polymorphisms (SNPs) and gene copy number variants [111, 112]. Thus, it is expected that the fish strains can be an important source of variation between behavioral assays. When comparing populations recently collected from the wild (so called 'wild populations') with laboratory strains, the behavioral differences are commonly associated with domestication (adaptation to lab living), such as higher growth rates, decreased startle, lesser shoaling, changes in boldness behaviors and stress coping [53, 113]. In terms of

Table 3 Comparison between D. Rerio strains in different social behavioral assays

Behavioral test	D. Rerio strains used	Results	Reference
Shoaling	AB, Santal (wild caught)	Shoal tendency: Santal>AB	[53]
	Wild strains: Nepal, Bangladesh, Santal, Canal	Nepal=Bangladesh=Santal=Canal	[113]
	AB, TU	Shoal cohesion at 7 dpf: AB>TU	[117]
		At 23 and 39 dpf: AB<TU	
		Older stages: AB=TU	
Shoal preference	AB, WIK, Golden, short-fin (SF)	Sociality: AB>WIK, Golden>SF	[27]
Shoal preference with computer animated images	AB, short-fin (SF)	Sociality: AB>SF	[54]
Aggression	TM1, Nadia, Scientific Hatecheries (SH)	Bites: TM1>Nadia>SH	[100]
Social learning	Wild strain (F3 of Assam wild caught), domesticated strain (from pet shops)	Wild fish less bold than domesticated fish; Wild fish increase their boldness after exposure to domesticated fish	[43]
Social recognition	AB, WIK, Golden, short-fin (SF)	Preference for social novelty	[27]
		AB=Golden>SF=WIK	

personality, laboratory strains typically are bolder [43, 53], have a shorter latency to feed, and are less sensitive to aversive stimuli (e.g., surface disturbance and alarm cue) [114], than the recently caught wild strains. A recent analysis of the brain transcriptome of two wild populations and two domesticated populations that differ in their predator avoidance behavior (e.g., domesticated strain individuals stay closer to the water surface than wild ones), identified a set of genes associated with behavioral domestication [115].

Behavioral variation among zebrafish strains have been observed in social preference and social recognition tests [27], anxiety-related behavior [116], shoal preference in response to alcohol [15] or D1R antagonists [54], and shoal development [117], among others (Table 3). These strain differences suggest a main genetic component in their behavioral traits. For example, the genetic basis for strain differences in boldness has been recently demonstrated [53, 113]. Therefore, strain-specific behavior should be carefully considered when designing experiments, to select the appropriate strains for studying zebrafish social and other behaviors.

4.2 *Social Behavior in Mutants and Transgenic Strains*

The development of genetic tools for zebrafish boosted exponentially the popularity of this animal model in developmental biology research [118]. In brief, the genetic toolbox available extends from morpholinos (gene knockdown [119]),

TILLING (targeting induced local lesions In genomes), N-ethyl-N-nitrosourea mutagenesis (ENU, [120] and zinc finger nucleases, CRISPR (clustered regularly interspaced short palindromic repeats) to TALLENS (transcription activator like effector nuclease) that allow for site-directed mutagenesis [121–123]. Another powerful genetic tool is the GAL4/upstream activating sequence system (UAS). There are already several GAL4 combined with enhancer trap zebrafish lines, that permit UAS-linked transgenes to be targeted to specific regions or specific cells of the brain [124]. Despite this wealth of genetic tools there are not yet many studies that link mutants/transgenes with a social behavioral phenotype. The knockout (null) mutant deficient in the fibroblast growth factor receptor 1a (fgfr1a) have been tested for mirror-elicited aggression [93]. The knockouts showed increased aggressive behavior, but also increased in boldness and exploratory behaviors, suggesting that this receptor is implicated in the aggression-boldness behavioral syndrome [93]. A mutant deficient in the glucocorticoid receptor shows freezing and reduced exploratory behavior, even after repeated-exposure to a "novel" tank [125]. However, when allowed visual interactions with other fish, this mutant spends less time freezing, suggesting that social interactions reversed freezing behavior in the mutant fish [125].

4.3 Sex Differences in Behavior

Sex-linked genes that influence behavior have been identified in zebrafish, but these genes are few in number and seem likely to indirectly influence behavior via the endocrine system, rather than directly control behavioral expression [126, 127]. While the number of such genes is small, differential hormone production is sufficient to enable sexing of fish using water from their holding tanks [128]. In an important early study, Moretz et al. [100] found superficial evidence for sex-differences in various contexts, although the importance of those sex differences was lesser than of strain differences. In other words, sex does explain a significant amount of variation in zebrafish behavior, but strain-differences affect behavior much more than sex-differences. This insight is likely to hold true in so far as the strains being tested do not experience selective pressure or epigenetic factors that favor sex-linked behaviors.

Sex-linked behaviors have been reported in multiple contexts, making sex an important consideration in experimental design, but the vast majority of studies identify no significant differences in behavior between the sexes. Sex effects aggression [99], activity levels [129, 130], boldness [82, 90, 131], and shoaling in zebrafish [4, 22, 61, 90]. The interoperability of these studies is tempered by a lack of consistency in the strain of zebrafish used (Table 4), making broader conclusions difficult to make. Therefore, although sex differences in zebrafish behavior reflect differences in boldness and risk-taking, there is no sufficient evidence yet to support this notion.

Table 4 Studies, strains, and behaviors in which a sex-linked difference in behavior has been identified

Authors	Year	D. Rerio Strain	Behavior	Result
Ariyomo and Watt	2015	Nacre	Aggression	Hungry females are less aggressive than hungry males
Conradsen and McGuigan	2015	WIK	Activity	Morphological differences between the sexes impact swimming performance; males can swim better/faster/farther all else equal
Way et al.	2015	Wild-type; lab stock	Shoaling and boldness	Bold males shoal more than shy males, but only with conspecifics; Bold females shoal more than shy females, but only with heterospecifics
Tran and Gerlai	2013	Gold; pet-trade	Activity	High activity females prefer top of holding tank, low activity females prefer bottom; pattern not held for males
Oswald et al.	2012	Wild-type; lab stock; artificially selected two generations for boldness traits	Boldness	Females bolder than males
Dahlbom et al.	2011	Wild, North Bengal, India	Boldness	Males bolder than females
Ruhl et al.	2009	Wild-type; pet-trade	Shoaling	Females prefer larger shoal regardless of sex; males prefer a larger shoal when males are presented and prefer a single female when females presented
Ruhl and McRobert	2005	Wild-type; pet-trade	Shoaling	Females prefer larger groups; males do not exhibit a group size preference

4.4 Intra-Strain Variation: Stress Coping

Individuals cope with stress in different ways, and stress coping style has been defined as "a coherent set of behavioral and physiological stress responses, which is consistent over time and which is characteristic to a certain group of individuals" [132].

Two divergent stress coping styles, referred to as proactive and reactive, were originally described in rodents but are now reported across the vertebrate subphylum [132–134]. In most cases, these two coping styles represent the extremes of a continuum. When challenged by a stressor, the proactive phenotype responds with active avoidance and aggression, whereas reactive animals are non-aggressive and respond by freezing and immobility. Proactive animals, being bold and aggressive, have a better chance of becoming socially dominant than reactive individuals [135, 136]. Moreover, proactive animals easily develop behavioral routines and are less sensitive to environmental cues than reactive individuals. The behavior of the reactive phenotype is highly responsive to changes in the environment. Thus, these animals are more flexible, rapidly adjusting their behavior to changing demands. It has been suggested that proactive animals may have an advantage in stable predictable environments, to develop routines and aggressively defend resources like food and shelter. In contrast, reactive animals may have an advantage in variable environments where fluctuations are non-predictable, and where resources in demand are difficult to monopolize [137–139]. Supporting this notion, Brelin et al. [140] observed that in brown trout (*Salmo trutta*) populations the frequency of proactive fish was higher in large rivers (than in small creeks) with non-predictive fluctuating environments. Proactive fish appear to have an advantage when reared in captivity and zebrafish of the AB strain display more proactive behaviors than wild-caught zebrafish [141]. In addition to behavioral differences, proactive and reactive animals also differ in autonomic and neuroendocrine stress responses [132–134], as proactive animals respond to stress with high plasma levels of catecholamines but a more modest increase in plasma glucocorticoids. The opposite is true for reactive animals, which usually respond to stress with high plasma glucocorticoids but a less pronounced activation of the sympathetic nervous system.

Little is known about stress coping in zebrafish, even though knowledge is rapidly growing [82]. For instance, boldness, a trait associated with proactive coping, could predict the outcome of fights for social dominance in size-matched pairs of wild-caught zebrafish. In this study, boldness was determined by screening zebrafish for behavior in a novel environment (open field test), tendency to seek shelter and willingness to approach a novel object. Using recently wild-derived zebrafish, Wong et al. [141] showed clearly divergent behavioral profiles resembling proactive and reactive coping styles. Although physiological traits were not included in these studies, the results clearly suggest that divergent stress coping styles described in other teleosts are also present in zebrafish. Confirming this suggestion, Tudorache et al. [142] reported divergent hypothalamic-pituitary-interrenal (HPI) axis reactivity in proactive and reactive zebrafish. When subjected to chronic netting stress, reactive zebrafish displayed higher whole body cortisol levels than proactive conspecifics. Moreover, the recovery to base line levels of cortisol following stress was slower in reactive fish. The divergent behavioral profile of proactive and reactive zebrafish is consistent over time and context [143]. Recently, stress coping style were related to temperature preference in zebrafish, with proactive fish prefereing higher temperatures than reactive fish [101]. There have also been studies focusing on molecular mechanisms potentially controlling the development of alternate stress coping styles in zebrafish. Analyses of the brain transcriptome of proactive and reactive zebrafish showed stress coping style accounts for 9 % of the individual variation in gene expression observed within the population [143]. Using brain transcriptomics, Wong et al. [144] identified 115 genes that differentiate zebrafish coping style, the majority of which are linked to neuronal metabolism.

5 Prospects for the Future: High-Throughput Phenotyping of Zebrafish Social Behavior

There are two key methodological limitations of high-throughput phenotyping of zebrafish social behavior that need to be overcome: (1) the automated quantification of social and collective behavior; and (2) the fine control of social stimuli to be used in the tests. As summarized below, both of these challenges are now being tackled, and major developments are therefore expected in the near future.

5.1 Automated Quantification of Social Behavior

Automated image-based tracking software has been available for some years now, and is currently used to quantify relevant behavior variables. It involves three general steps: (1) video-recordings of fish behavior; (2) detection of individuals in each frame of the image (i.e., computer digitized images are converted into x, y coordinate data) and linking of these detections on consecutive frames to create tracks through time for each individual; and (3) analysis of trajectory and behavioral data [145].

There are multiple versions of automated image-based tracking software available including commercial versions (e.g., Ethovision developed by Noldus IT (Netherlands); LocoScan by CleverSys, Inc. (USA); ZebraLab by ViewPoint (France)) and custom-made systems. The former are generally more user-friendly, standardized, but more expensive. Custom-made systems are usually freely available from several research laboratories, such as the software developed in Python (pythonTM) by the Oliveira laboratory (https://github.com/joseaccruz/fishtracker), which determines and extracts into data files, the pixel coordinates of the head, centroid and tail of fish for each frame, hence allowing to track not only the position but also the orientation of the focal fish. This is particularly useful for getting a measure of the attentional engagement of the focal fish, such as in the social attention test. When only registering the position most video-tracking packages are adequate to characterize 2D trajectories of individual fish (e.g., traveled distance, average speed, meandering score of the trajectory) and/or its behavior in relation to a region of interest (ROI) in the test tank (e.g., latencies to enter ROI, time spend in ROI, frequency of entries in ROI). Thus, this approach is very useful for most of the social behavior tests described above, including social recognition, social learning, and social preference. However, there are three common limitations to most readily available video-tracking packages: (1) they are error prone when multiple individuals need to be tracked simultaneously; (2) they mostly track 2D coordinates; and (3) they do not allow the automatic recognition of specific behavioral patterns (e.g., courtship or aggression). Some recent developments in image-based tracking have addressed these issues, enabling the expansion of this technology to other social behavior paradigms.

For example, Cachat et al.[146] generated temporal and spatial 3D reconstructions of adult zebrafish behavior by combining video acquisition/tracking using Ethovision software with track integration using RapidMiner 5.0 software and temporal reconstructions using Scatter 3D Color plot software. The advantage of the spatiotemporal 3D swim paths reconstructions is in more realistic representations of the zebrafish swimming activity, minimizing errors and providing valuable visualizations of general behavioral patterns. Miller and Gerlai [147] generated an automated tracking of zebrafish shoals to acquire and analyse detailed trajectory data from shoals of zebrafish. A multi-tracking algorithm (idTracker) has also been recently developed to enable simultaneous tracking of multiple unmarked individuals without error propagation [72]—a key methodological advancement since other currently available video-tracking methods frequently switch identities of unmarked individuals, resulting in errors that propagate over the tracking period, unless manually corrected. Thus, idTracker enables tracking of individuals in groups without the need of tagging them, hence prompting high-throughput analysis of collective behavior, such as shoaling. IdTracker also allows the quantification of some social behaviors that only rely on positional data. Classification algorithms that detect behavioral patterns in sequences of social behavior have also been developed recently, and are already in use for some model organisms (e.g., Drosophila), presenting data as plots of the time course of behaviors displayed by each individual in a given period [148]. The development of similar classification algorithms for zebrafish will foster the automatization of other social behavior tests, including models of aggressive or mating behaviors.

5.2 Video Playbacks

Video playbacks are a powerful tool to study visual communication in animals, and have been widely used in fish. Video playbacks generate more controllable and repeatable visual stimuli than the use of live animals, and hence allow standardization of the stimulus presented. They are suitable for studying motion, shape, texture, size, and brightness. However, studying color using video playbacks is problematic because video systems are specifically designed for human vision [149]. Another limitation of video images is that they lack depth cues. Nevertheless, zebrafish are responsive to video playbacks of conspecific stimuli (e.g., [5, 52]). Two different types of techniques are used to produce video-clips: (1) video-recordings of live fish that are subsequently edited; and (2) computer-animated images. The first technique may render more realistic images, but is less flexible for editing the relevant characteristics of the stimulus. For example, this approach has been used recently to test the attentional response of zebrafish to the motion pattern of conspecifics involved in social interactions without the presence of the form features of the fish. For this purpose, a video-sequence of live fish fighting was edited and the two opponents were replaced by dots of similar sizes on each frame [24]. Computer animated images allow to alter specific features of the conspecifics to give insight in the specific mechanisms underlying social behavior in zebrafish. This approach has also been applied in different social behavior paradigms in zebrafish, such as social preference [52] and mate-choice [105]. However, animations are difficult to program, thus limiting their usage in behavioral research. Recently, a user-friendly platform for creating realistic animated 3D fish for several fish model species (e.g. sticklebacks, poeciliids, and zebrafish) has been released (anyFish 2.0), allowing researchers to easily manipulate the visual appearance and behaviors of the stimulus fish [150] and thereby markedly enhancing the study of social behavior in zebrafish.

Taken together, these recent methodological advances in the automatization of behavior quantification paired with the high-control of stimulus manipulation, open a new era for the characterization of social phenotypes in zebrafish, rapidly emerging as a golden standard for the study of social behavior.

Acknowledgements During the preparation of this manuscript ARN and RFO were supported by Fundação para a Ciência e a Tecnologia (grants SFRH/BPD/93317/2013 and EXCL/BIA-ANM/0549/2012, respectively). SW was supported by the Swedish Research Council (VR) and the Swedish research council FORMAS.

References

1. Grutter A. Parasite removal rates by the cleaner wrasse Labroides dimidiatus. Mar Ecol Prog Ser. 1996;130:61–70. doi:10.3354/meps130061.
2. Soares MC, Bshary R, Mendonça R, Grutter AS, Oliveira RF. Arginine vasotocin regulation of interspecific cooperative behaviour in a cleaner fish. PLoS One. 2012;7, e39583. doi:10.1371/journal.pone.0039583.
3. Ros AFH, Vullioud P, Bruintjes R, Vallat A, Bshary R. Intra- and interspecific challenges modulate cortisol but not androgen levels in a year-round territorial damselfish. J Exp Biol. 2014;217:1768–74. doi:10.1242/jeb.093666.
4. Engeszer RE, Wang G, Ryan MJ, Parichy DM. Sex-specific perceptual spaces for a vertebrate basal social aggregative behavior. Proc Natl Acad Sci U S A. 2008;105:929–33. doi:10.1073/pnas.0708778105.
5. Saverino C, Gerlai R. The social zebrafish: behavioral responses to conspecific, heterospecific, and computer animated fish. Behav Brain Res. 2008;191:77–87. doi:10.1016/j.bbr.2008.03.013.
6. Snekser JL, Ruhl N, Bauer K, McRobert SP. The influence of sex and phenotype on shoaling decisions in zebrafish. Int J Comp Psychol. 2010;23:70–81.
7. Engeszer RE, Patterson LB, Rao AA, Parichy DM. Zebrafish in the wild: a review of natural history and new notes from the field. Zebrafish. 2007;4:21–40. doi:10.1089/zeb.2006.9997.
8. Parichy DM. Advancing biology through a deeper understanding of zebrafish ecology and evolution. Elife. 2015;4:e05635. doi:10.7554/eLife.05635.
9. Bhat A, Greulich MM, Martins EP. Behavioral plasticity in response to environmental manipulation among zebrafish (*Danio rerio*) populations. PLoS One. 2015;10, e0125097. doi:10.1371/journal.pone.0125097.
10. Carfagnini AG, Rodd FH, Jeffers KB, Bruce AEE. The effects of habitat complexity on aggression and fecundity in zebrafish (*Danio rerio*). Environ Biol Fishes. 2009;86:403–9. doi:10.1007/s10641-009-9539-7.
11. Collymore C, Tolwani RJ, Rasmussen S. The behavioral effects of single housing and environmental enrichment on adult zebrafish (*Danio rerio*). J Am Assoc Lab Anim Sci. 2015;54:280–5.

12. Spence R, Gerlach G, Lawrence C, Smith C. The behaviour and ecology of the zebrafish, *Danio rerio*. Biol Rev Camb Philos Soc. 2008;83:13–34. doi:10.1111/j.1469-185X.2007.00030.x.
13. Gerlai R. Social behavior of zebrafish: from synthetic images to biological mechanisms of shoaling. J Neurosci Methods. 2014;234:59–65. doi:10.1016/j.jneumeth.2014.04.028.
14. Bass SLS, Gerlai R. Zebrafish (*Danio rerio*) responds differentially to stimulus fish: the effects of sympatric and allopatric predators and harmless fish. Behav Brain Res. 2008;186:107–17. doi:10.1016/j.bbr.2007.07.037.
15. Gerlai R, Fernandes Y, Pereira T. Zebrafish (*Danio rerio*) responds to the animated image of a predator: towards the development of an automated aversive task. Behav Brain Res. 2009;201:318–24. doi:10.1016/j.bbr.2009.03.003.
16. Ahmed TS, Gerlai R, Fernandes Y. Effects of animated images of sympatric predators and abstract shapes on fear responses in zebrafish. Behaviour. 2012;149:1125–53. doi:10.1163/1568539X-00003011.
17. Barcellos LJG, Ritter F, Kreutz LC, Quevedo RM, da Silva LB, Bedin AC, Finco J, Cericato L. Whole-body cortisol increases after direct and visual contact with a predator in zebrafish, *Danio rerio*. Aquaculture. 2007;272:774–8. doi:10.1016/j.aquaculture.2007.09.002.
18. Spence R, Ashton R, Smith C. Oviposition decisions are mediated by spawning site quality in wild and domesticated zebrafish, *Danio rerio*. Behaviour. 2007;144:953–66. doi:10.1163/156853907781492726.
19. Spence R, Smith C. Mating preference of female zebrafish, *Danio rerio*, in relation to male dominance. Behav Ecol. 2006. doi:10.1093/beheco/arl016.
20. Spence ROW, Smith C. Male territoriality mediates density and sex ratio effects on oviposition in the zebrafish, *Danio rerio*. Anim Behav. 2005;69:1317–23. doi:10.1016/j.anbehav.2004.10.010.
21. Ariyomo TO, Watt PJ. The effect of variation in boldness and aggressiveness on the reproductive success of zebrafish. Anim Behav. 2012;83:41–6. doi:10.1016/j.anbehav.2011.10.004.
22. Ruhl N, McRobert SP, Currie WJS. Shoaling preferences and the effects of sex ratio on spawning and aggression in small laboratory populations of zebrafish (*Danio rerio*). Lab Anim (NY). 2009;38:264–9. doi:10.1038/laban0809-264.
23. Oliveira RF. Mind the fish: zebrafish as a model in cognitive social neuroscience. Front Neural Circuits. 2013;7:131. doi:10.3389/fncir.2013.00131.
24. Abril-de-Abreu R, Cruz J, Oliveira RF. Social eavesdropping in zebrafish: tuning of attention to social interactions. Sci Rep. 2015;5:12678. doi:10.1038/srep12678.
25. Engeszer RE, DA Barbiano LA, Ryan MJ, Parichy DM. Timing and plasticity of shoaling behaviour in the zebrafish, *Danio rerio*. Anim Behav. 2007;74:1269–75. doi:10.1016/j.anbehav.2007.01.032.
26. Silverman JL, Yang M, Lord C, Crawley JN. Behavioural phenotyping assays for mouse models of autism. Nat Publ Gr. 2010;11:490–502. doi:10.1038/nrn2851.
27. Barba-Escobedo PA, Gould GG. Visual social preferences of lone zebrafish in a novel environment: strain and anxiolytic effects. Genes Brain Behav. 2012;11:366–73. doi:10.1111/j.1601-183X.2012.00770.x.
28. Gerlach G, Lysiak N. Kin recognition and inbreeding avoidance in zebrafish, *Danio rerio*, is based on phenotype matching. Anim Behav. 2006;71:1371–7. doi:10.1016/j.anbehav.2005.10.010.
29. Gerlach G, Hodgins-Davis A, Avolio C, Schunter C. Kin recognition in zebrafish: a 24-hour window for olfactory imprinting. Proc Biol Sci. 2008;275:2165–70. doi:10.1098/rspb.2008.0647.
30. Hinz C, Kobbenbring S, Kress S, Sigman L, Müller A, Gerlach G. Kin recognition in zebrafish, *Danio rerio*, is based on imprinting on olfactory and visual stimuli. Anim Behav. 2013;85:925–30. doi:10.1016/j.anbehav.2013.02.010.
31. Heyes CM. Social learning in animals: categories and mechanisms. Biol Rev Camb Philos Soc. 1994;69:207–31.
32. Abril-de-Abreu R, Cruz AS, Oliveira RF. Social dominance modulates eavesdropping in zebrafish. R Soc Open Sci. 2015;2:150220. doi:10.1098/rsos.150220.
33. Oliveira RF, McGregor PK, Latruffe C. Know thine enemy: fighting fish gather information from observing conspecific interactions. Proc R Soc B Biol Sci. 1998;265:1045–9. doi:10.1098/rspb.1998.0397.

34. Peake TM. Eavesdropping in communication networks. In: McGregor PK, editor, Animal communication networks. Cambridge: Cambridge University Press; 2005. doi:10.1017/CBO9780511610363.
35. von Frisch K. Zur Psychologie des Fisch-Schwarmes. Naturwissenschaften. 1938;26:601–6. doi:10.1007/BF01590598.
36. Speedie N, Gerlai R. Alarm substance induced behavioral responses in zebrafish (*Danio rerio*). Behav Brain Res. 2008;188:168–77. doi:10.1016/j.bbr.2007.10.031.
37. Parra KV, Adrian JC, Gerlai R. The synthetic substance hypoxanthine 3-N-oxide elicits alarm reactions in zebrafish (*Danio rerio*). Behav Brain Res. 2009;205:336–41. doi:10.1016/j.bbr.2009.06.037.
38. Mathuru AS, Kibat C, Cheong WF, Shui G, Wenk MR, Friedrich RW, Jesuthasan S. Chondroitin fragments are odorants that trigger fear behavior in fish. Curr Biol. 2012;22:538–44. doi:10.1016/j.cub.2012.01.061.
39. Suboski MD, Bain S, Carty AE, McQuoid LM, Seelen MI, Seifert M. Alarm reaction in acquisition and social transmission of simulated-predator recognition by zebra danio fish (*Brachydanio rerio*). Comp Biochem Physiol A Mol Integr Physiol. 1990;104:101–12.
40. Hall D, Suboski MD. Visual and olfactory stimuli in learned release of alarm reactions by zebra danio fish (*Brachydanio rerio*). Neurobiol Learn Mem. 1995;63:229–40. doi:10.1006/nlme.1995.1027.
41. Gleason PE, Weber PG, Weber SP. Effect of group size on avoidance learning in zebra fish, *Brachydanio rerio* (Pisces: Cyprinidae). Anim Learn Behav. 1977;5:213–6. doi:10.3758/BF03214081.
42. Lindeyer CM, Reader SM. Social learning of escape routes in zebrafish and the stability of behavioural traditions. Anim Behav. 2010;79:827–34. doi:10.1016/j.anbehav.2009.12.024.
43. Zala SM, Määttänen I, Penn DJ. Different social-learning strategies in wild and domesticated zebra fish, *Danio rerio*. Anim Behav. 2012;83:1519–25. doi:10.1016/j.anbehav.2012.03.029.
44. Zala SM, Määttänen I. Social learning of an associative foraging task in zebrafish. Naturwissenschaften. 2013;100:469–72. doi:10.1007/s00114-013-1017-6.
45. Parker MO, Gaviria J, Haigh A, Millington ME, Verity J. Discrimination reversal and attentional sets in zebrafish (*Danio rerio*). Behav Brain Res. 2014;232:264–8. doi:10.1016/j.bbr.2012.04.035.Discrimination.
46. Buske C, Gerlai R. Shoaling develops with age in zebra fish (*Danio rerio*). Prog Neuropsychopharmacol Biol Psychiatry. 2011;35:1409–15. doi:10.1016/j.pnpbp.2010.09.003.
47. Valente A, Huang K-H, Portugues R, Engert F. Ontogeny of classical and operant learning behaviors in zebrafish. Learn Mem. 2012;19:170–7. doi:10.1101/lm.025668.112.
48. Hinz FI, Aizenberg M, Tushev G, Schuman EM. Protein synthesis-dependent associative long-term memory in larval zebrafish. J Neurosci. 2013;33:15382–7. doi:10.1523/JNEUROSCI.0560-13.2013.
49. Miller NY, Gerlai R. Shoaling in zebrafish: what we don't know. Rev Neurosci. 2011;22:17–25. doi:10.1515/RNS.2011.004.
50. Wright D, Krause J. Repeated measures of shoaling tendency in zebrafish (*Danio rerio*) and other small teleost fishes. Nat Protoc. 2006;1:1828–31. doi:10.1038/nprot.2006.287.
51. Braida D, Donzelli A, Martucci R, Capurro V, Busnelli M, Chini B, Sala M. Neurohypophyseal hormones manipulation modulate social and anxiety-related behavior in zebrafish. Psychopharmacology (Berl). 2012;220:319–30. doi:10.1007/s00213-011-2482-2.
52. Qin M, Wong A, Seguin D, Gerlai R. Induction of social behavior in zebrafish: live versus computer animated fish as stimuli. Zebrafish. 2014;11:185–97. doi:10.1089/zeb.2013.0969.
53. Wright D, Nakamichi R, Krause J, Butlin RK. QTL analysis of behavioral and morphological differentiation between wild and laboratory zebrafish (Danio rerio). Behav Genet. 2006;36. doi:10.1007/s10519-005-9029-4.
54. Scerbina T, Chatterjee D, Gerlai R. Dopamine receptor antagonism disrupts social preference in zebrafish: a strain comparison study. Amino Acids. 2012;43:2059–72. doi:10.1007/s00726-012-1284-0.

55. Fernandes Y, Gerlai R. Long-term behavioral changes in response to early developmental exposure to ethanol in zebrafish. Alcohol Clin Exp Res. 2009;33:601–9. doi:10.1111/j.1530-0277.2008.00874.x.
56. Gerlai R. Zebra fish: an uncharted behavior genetic model. Behav Genet. 2003;33:461–8.
57. Ladu F, Mwaffo V, Li J, Macrì S, Porfiri M. Acute caffeine administration affects zebrafish response to a robotic stimulus. Behav Brain Res. 2015;289:48–54. doi:10.1016/j.bbr.2015.04.020.
58. Grossman L, Utterback E, Stewart A, Gaikwad S, Chung KM, Suciu C, Wong K, Elegante M, Elkhayat S, Tan J, et al. Characterization of behavioral and endocrine effects of LSD on zebrafish. Behav Brain Res. 2010;214:277–84. doi:10.1016/j.bbr.2010.05.039.
59. Fernandes Y, Rampersad M, Jia J, Gerlai R. The effect of the number and size of animated conspecific images on shoaling responses of zebrafish. Pharmacol Biochem Behav. 2015;139(Pt B):94–102. doi:10.1016/j.pbb.2015.01.011.
60. Dreosti E, Lopes G, Kampff AR, Wilson SW. Development of social behavior in young zebrafish. Front Neural Circuits. 2015;9:39. doi:10.3389/fncir.2015.00039.
61. Ruhl N, McRobert SP. The effect of sex and shoal size on shoaling behaviour in *Danio rerio*. J Fish Biol. 2005;67:1318–26. doi:10.1111/j.0022-1112.2005.00826.x.
62. Engeszer RE, Ryan MJ, Parichy DM. Learned social preference in zebrafish. Curr Biol. 2004;14:881–4. doi:10.1016/j.
63. Al-imari L, Gerlai R. Sight of conspecifics as reward in associative learning in zebrafish (*Danio rerio*). Behav Brain Res. 2008;189:216–9. doi:10.1016/j.bbr.2007.12.007.
64. Saif M, Chatterjee D, Buske C, Gerlai R. Sight of conspecific images induces changes in neurochemistry in zebrafish. Behav Brain Res. 2013;243:294–9. doi:10.1016/j.bbr.2013.01.020.
65. Spinello C, Macrì S, Porfiri M. Acute ethanol administration affects zebrafish preference for a biologically inspired robot. Alcohol. 2013;47:391–8. doi:10.1016/j.alcohol.2013.04.003.
66. Pitcher TJ. Functions of shoaling behaviour in teleost. In: Pitcher TJ, editor, The Behaviour of Teleost Fishes, 2nd edition, Springer US, 1993. doi:10.1007/978-1-4684-8261-4_12.
67. Miller N, Gerlai R. Quantification of shoaling behaviour in zebrafish (*Danio rerio*). Behav Brain Res. 2007;184:157–66. doi:10.1016/j.bbr.2007.07.007.
68. Pham M, Raymond J, Hester J, Kyzar E, Gaikwad S, Bruce I, Fryar C, Chanin S, Enriquez J, Bagawandoss S, Zapolsky I, Green J, Michael Stewart A, Robison B, Kalueff AV. Assessing Social Behavior Phenotypes in Adult Zebrafish: Shoaling, Social preference and Mirror Biting tests. In: Klueff AV, Stewart AM, editors. Zebrafish Protocols for Neurobehavioral Research, 2012, Vol.66, p 231-46, New York: Humana Press. doi:10.1007/978-1-61779-597-8.
69. Green J, Collins C, Kyzar EJ, Pham M, Roth A, Gaikwad S, Cachat J, Stewart AM, Landsman S, Grieco F, et al. Automated high-throughput neurophenotyping of zebrafish social behavior. J Neurosci Methods. 2012;210:266–71. doi:10.1016/j.jneumeth.2012.07.017.
70. Echevarria DJ, Hammack CM, Pratt DW, Hosemann JD. A novel behavioral test battery to assess global drug effects using the zebrafish. Int J Comp Psychol. 2008;21:19–34.
71. Gebauer DL, Pagnussat N, Piato ÂL, Schaefer IC, Bonan CD, Lara DR. Pharmacology, biochemistry and behavior effects of anxiolytics in zebra fish: similarities and differences between benzodiazepines, buspirone and ethanol. Pharmacol Biochem Behav. 2011;99:480–6. doi:10.1016/j.pbb.2011.04.021.
72. Pérez-Escudero A, Vicente-Page J, Hinz RC, Arganda S, de Polavieja GG. idTracker: tracking individuals in a group by automatic identification of unmarked animals. Nat Methods. 2014;11:743–8. doi:10.1038/nmeth.2994.
73. Buske C, Gerlai R. Maturation of shoaling behavior is accompanied by changes in the dopaminergic and serotoninergic systems in zebrafish. Dev Psychobiol. 2012;54:28–35. doi:10.1002/dev.20571.
74. Rhee SH, Waldman ID. Genetic and environmental influences on antisocial behavior: a meta-analysis of twin and adoption studies. Psychol Bull. 2002;128:490–529.

75. Jones LJ, Norton WHJ. Using zebrafish to uncover the genetic and neural basis of aggression, a frequent comorbid symptom of psychiatric disorders. Behav Brain Res. 2015;276:171–80. doi:10.1016/j.bbr.2014.05.055.
76. Stewart AM, Ullmann JFP, Norton WHJ, Parker MO, Brennan CH, Gerlai R, Kalueff AV. Molecular psychiatry of zebrafish. Mol Psychiatry. 2015;20:2–17. doi:10.1038/mp.2014.128.
77. Filby AL, Paull GC, Hickmore TF, Tyler CR. Unravelling the neurophysiological basis of aggression in a fish model. BMC Genomics. 2010;11:498. doi:10.1186/1471-2164-11-498.
78. Norton W, Bally-Cuif L. Adult zebrafish as a model organism for behavioural genetics. BMC Neurosci. 2010;11:90. doi:10.1186/1471-2202-11-90.
79. Ricci L, Summers CH, Larson ET, O'Malley D, Melloni RH. Development of aggressive phenotypes in zebrafish: interactions of age, experience and social status. Anim Behav. 2013;86:245–52. doi:10.1016/j.anbehav.2013.04.011.
80. Weber DN, Ghorai JK. Experimental design affects social behavior outcomes in adult zebrafish developmentally exposed to lead. Zebrafish. 2013;10:294–302. doi:10.1089/zeb.2012.0780.
81. Weber DN, Hoffmann RG, Hoke ES, Tanguay RL. Bisphenol A exposure during early development induces sex-specific changes in adult zebrafish social interactions. J Toxicol Environ Health A. 2015;78:50–66. doi:10.1080/15287394.2015.958419.
82. Dahlbom SJ, Lagman D, Lundstedt-Enkel K, Sundström LF, Winberg S. Boldness predicts social status in zebrafish (*Danio rerio*). PLoS One. 2011;6, e23565. doi:10.1371/journal.pone.0023565.
83. Howard RD, Rohrer K, Liu Y, Muir WM. Mate competition and evolutionary outcomes in genetically modified zebrafish (*Danio rerio*). Evolution. 2015;69:1143–57. doi:10.1111/evo.12662.
84. Paull GC, Filby AL, Giddins HG, Coe TS, Hamilton PB, Tyler CR. Dominance hierarchies in zebrafish (*Danio rerio*) and their relationship with reproductive success. Zebrafish. 2010;7:109–17. doi:10.1089/zeb.2009.0618.
85. Basquill SP, Grant JW. An increase in habitat complexity reduces aggression and monopolization of food by zebra fish (*Danio rerio*). Can J Zool. 1998;76:770–2. doi:10.1139/z97-232.
86. Oliveira RF, Silva JF, Simões JM. Fighting zebrafish: characterization of aggressive behavior and winner-loser effects. Zebrafish. 2011;8:73–81. doi:10.1089/zeb.2011.0690.
87. Way GP, Ruhl N, Snekser JL, Kiesel AL, McRobert SP. A comparison of methodologies to test aggression in zebrafish. Zebrafish. 2015;12:144–51. doi:10.1089/zeb.2014.1025.
88. Toms CN, Echevarria DJ. Back to basics: searching for a comprehensive framework for exploring individual differences in zebrafish (*Danio rerio*) behavior. Zebrafish. 2014;11:325–40. doi:10.1089/zeb.2013.0952.
89. Gerlai R, Lahav M, Guo S, Rosenthal A. Drinks like a fish: zebra fish (*Danio rerio*) as a behavior genetic model to study alcohol effects. Pharmacol Biochem Behav. 2000;67:773–82.
90. Way GP, Kiesel AL, Ruhl N, Snekser JL, McRobert SP. Sex differences in a shoaling-boldness behavioral syndrome, but no link with aggression. Behav Processes. 2015;113:7–12. doi:10.1016/j.beproc.2014.12.014.
91. Ariyomo TO, Carter M, Watt PJ. Heritability of boldness and aggressiveness in the zebrafish. Behav Genet. 2013;43:161–7. doi:10.1007/s10519-013-9585-y.
92. McRobert SP, Kiesel AL, Snekser JL, Ruhl N. Behavioural syndromes and shoaling: connections between aggression, boldness and social behaviour in three different Danios. Behaviour. 2012;149:1155–75. doi:10.1163/1568539X-00003015.
93. Norton WHJ, Stumpenhorst K, Faus-Kessler T, Folchert A, Rohner N, Harris MP, Callebert J, Bally-Cuif L. Modulation of Fgfr1a signaling in zebrafish reveals a genetic basis for the aggression-boldness syndrome. J Neurosci. 2011;31:13796–807. doi:10.1523/JNEUROSCI.2892-11.2011.
94. Martins EP, Bhat A. Population-level personalities in zebrafish: aggression-boldness across but not within populations. Behav Ecol. 2014;25:368–73. doi:10.1093/beheco/aru007.

95. Norton WHJ, Bally-Cuif L. Unravelling the proximate causes of the aggression-boldness behavioural syndrome in zebrafish. Behaviour. 2012;149:1063–79. doi:10.1163/1568539X-00003012.
96. Dahlbom SJ, Backström T, Lundstedt-Enkel K, Winberg S. Aggression and monoamines: effects of sex and social rank in zebrafish (*Danio rerio*). Behav Brain Res. 2012;228:333–8. doi:10.1016/j.bbr.2011.12.011.
97. Gronquist D, Berges JA. Effects of aquarium-related stressors on the zebrafish: a comparison of behavioral, physiological, and biochemical indicators. J Aquat Anim Health. 2013;25:53–65. doi:10.1080/08997659.2012.747450.
98. Hamilton IM, Dill LM. Monopolization of food by zebrafish (*Danio rerio*) increases in risky habitats. Can J Zool. 2002;80:2164–9. doi:10.1139/z02-199.
99. Ariyomo TO, Watt PJ. Effect of hunger level and time of day on boldness and aggression in the zebrafish *Danio rerio*. J Fish Biol. 2015;86:1852–9. doi:10.1111/jfb.12674.
100. Moretz JA, Martins EP, Robison BD. Behavioral syndromes and the evolution of correlated behavior in zebrafish. Behav Ecol. 2007;18:556–62. doi:10.1093/beheco/arm011.
101. Rey S, Digka N, MacKenzie S. Animal personality relates to thermal preference in wild-type zebrafish, *Danio rerio*. Zebrafish. 2015;12:243–9. doi:10.1089/zeb.2014.1076.
102. Darrow KO, Harris WA. Characterization and development of courtship in zebrafish, *Danio rerio*. Zebrafish. 2004;1:40–5. doi:10.1089/154585404774101662.
103. Owen MA, Rohrer K, Howard RD. Mate choice for a novel male phenotype in zebrafish, *Danio rerio*. Anim Behav. 2012;83:811–20. doi:10.1016/j.anbehav.2011.12.029.
104. Gumm JM, Snekser JL, Iovine MK. Fin-mutant female zebrafish (*Danio rerio*) exhibit differences in association preferences for male fin length. Behav Processes. 2009;80:35–8. doi:10.1016/j.beproc.2008.09.004.
105. Turnell ER, Mann KD, Rosenthal GG, Gerlach G. Mate choice in zebrafish (Danio rerio) analyzed with video-stimulus techniques. Biol Bull. 2003;205:225–6.
106. Skinner AMJ, Watt PJ. Strategic egg allocation in the zebra fish, Danio rerio. Behav Ecol. 2007;18:905–9. doi:10.1093/beheco/arm059.
107. Pyron M. Female preferences and male-male interactions in zebrafish (Danio rerio). Can J Zool. 2003;81(1):122–5. doi:10.1139/z02-229.
108. Kitevski B, Pyron M. Female zebrafish (*Danio rerio*) do not prefer mutant longfin males. J Freshw Ecol. 2003;18:501–2. doi:10.1080/02705060.2003.9663988.
109. Hutter S, Zala SM, Penn DJ. Sex recognition in zebrafish (*Danio rerio*). J Ethol. 2010;29:55–61. doi:10.1007/s10164-010-0221-5.
110. van den Hurk R, Lambert JGD. Ovarian steroid glucuronides function as sex pheromones for male zebrafish, *Brachydanio rerio*. Can J Zool. 1983;61:2381–7. doi:10.1139/z83-317.
111. Coe TS, Hamilton PB, Griffiths AM, Hodgson DJ, Wahab MA, Tyler CR. Genetic variation in strains of zebrafish (*Danio rerio*) and the implications for ecotoxicology studies. Ecotoxicology. 2009;18:144–50. doi:10.1007/s10646-008-0267-0.
112. Whiteley AR, Bhat A, Martins EP, Mayden RL, Arunachalam M, Uusi-Heikkilä S, Ahmed ATA, Shrestha J, Clark M, Stemple D, et al. Population genomics of wild and laboratory zebrafish (*Danio rerio*). Mol Ecol. 2011;20:4259–76. doi:10.1111/j.1365-294X.2011.05272.x.
113. Wright D, Rimmer LB, Pritchard VL, Krause J, Butlin RK. Inter and intra-population variation in shoaling and boldness in the zebrafish (*Danio rerio*). Naturwissenschaften. 2003;374–7. doi:10.1007/s00114-003-0443-2.
114. Oswald M, Robison BD. Strain-specific alteration of zebrafish feeding behavior in response to aversive stimuli. Can J Zool. 2008;86:1085–94. doi:10.1139/Z08-085.
115. Drew RE, Settles ML, Churchill EJ, Williams SM, Balli S, Robison BD. Brain transcriptome variation among behaviorally distinct strains of zebrafish (*Danio rerio*). BMC Genomics. 2012;13:323. doi:10.1186/1471-2164-13-323.
116. Egan RJ, Bergner CL, Hart PC, Cachat JM, Canavello PR, Elegante MF, Elkhayat SI, Bartels BK, Tien AK, Tien DH, et al. Understanding behavioral and physiological phenotypes of stress and anxiety in zebrafish. Behav Brain Res. 2009;205:38–44. doi:10.1016/j.bbr.2009.06.022.

117. Mahabir S, Chatterjee D, Buske C, Gerlai R. Maturation of shoaling in two zebrafish strains: a behavioral and neurochemical analysis. Behav Brain Res. 2013;247:1–8. doi:10.1016/j.bbr.2013.03.013.
118. Sertori R, Trengove M, Basheer F, Ward AC, Liongue C. Genome editing in zebrafish: a practical overview. Brief Funct Genomics. 2015. doi:10.1093/bfgp/elv051.
119. Bill BR, Petzold AM, Clark KJ, Schimmenti LA, Ekker SC. A primer for morpholino use in zebrafish. Zebrafish. 2009;6:69–77. doi:10.1089/zeb.2008.0555.
120. Wienholds E, van Eeden F, Kosters M, Mudde J, Plasterk RHA, Cuppen E. Efficient target-selected mutagenesis in zebrafish. Genome Res. 2003;13:2700–7. doi:10.1101/gr.1725103.
121. Doyon Y, McCammon JM, Miller JC, Faraji F, Ngo C, Katibah GE, Amora R, Hocking TD, Zhang L, Rebar EJ, et al. Heritable targeted gene disruption in zebrafish using designed zinc-finger nucleases. Nat Biotechnol. 2008;26:702–8. doi:10.1038/nbt1409.
122. Hwang WY, Fu Y, Reyon D, Maeder ML, Tsai SQ, Sander JD, Peterson RT, Yeh J-RJ, Joung JK. Efficient genome editing in zebrafish using a CRISPR-Cas system. Nat Biotechnol. 2013;31:227–9. doi:10.1038/nbt.2501.
123. Zu Y, Tong X, Wang Z, Liu D, Pan R, Li Z, Hu Y, Luo Z, Huang P, Wu Q, et al. TALEN-mediated precise genome modification by homologous recombination in zebrafish. Nat Methods. 2013;10:329–31. doi:10.1038/nmeth.2374.
124. Asakawa K, Kawakami K. Targeted gene expression by the Gal4-UAS system in zebrafish. Dev Growth Differ. 2008;50:391–9. doi:10.1111/j.1440-169X.2008.01044.x.
125. Ziv L, Muto A, Schoonheim PJ, Meijsing SH, Strasser D, Ingraham HA, Schaaf MJM, Yamamoto KR, Baier H. An affective disorder in zebrafish with mutation of the glucocorticoid receptor. Mol Psychiatry. 2013;18:681–91. doi:10.1038/mp.2012.64.
126. Santos EM, Kille P, Workman VL, Paull GC, Tyler CR. Sexually dimorphic gene expression in the brains of mature zebrafish. Comp Biochem Physiol A Mol Integr Physiol. 2008;149:314–24. doi:10.1016/j.cbpa.2008.01.010.
127. Wong RY, McLeod MM, Godwin J. Limited sex-biased neural gene expression patterns across strains in Zebrafish (*Danio rerio*). BMC Genomics. 2014;15:905. doi:10.1186/1471-2164-15-905.
128. Félix AS, Faustino AI, Cabral EM, Oliveira RF. Noninvasive measurement of steroid hormones in zebrafish holding-water. Zebrafish. 2013;10:110–5. doi:10.1089/zeb.2012.0792.
129. Conradsen C, McGuigan K. Sexually dimorphic morphology and swimming performance relationships in wild-type zebrafish *Danio rerio*. J Fish Biol. 2015;87:1219–33. doi:10.1111/jfb.12784.
130. Tran S, Gerlai R. Individual differences in activity levels in zebrafish (*Danio rerio*). Behav Brain Res. 2013;257:224–9. doi:10.1016/j.bbr.2013.09.040.
131. Oswald ME, Drew RE, Racine M, Murdoch GK, Robison BD. Is behavioral variation along the bold-shy continuum associated with variation in the stress axis in zebrafish? Physiol Biochem Zool. 2012;85:718–28. doi:10.1086/668203.
132. Koolhaas JM, Korte SM, De Boer SF, Van Der Vegt BJ, Van Reenen CG, Hopster H, De Jong IC, Ruis MA, Blokhuis HJ. Coping styles in animals: current status in behavior and stress-physiology. Neurosci Biobehav Rev. 1999;23:925–35.
133. Koolhaas JM, de Boer SF, Coppens CM, Buwalda B. Neuroendocrinology of coping styles: towards understanding the biology of individual variation. Front Neuroendocrinol. 2010;31:307–21. doi:10.1016/j.yfrne.2010.04.001.
134. Øverli Ø, Sørensen C, Pulman KGT, Pottinger TG, Korzan W, Summers CH, Nilsson GE. Evolutionary background for stress-coping styles: relationships between physiological, behavioral, and cognitive traits in non-mammalian vertebrates. Neurosci Biobehav Rev. 2007;31:396–412. doi:10.1016/j.neubiorev.2006.10.006.
135. Pottinger TG, Carrick TR. Modification of the plasma cortisol response to stress in rainbow trout by selective breeding. Gen Comp Endocrinol. 1999;116:122–32. doi:10.1006/gcen.1999.7355.
136. Schjolden J, Stoskhus A, Winberg S. Does individual variation in stress responses and agonistic behavior reflect divergent stress coping strategies in juvenile rainbow trout? Physiol Biochem Zool. 2004;78:715–23. doi:10.1086/432153.

137. Bolhuis JE, Schouten WGP, de Leeuw JA, Schrama JW, Wiegant VM. Individual coping characteristics, rearing conditions and behavioural flexibility in pigs. Behav Brain Res. 2004;152:351–60. doi:10.1016/j.bbr.2003.10.024.
138. Koolhaas JM, van Oortmerssen GA, den Daas S, Benus RF. Routine formation and flexibility in social and non-social behaviour of aggressive and non-aggressive male mice. Behaviour. 1990;112:176–93. doi:10.1163/156853990X00185.
139. Ruiz-Gomez Mde L, Huntingford FA, Øverli Ø, Thörnqvist P-O, Höglund E. Response to environmental change in rainbow trout selected for divergent stress coping styles. Physiol Behav. 2011;102:317–22. doi:10.1016/j.physbeh.2010.11.023.
140. Brelin D, Petersson E, Dannewitz J, Dahl J, Winberg S. Frequency distribution of coping strategies in four populations of brown trout (*Salmo trutta*). Horm Behav. 2008;53:546–56. doi:10.1016/j.yhbeh.2007.12.011.
141. Wong RY, Dereje S, Sawyer S, Oxendine SE, Zhou L, Kezios ZD, Godwin J, Perrin F. Comparing behavioral responses across multiple assays of stress and anxiety in zebrafish (*Danio rerio*). Behaviour. 2012;149:1205–40. doi:10.1163/1568539X-00003018.
142. Tudorache C, Schaaf MJM, Slabbekoorn H. Covariation between behaviour and physiology indicators of coping style in zebrafish (*Danio rerio*). J Endocrinol. 2013;219:251–8. doi:10.1530/JOE-13-0225.
143. Rey S, Boltana S, Vargas R, Roher N, Mackenzie S. Combining animal personalities with transcriptomics resolves individual variation within a wild-type zebrafish population and identifies underpinning molecular differences in brain function. Mol Ecol. 2013;22:6100–15. doi:10.1111/mec.12556.
144. Wong RY, Lamm MS, Godwin J. Characterizing the neurotranscriptomic states in alternative stress coping styles. BMC Genomics. 2015;16:425. doi:10.1186/s12864-015-1626-x.
145. Dell AI, Bender JA, Branson K, Couzin ID, de Polavieja GG, Noldus LPJJ, Pérez-Escudero A, Perona P, Straw AD, Wikelski M, et al. Automated image-based tracking and its application in ecology. Trends Ecol Evol. 2014;29:417–28. doi:10.1016/j.tree.2014.05.004.
146. Cachat J, Stewart A, Utterback E, Hart P, Gaikwad S, Wong K, Kyzar E, Wu N, Kalueff AV. Three-dimensional neurophenotyping of adult zebrafish behavior. PLoS One. 2011;6, e17597. doi:10.1371/journal.pone.0017597.
147. Miller N, Gerlai R. Automated tracking of zebrafish shoals and the analysis of shoaling behavior. In: Kalueff AV, Stewart AM, editors. Zebrafish Protocols for Neurobehavioral Research, 2012, Vol. 66, pp. 217–230. New York: Humana Press doi:10.1007/978-1-61779-597-8.
148. Branson K, Robie AA, Bender J, Perona P, Dickinson MH. High-throughput ethomics in large groups of Drosophila. Nat Methods. 2009;6:451–7. doi:10.1038/nmeth.1328.
149. Oliveira RF, Rosenthal GG, Schlupp I, McGregor PK, Cuthill IC, Endler JA, Fleishman LJ, Zeil J, Barata E, Burford F, et al. Considerations on the use of video playbacks as visual stimuli: the Lisbon workshop consensus. Acta Ethol. 2000;3:61–5. doi:10.1007/s102110000019.
150. Ingley SJ, Rahmani Asl M, Wu C, Cui R, Gadelhak M, Li W, Zhang J, Simpson J, Hash C, Butkowski T, et al. anyFish 2.0: an open-source software platform to generate and share animated fish models to study behavior. SoftwareX. 2015;3–4:13–21. doi:10.1016/j.softx.2015.10.001.

Modeling OCD Endophenotypes in Zebrafish

Matthew Parker

Abstract Obsessive compulsive disorder (OCD) is a pervasive, debilitating neuropsychiatric disorder. Despite over half a century of effort, OCD has remained remarkably resistant to treatment, partly owing to a lack of understanding of the underlying biology. Recently, there has been a growing consensus that in order to understand the basis of neuropsychiatric disorders such as OCD, we should focus on transdiagnostic, observable, measurable behavioral or neural elements, *endophenotypes*. Zebrafish have the well-characterized neural development and available cutting-edge genetic tools that make them the ideal species for studying psychiatric disorders. In addition, a number of endophenotypes linked to OCD have been observed, and can be objectively measured, in zebrafish. In this chapter, some key behavioral tests of relevance to OCD will be outlined. If the neural substrates underlying these behaviors are elucidated, this may represent significant progress in understanding the biological underpinnings of OCD. This will ultimately lead to increased specificity for drug discovery, as well as providing targets for personalized treatments for one of the most common neuropsychiatric disorders.

Keywords OCD • Zebrafish • Behavior • Personalized medicine • Neuropsychiatric

1 Introduction

Considerable progress has been made in basic biomedical science, with genetics and genomics providing the tools by which we can understand and characterize disease. Psychiatry, however, is falling somewhat behind. Part of the reason for this may be that disease classifications, as defined within the Diagnostic and Statistical Manual of Mental Disorder (DSM-IV; DSM-V), are not sufficiently rooted in the biology of the disorders [1]. This presents a challenge in terms of both diagnostic precision and treatment efficacy of what are widely recognized as polygenic

M. Parker (✉)
School of Health Sciences and Social Work, University of Portsmouth,
King Richard 1st Road, Portsmouth, UK
e-mail: matthew.parker@port.ac.uk

A.V. Kalueff (ed.), *The rights and wrongs of zebrafish: Behavioral phenotyping of zebrafish*, DOI 10.1007/978-3-319-33774-6_6

disorders. Taking a biological perspective of neuropsychiatric disorder is particularly important for those involved in translational research, as an understanding of the underlying mechanisms of heterogeneous psychiatric disorders may help to develop appropriate models with high external validity. One answer is to adopt a dimensional approach to psychiatry, examining diseases in terms of their (often trans-diagnositc) constituent behavioral correlates [2]. Endophenotypes represent observable, heritable, measureable components of neuropsychiatric disease [3]. These can be characterized either as neurophysiological endophenotypes (for example, alterations in brain structure or morphology) or neurocognitive endophenotypes (for example, impulsivity or compulsivity) [4]. Not only do endophenotypes benefit from being objective and quantifiable but they are observed in both patients and first-degree relatives demonstrating their allelic links [2, 3].

Obsessive Compulsive Disorder (OCD) is a pervasive, disabling, neuropsychiatric disorder characterized both by persistent and unwanted thoughts (obsessions) about potential harm, for example through contamination; and by checking and monitoring (compulsions) that directly relate to the obsessions, for example by repetitive washing of the hands [5]. OCD affects approximately 2–3 % of people over the life course [6], and is remarkably resistant to treatment in many patients [7–9]. Gaining a better understanding of the molecular and cellular basis of OCD will help in the search for effective pharmacological and behavioral treatments by allowing clinicians to personalize treatment [10, 11]. Neurocognitive endophenotypes for OCD include stereotypic behavior (inappropriate repetition of behavioral sequences) [12, 13], persistence of behavioral habits despite devaluation or adverse consequences [14], rigid attentional sets and reversal learning deficits [15, 16] and resistance to extinction [2].

In recent years, zebrafish have demonstrated their ability to perform well in a number of behavioral tasks relevant to these neurocognitive endophenotypes [17, 18]. Forward-genetic mutational screening is a powerful approach for identifying novel candidate genes. Individuals demonstrating the phenotype of interest are selected from a large-scale loss-of-function (LoF) mutagenized population, thus allowing identification of the functional mutation underlying the phenotype [19]. Although forward genetic screens are possible in mammals, extensive infrastructure and personnel requirements limit their utility to all but a few large-scale projects. For this reason, zebrafish are fast becoming the vertebrate of choice for forward-genetic screens. High fecundity, as well as high degree of homology with mammals and extensive availability of genomic tools together facilitate simple, large-scale, translationally relevant phenotype screens [20]. Zebrafish encode orthologs for the majority of characterized human compulsivity-related gene including all members of the dopamine [21], serotonin [22], and glutamatergic [23] receptor gene families. Zebrafish are thus uniquely positioned for efficient whole-genome functional assessments of genetic factors that mediate the variability in OCD-related endophenotypes. In this regard, many behavioral assays relevant to psychiatric endophenotypes are beginning to be pharmacologically or genetically validated in a bid to demonstrate their translational relevance to understanding human behavioral conditions. This chapter will outline behavioral assays pertinent to OCD endophenotypes that have been designed and validated in zebrafish. If these tasks are used for high-throughput forward-genetic screening, this will expedite the search for effective treatment for OCD.

2 Behavioral Phenotypic Tests in Zebrafish and Observed Phenotypes

2.1 Stereotypic (Repetitive) Behavior

Stereotypic behavior is defined as repeated and invariant response patterns [24], is an endophenotype associated with OCD in humans [4], and is observed in animal models of OCD [25]. In zebrafish, several attempts have been made to characterize stereotypic behavior in the laboratory, typically in terms of quantifying locomotor responses to drug challenges. For example, Rhehl and colleagues [26] described stereotypic behavior following administration of the non-competitive NMDA-receptor antagonist, ketamine. They operationalized stereotypic behavior in terms of 'circling', or repetitive rotations, in an open-field test following analysis of route patterns and locomotion in automated, commercially available software (Ethovision, Noldus; Fig. 1a) [26]. This same group also characterized stereotypic behavior following administration of the hallucinogen, ibogaine, in a similar fashion [27]. However, following administration of ibogaine, they observed an idiosyncratic pattern of stereotypic exploration in the novel tank test. Specifically, the fish would descend to the lower portion of the tank into a corner, and subsequently move rapidly to another corner. This sequence would then be repeated a number of times. This may be a useful operational measure of stereotypic behavior as an OCD endophenotype as it is objective, quantifiable, automated and has the potential for high-throughput analysis. However, it is important to consider that these measures of stereotypic behavior may be specific to drug-induced stereotypy, thus potentially limiting the construct validity in terms of OCD models [28].

López-Patiño and colleagues [29] described stereotypic behavior that increases during cocaine withdrawal in zebrafish, an effect rescued by administration of cocaine or diazepam. Administration of FG-7142, a partial inverse agonist at the benzodiazepine allosteric site of the $GABA_A$ receptor, also increases stereotypy. This group operationally defined stereotypic behavior as any repetitive, unvarying behavior, with no obvious goal or function, that lasted >1-min [29]. According to their observations, this was typically characterized as the fish swimming continuously, at a fast pace, back-and-forth along the side of the tank [29]. López-Patiño and colleagues [29] were able to collect these data via continuous video recording and computer algorithms, allowing a degree of objectivity in the observations. The fact that these behavioral sequences were observed in withdrawal from cocaine indicates that this stereotypy was based on complex interactions between the drug and anxiety associated with its removal, a process linked to OCD-like patterns of stereotypy [30–32].

<table>
<tr><th>Assay and relevance to OCD</th><th>Schematic and apparatus</th><th>Refs</th></tr>
<tr><td>A. Open field test

Can measure stereotypic behavior in fish either by analysing circling behavior or other repetitive, invariant swimming patterns</td><td>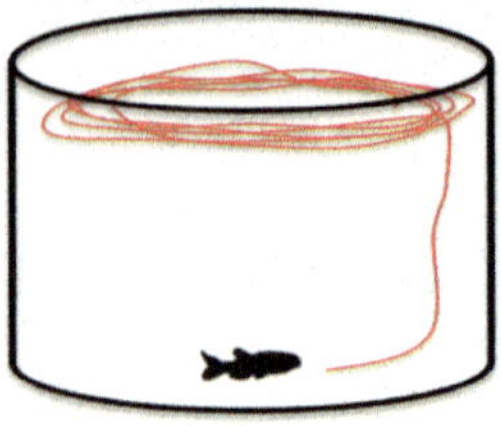</td><td>[26, 27, 29]</td></tr>
<tr><td>B. Place-response test

Can measure habit formation in fish by examining the rate at which they make the locomotor 'response' during the probe trial, as opposed to the 'place' where the reinforcer was delivered</td><td>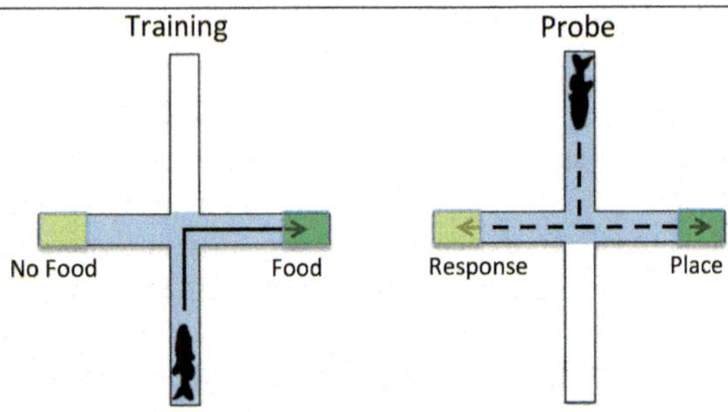
</td><td>[39]</td></tr>
<tr><td>C. Two choice discrimination

Can measure reversal learning by examining the trials to criterion (6-correct trials in a row) on the simple discrimination, the first reversal, and the intra-dimensional set shit.</td><td>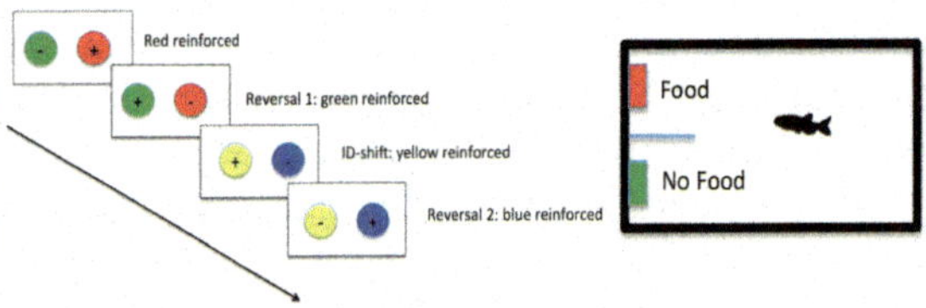
</td><td>[37]</td></tr>
<tr><td>D. Resistance to extinction

Can measure the resistance to extinguishing a Pavlovian response following training and extinction trials.</td><td>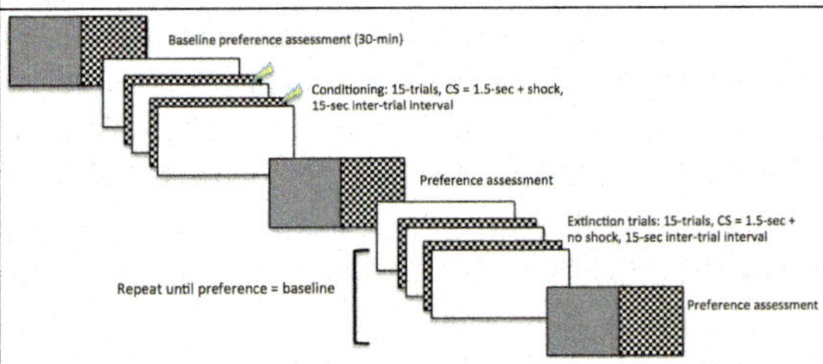
</td><td>[44, 45]</td></tr>
</table>

Fig. 1 Assays that can be used to study OCD endophenotypes in zebrafish

2.2 Stimulus-Response/Habit Formation

During the process of acquiring an operant response, behavior is initially goal-directed and purposeful, with a direct associative relationship between the discriminative stimulus, the response, and its outcome (response-stimulus; R-S learning). However, after prolonged learning, the response to the discriminative stimulus becomes automatic or 'habitual', operationally defined as the point at which the probability of the outcome given the response (P(O|R)) becomes impervious to changes in the contingency, or to devaluation of the outcome (stimulus-response; S-R learning [33, 34]). Variability in the rate at which S-R learning is acquired during training has been reliably observed in OCD patients, suggesting that it may be an endophenotype for the disorder [35, 36].

The study of habit formation in model species has a long and rich tradition [14], and has afforded rodents high status as translational models for habit-related neuropsychiatric disorders. Using fish to study habit is relatively new. However, recent advances in our understanding of the neurocognitive profiles of zebrafish, in particular with respect to their performance on tests of executive function [17, 18, 37, 38], suggest that this species may be a useful addition to translational work on habit and OCD. There is currently only one test in the literature relating to habit formation in zebrafish. Parker and colleagues [39] examined the rate at which zebrafish exposed developmental to ethanol acquired a habit following training in a T-maze, using an assay known as the 'place-response' task (Fig. 1b). The procedure was based on previous work with a variety of species where the animal is over-trained to locate food in a particular location in a T-maze, and subsequently challenged to locate the food from a novel entry point to assess what has been learnt [40–42]. Specifically, the animal initially is trained (starting in the South arm of the maze) that the food will always be located in the East arm. During this time, the North arm is blocked and inaccessible. Following a period of training, the animal is exposed to a probe trial where they are placed in the maze from the North arm (this time with the South arm inaccessible) and their choice of arms (East or West) is noted. Extensive research with rodents has shown that initially, the animals show R-S learning, choosing the 'place' where the food was located (in our example, the East arm). However, following extensive training (over-training) the animals show S-R learning, choosing the West arm having performed an automatic locomotor 'response' [43]. This effect was observed in zebrafish [39].

There may be some potential for habit formation to be tested based on a negative reinforcement stimulus-response-outcome procedure shown to be effective in determining differences in S-R habit in human patients with OCD [35]. In this assay, the fish ostensibly would be trained on an active avoidance procedure, where they were required to avoid a CS previously paired with a shock US [44, 45]. The procedure would be organized into four discrete blocks. In Block 1, the shock US would be repeatedly paired with a visual CS. In Block 2, the fish would be trained to actively avoid the shock-paired CS. In Block 3, devaluation of the CS can be achieved through extinction, by extinguishing the CS-US pairing in a series of Pavlovian discrete trials. In Block 4, habit can be assessed by testing outcome devaluation. The outcome variable would be the distribution of responses to the devalued CS as compared to a non-devalued control group.

2.3 *Behavioral Flexibility and Attentional Set Formation*

Behavioral flexibility describes the potential for an animal to actively adjust an established behavior following situational changes in the response requirement [46]. One empirical definition of behavioral flexibility is the ability to form an attentional set, whereby the 'rule' that governs within-session steady-state responding can be altered according to negative feedback from preceding non-reinforced trials [46, 47]. A marked reduction in behavioral flexibility is seen in patients with OCD and their first-degree relatives (i.e., below that which would be expected in the general population), suggesting it is an important endophenotype for OCD [48, 49]. Behavioral flexibility can be operationally defined in terms of reversal learning and set shifting on two choice discrimination procedures (Fig. 1c) [46]. Specifically, an animal is initially trained to discriminate between two stimuli (e.g., red + vs green −). Once they have reached criterion (e.g., six correct choices in a row [46]) the correct choice is reversed such that the previously unreinforced alternative is now correct, and the previously reinforced alternative, incorrect (i.e., green +, red −). The animal is then trained again to criterion, and the discrimination is then switched again, this time to two novel colors (intra-dimensional set shift; e.g., yellow +, blue −). Once it reaches criterion, the choice is reversed a further time (i.e., blue +, yellow −). If the animal is displaying behavioral flexibility, it would be predicted that the first stages of learning (the initial discrimination and first reversal) would take relatively more trials to reach criterion than latter conditions, as the animal will gradually acquire a 'rule'; i.e., instead of simply learning color A = reinforcement; color B = non-reinforcement, the animal learns to respond according to negative feedback from previous non-reinforced trials [50]. In rodents with intact orbitofrontal cortex, reversal learning and intra-dimensional set shifting results in this pattern of responding [47].

This procedure has been developed for zebrafish, and it has been variously demonstrated that the fish can develop and maintain an attentional set [37]. Briefly, the fish were trained initially to discriminate between two visual stimuli (blue and green cards) in order to receive a food reinforcer. Upon reaching criterion (six correct responses in a row), the contingency was reversed, such that the previously incorrect stimulus was now correct. This was followed by an intra-dimensional set shift and subsequent reversal (as outlined above). We found that, similar to mammalian models, the zebrafish showed a gradual reduction in trials to criterion through the process of reversals and set-shifts, suggesting that they were capable of forming and maintaining an attentional set [37] suggesting that zebrafish may be a useful model organism for investigating behavioral flexibility.

2.4 *Resistance to Extinction*

Following acquisition of an operant or Pavlovian response, if the contingency is removed (i.e., the reinforcer withheld in the presence of the discriminative stimulus following an operant response, or the conditioned stimulus [CS] is repeatedly presented in the absence of the unconditioned stimulus [US] during Pavlovian conditioning) the response-reinforcer or CS-US association reduces (as a function of trials in extinction) [51]. In OCD, patients have been shown to have deficits in the extinction of fear-eliciting responses [52, 53].

In zebrafish, Pavolvian and operant fear responses can be measured using an avoidance paradigm (Fig. 1d) [44, 45]. During this procedure, baseline preference for the CS is determined over a 30-min period. Following this, the CS is presented 15 times for 1.5-s, and during each presentation the fish is given an electric shock US (9v), interspersed with 30-s inter-trial intervals (ITIs). The fish is then probed for preference again, to ascertain avoidance of the CS [45]. To examine extinction, following this probe trial, the CS-US pairing would be extinguished over a series of CS presentations in the absence of the US. Resistance to extinction would be characterized as the rate at which the fish showed no active avoidance of the CS.

3 Pitfalls and Problems with Behavioral Testing in Zebrafish

As well as their numerous benefits as a model system, there are a number of challenges facing researchers wishing to carry out neurocognitive assessments in zebrafish. These include: (1) Ensuring within and between laboratory reliability; (2) Low numbers of fish reaching learning criterion; (3) Satiety during testing. Below, the chapter will outline the problems in more detail, as well as some potential solutions, or safeguards that can be employed to mitigate the effects.

1. Ensuring high levels of within- and between-laboratory reliability is essential for any model species if it is to be considered translationally relevant. This invariably involves a complex interplay between increasing standardization of procedures while at the same time minimizing stress owing to handling, housing, and husbandry [54]. One solution to this is to increase the translational relevance of zebrafish as a model system for exploring neurocognitive phenotypes by maximizing the species-specificity of the assay [55]. For example, naturalistic experiments that utilize the species' anti-predatory or escape-avoidance response may be prudent in order to ensure reliability, and ultimately, construct validity of the behavior [18, 55, 56]. In addition, care should be taken to ensure that handling of the fish is minimal and that automation is used wherever possible [18].
2. During some of the long-term studies of learning and steady-state performance in zebrafish, there have been consistent reports that low numbers of fish reach learning criterion and have to be excluded from the final study (e.g., [37, 38, 57, 58]). This clearly raises a number of problems in terms of time and resource

allocation to projects, the potential for selection bias in the final sample (i.e., selecting only the best performers could lead to ceiling effects, or mask marginal effects), and ultimately for the role of this species in neurocognitive testing. One potential solution may be to use aversive learning (fear conditioning) protocols such as those discussed above [45], as this may control inherent differences in appetitive motivation [59]. However, shock-based (aversive) learning is not yet a perfect solution, as the potential for this to be used in long-term training programs is not yet established, and in some cases may not be a relevant behavioral response for the particular neurocognitive endophenotype. In addition, using these procedures raises ethical difficulties with using this species.

3. During protocols in which the subject is required to perform many trials during training sessions the fish may reach satiety and motivation (and hence task performance) may drop off during the latter part of a session [57]. This can become problematic for some procedures; for example, in protocols that require steady state responding such as reversal learning or tasks of attention [37]. There are theoretical solutions, such as employing schedules of reinforcement, but these are as yet, untested in fish. It may be possible to mitigate this by using aversive learning procedures but, for the reasons outlined above, this is not without problems.

4 Remaining Challenges

Despite their clear advantages as a model system, there are also some challenges facing those wishing to use zebrafish in OCD research, or in neuropsychiatric research in general. Many of these have been outlined elsewhere in detail [18, 60]. However, three challenges that are particularly applicable here are modeling the heterogeneity of OCD, finding consistency in underlying neural circuits for the behaviors of interest and ensuring inter-species validity of behavioral tools.

4.1 Modeling Etiological Heterogeneity

OCD is a heterogenous disorder and the etiology is multifaceted, involving as it does the interaction of multiple genetic and environmental factors. For example, dysregulation of the hypothalamic-pituitary adrenocortical axis (HPA), which is characterised by an impairment in cortisol regulation, increases in stress hormones (e.g., corticotrophin-releasing hormone [CRH], adrenocorticotrophic hormone [ACTH]) and impairment of the function of glucocorticoid receptors [GR]), have been consistently linked to neuropsychiatric disorders [61]. However, neither the developmental biology nor the underlying cellular/molecular processes that mediate the links between stress and neuropsychiatric disorder phenotypes are currently clear. There is, for example, a heritable component to OCD with genome-wide association studies (GWAS) identifying potential candidate genes [62]. Also, not only is

OCD likely to be polygenic and relate to alleles of variable (usually very low) penetrance, but genetic variants do not operate in isolation. Instead, they are modulated by the extensive and fluctuating array of environmental influences, including a range of early life stressors [63]. Thus it is critical that we understand the effect of stressors (and other life events) on gene expression. At present, there is no evidence of how gene-environment interactions may manifest in zebrafish. There is a rich literature in rodent models [64], and it may be prudent for zebrafish researchers to initiate similar investigations.

4.2 *Consistency in Neural Circuits Underlying Behavior*

Although there is considerable variance between the brain topography of fish and mammals, there exist a number of putative functional homologies between neural circuits [17, 18]. For example, zebrafish lack the expanded telecephalon that has formed the laminar cortex mammals. Instead, zebrafish telencephalic cells have formed the pallium [65], and this has been functionally compared to the mammalian cortex on account of the homologies in connectivity (e.g., ascending dopamine pathways [66]). Both the mammalian and the zebrafish thalamic nuclei are located in the diencephalon [21], and similar structures of the mammalian striatum (an area of the midbrain linked to OCD [67]) such as the ventral tegmental area (in fish, posterior tuberal nucleus) and nucleus accumbens (in fish, ventral and dorsal telencephalic nuclei) [68]. In addition, the central region of the dorsal pallium (termed area Dc) has been suggested as a homologue of the isocortex [68]. In summary, the extent to which differences in morphology challenge the validity of zebrafish models of neuropsychiatric disorder is not yet clear. Until such time as this is clarified, differences merit attention and should be carefully considered by any researcher attempting to extrapolate translational benefit. More data are required on functional homologies using validated behavioral tools and biomarkers where possible to ensure our ultimate goals are achievable.

4.3 *Inter-Species Validity of Behavioral Tools*

While developing behavioral tests for zebrafish, it is crucial that we are clear that the observed behaviors are comparable with those observed in mammals if we are to argue that zebrafish have translational relevance to human neuropsychiatric conditions such as OCD. In the first instance we must ensure that the behaviors have high face-validity (i.e., they subjectively appear to be measuring what they are purported to be measuring) and high construct validity (i.e., that the behavior truly is related to what it is purported to be related to) [69]. Ensuring construct validity will require that all behavioral tests are validated either with pharmacological and/or genetic evidence. Until this is the case, researchers must exercise caution when extrapolating findings from zebrafish to mammals.

5 Conclusions

This chapter outlines the potential utility for zebrafish as a model for exploring neurocognitive endophenotypes relating to OCD and repetitive behavior. Some of the behavioral tests discussed in the chapter have been validated in zebrafish, with neural circuits known to mediate the behavior in mammals having similar controls in the fish. This further highlights the potential for this species to be used in drug discovery for OCD, as well as in other related conditions. However, the chapter also highlighted some potential limitations that should be considered prior to pursuing such projects.

Gaining a clear understanding of the biological basis of psychiatric disorder is critical as we enter a time in which the realization that personalized, individualized medicine is the most effective treatment option [3]. Zebrafish are uniquely positioned in terms of their potential utility for expediting the characterization of the molecular basis of disease phenotypes, and now that we are developing tools to probe behavioral endophenotypes, their potential for modeling aberrant neural circuits underlying devastating psychiatric conditions could soon be realized. Therefore, it is essential that researchers continue to develop more valid and reliable assays, which are transferable to different laboratories, in order to move forward.

References

1. McHugh PR. Striving for coherence: psychiatry's efforts over classification. JAMA. 2005;293:2526–8.
2. Robbins TW, Gillan CM, Smith DG, de Wit S, Ersche KD. Neurocognitive endophenotypes of impulsivity and compulsivity: towards dimensional psychiatry. Trends Cogn Sci. 2012;16:81–91.
3. Gottesman II, Gould TD. The endophenotype concept in psychiatry: etymology and strategic intentions. Am J Psychiatry. 2003;160:636–45.
4. Menzies L, Achard S, Chamberlain SR, Fineberg N, Chen C-H, Del Campo N, et al. Neurocognitive endophenotypes of obsessive-compulsive disorder. Brain. 2007;130:3223–36.
5. Stein DJ. Obsessive-compulsive disorder. Lancet. 2002;360:397–405.
6. Ruscio A, Stein D, Chiu W, Kessler R. The epidemiology of obsessive-compulsive disorder in the National Comorbidity Survey Replication. Mol Psychiatry. 2010;15:53–63.
7. Abramowitz JS. Effectiveness of psychological and pharmacological treatments for obsessive-compulsive disorder: a quantitative review. J Consult Clin Psychol. 1997;65:44.
8. Mataix-Cols D, Rauch SL, Baer L, Eisen JL, Shera DM, Goodman WK, et al. Symptom stability in adult obsessive-compulsive disorder: data from a naturalistic two-year follow-up study. Am J Psychiatry. 2002;159:263–268.
9. Koran LM, Hackett E, Rubin A, Wolkow R, Robinson D. Efficacy of sertraline in the long-term treatment of obsessive-compulsive disorder. Am J Psychiatry. 2002;159:88–95.
10. Woodcock J. The prospects for "personalized medicine" in drug development and drug therapy. Clin Pharmacol Ther. 2007;81:164–9.
11. Campbell P. A decade for psychiatric disorders. Nature. 2010;463:9.
12. Berridge KC, Aldridge JW, Houchard KR, Zhuang X. Sequential super-stereotypy of an instinctive fixed action pattern in hyper-dopaminergic mutant mice: a model of obsessive compulsive disorder and Tourette's. BMC Biol. 2005;3:4.

13. Chamberlain SR, Blackwell AD, Fineberg NA, Robbins TW, Sahakian BJ. The neuropsychology of obsessive compulsive disorder: the importance of failures in cognitive and behavioural inhibition as candidate endophenotypic markers. Neurosci Biobehav Rev. 2005;29:399–419.
14. Burguière E, Monteiro P, Mallet L, Feng G, Graybiel AM. Striatal circuits, habits, and implications for obsessive–compulsive disorder. Curr Opin Neurobiol. 2015;30:59–65.
15. Laughlin RE, Grant TL, Williams RW, Jentsch JD. Genetic dissection of behavioral flexibility: reversal learning in mice. Biol Psychiatry. 2011;69:1109–16.
16. Veale D, Sahakian B, Owen A, Marks I. Specific cognitive deficits in tests sensitive to frontal lobe dysfunction in obsessive–compulsive disorder. Psychol Med. 1996;26:1261–9.
17. Parker MO, Brock AJ, Walton RT, Brennan CH. The role of zebrafish (*Danio rerio*) in dissecting the genetics and neural circuits of executive function. Front Neural Circuits. 2013;7:63.
18. Stewart AM, Ullmann JF, Norton W, Parker M, Brennan C, Gerlai R, et al. Molecular psychiatry of zebrafish. Mol Psychiatry. 2015;20:2–17.
19. Takahashi JS, Pinto LH, Vitaterna MH. Forward and reverse genetic approaches to behavior in the mouse. Science. 1994;264:1724.
20. Lieschke GJ, Currie PD. Animal models of human disease: zebrafish swim into view. Nat Rev Genet. 2007;8:353–67.
21. Rink E, Wullimann MF. The teleostean (zebrafish) dopaminergic system ascending to the subpallium (striatum) is located in the basal diencephalon (posterior tuberculum). Brain Res. 2001;889:316–30.
22. Norton WH, Folchert A, Bally-Cuif L. Comparative analysis of serotonin receptor (HTR1A/HTR1B families) and transporter (*slc6a4a/b*) gene expression in the zebrafish brain. J Comp Neurol. 2008;511:521–42.
23. Edwards J, Michel W. Pharmacological characterization of ionotropic glutamate receptors in the zebrafish olfactory bulb. Neuroscience. 2003;122:1037–47.
24. Ridley R. The psychology of perseverative and stereotyped behaviour. Prog Neurobiol. 1994;44:221–31.
25. Korff S, Stein DJ, Harvey BH. Stereotypic behaviour in the deer mouse: pharmacological validation and relevance for obsessive compulsive disorder. Prog Neuropsychopharmacol Biol Psychiatry. 2008;32:348–55.
26. Riehl R, Kyzar E, Allain A, Green J, Hook M, Monnig L, et al. Behavioral and physiological effects of acute ketamine exposure in adult zebrafish. Neurotoxicol Teratol. 2011;33:658–67.
27. Cachat J, Kyzar EJ, Collins C, Gaikwad S, Green J, Roth A, et al. Unique and potent effects of acute ibogaine on zebrafish: the developing utility of novel aquatic models for hallucinogenic drug research. Behav Brain Res. 2013;236:258–69.
28. McBride SD, Parker MO. The disrupted basal ganglia and behavioural control: an integrative cross-domain perspective of spontaneous stereotypy. Behav Brain Res. 2015;276:45–58.
29. López-Patiño MA, Yu L, Cabral H, Zhdanova IV. Anxiogenic effects of cocaine withdrawal in zebrafish. Physiol Behav. 2008;93:160–71.
30. Richter RM, Weiss F. In vivo CRF release in rat amygdala is increased during cocaine withdrawal in self-administering rats. Synapse. 1999;32:254–61.
31. Sarnyai Z, Bíró É, Gardi J, Vecsernyés M, Julesz J, Telegdy G. Brain corticotropin-releasing factor mediates 'anxiety-like' behavior induced by cocaine withdrawal in rats. Brain Res. 1995;675:89–97.
32. Metaxas A, Keyworth H, Yoo J, Chen Y, Kitchen I, Bailey A. The stereotypy-inducing and OCD-like effects of chronic 'binge' cocaine are modulated by distinct subtypes of nicotinic acetylcholine receptors. Br J Pharmacol. 2012;167:450–64.
33. Dickinson A. Actions and habits: the development of behavioural autonomy. Philos Trans R Soc B. 1985;308:67–78.
34. Adams C, Dickinson A. Actions and habits: variations in associative representations during instrumental learning. In: Information processing in animals: memory mechanisms. Eds: Ralph R. Miller, Norman E. Spear 1981:143–65.
35. Gillan CM, Morein-Zamir S, Urcelay GP, Sule A, Voon V, Apergis-Schoute AM, et al. Enhanced avoidance habits in obsessive-compulsive disorder. Biol Psychiatry. 2014;75:631–8.

36. Gillan CM, Papmeyer M, Morein-Zamir S, Sahakian BJ, Fineberg NA, Robbins TW, et al. Disruption in the balance between goal-directed behavior and habit learning in obsessive-compulsive disorder. Am J Psychiatry. 2011;168:718–26.
37. Parker MO, Gaviria J, Haigh A, Millington ME, Brown VJ, Combe FJ, et al. Discrimination reversal and attentional sets in zebrafish (*Danio rerio*). Behav Brain Res. 2012;232:264–8.
38. Parker MO, Ife D, Ma J, Pancholi M, Smeraldi F, Straw C, et al. Development and automation of a test of impulse control in zebrafish. Front Syst Neurosci. 2013;7.
39. Parker MO, Evans AMD, Brock AJ, Combe FJ, Teh MT, Brennan CH. Moderate alcohol exposure during early brain development increases stimulus-response habits in adulthood. Addict Biol. 2016;21:49–60.
40. Parker M, McBride SD, Redhead ES, Goodwin D. Differential place and response learning in horses displaying an oral stereotypy. Behav Brain Res. 2009;200:100–5.
41. Faure A, Haberland U, Condé F, El Massioui N. Lesion to the nigrostriatal dopamine system disrupts stimulus-response habit formation. J Neurosci. 2005;25:2771–80.
42. Yu C, Gupta J, Chen J-F, Yin HH. Genetic deletion of A2A adenosine receptors in the striatum selectively impairs habit formation. J Neurosci. 2009;29:15100–3.
43. Dickinson A, Nicholas D, Adams CD. The effect of the instrumental training contingency on susceptibility to reinforcer devaluation. Q J Exp Psychol. 1983;35:35–51.
44. Xu X, Scott-Scheiern T, Kempker L, Simons K. Active avoidance conditioning in zebrafish (*Danio rerio*). Neurobiol Learn Mem. 2007;87:72–7.
45. Valente A, Huang K-H, Portugues R, Engert F. Ontogeny of classical and operant learning behaviors in zebrafish. Learn Mem. 2012;19:170–7.
46. Birrell JM, Brown VJ. Medial frontal cortex mediates perceptual attentional set shifting in the rat. J Neurosci. 2000;20:4320–4.
47. McAlonan K, Brown VJ. Orbital prefrontal cortex mediates reversal learning and not attentional set shifting in the rat. Behav Brain Res. 2003;146:97–103.
48. Chamberlain SR, Menzies L, Hampshire A, Suckling J, Fineberg NA, del Campo N, et al. Orbitofrontal dysfunction in patients with obsessive-compulsive disorder and their unaffected relatives. Science. 2008;321:421–2.
49. Chamberlain SR, Fineberg NA, Menzies LA, Blackwell AD, Bullmore ET, Robbins TW, et al. Impaired cognitive flexibility and motor inhibition in unaffected first-degree relatives of patients with obsessive-compulsive disorder. Am J Psychiatry. 2007;164:335–8.
50. Clarke H, Dalley J, Crofts H, Robbins T, Roberts A. Cognitive inflexibility after prefrontal serotonin depletion. Science. 2004;304:878–80.
51. Rescorla RA. Probability of shock in the presence and absence of CS in fear conditioning. J Comp Physiol Psychol. 1968;66:1–5.
52. Milad MR, Furtak SC, Greenberg JL, Keshaviah A, Im JJ, Falkenstein MJ, et al. Deficits in conditioned fear extinction in obsessive-compulsive disorder and neurobiological changes in the fear circuit. JAMA Psychiatry. 2013;70:608–18.
53. Milad MR, Rauch SL. Obsessive-compulsive disorder: beyond segregated cortico-striatal pathways. Trends Cogn Sci. 2012;16:43–51.
54. Richter SH, Garner JP, Würbel H. Environmental standardization: cure or cause of poor reproducibility in animal experiments? Nat Methods. 2009;6:257–61.
55. Gerlai R. Zebrafish antipredatory responses: a future for translational research? Behav Brain Res. 2010;207:223–31.
56. Parker MO, Millington ME, Combe FJ, Brennan CH. Housing conditions differentially affect physiological and behavioural stress responses of zebrafish, as well as the response to anxiolytics. PLoS One. 2012;7:e34992-e.
57. Parker MO, Millington ME, Combe FJ, Brennan CH. Development and implementation of a three-choice serial reaction time task for zebrafish (*Danio rerio*). Behav Brain Res. 2012;227:73–80.
58. Parker MO, Brock AJ, Sudwarts A, Brennan CH. Atomoxetine reduces anticipatory responding in a 5-choice serial reaction time task for adult zebrafish. Psychopharmacology (Berl). 2014;231(13):2671–9.

59. Blaser R, Vira D. Experiments on learning in zebrafish (*Danio rerio*): a promising model of neurocognitive function. Neurosci Biobehav Rev. 2014;42:224–31.
60. Stewart AM, Braubach O, Spitsbergen J, Gerlai R, Kalueff AV. Zebrafish models for translational neuroscience research: from tank to bedside. Trends Neurosci. 2014;37:264–78.
61. Kluge M, Schüssler P, Künzel HE, Dresler M, Yassouridis A, Steiger A. Increased nocturnal secretion of ACTH and cortisol in obsessive compulsive disorder. J Psychiatr Res. 2007;41:928–33.
62. Yu D, Mathews CA, Scharf JM, Neale BM, Davis LK, Gamazon ER, et al. Cross-disorder genome-wide analyses suggest a complex genetic relationship between Tourette's syndrome and OCD. Am J Psychiatry. 2015;172:82–93.
63. van Winkel R, Stefanis NC, Myin-Germeys I. Psychosocial stress and psychosis. A review of the neurobiological mechanisms and the evidence for gene-stress interaction. Schizophr Bull. 2008;34:1095–105.
64. Gutman DA, Nemeroff CB. Neurobiology of early life stress: rodent studies. Semin Clin Neuropsychiatry. 2002;7:89–95.
65. Ito H, Yamamoto N. Non-laminar cerebral cortex in teleost fishes? Biol Lett. 2009;5:117–21.
66. Tay TL, Ronneberger O, Ryu S, Nitschke R, Driever W. Comprehensive catecholaminergic projectome analysis reveals single-neuron integration of zebrafish ascending and descending dopaminergic systems. Nat Commun. 2011;2:171.
67. Welch JM, Lu J, Rodriguiz RM, Trotta NC, Peca J, Ding J-D, et al. Cortico-striatal synaptic defects and OCD-like behaviours in Sapap3-mutant mice. Nature. 2007;448:894–900.
68. Mueller T, Dong Z, Berberoglu MA, Guo S. The dorsal pallium in zebrafish, *Danio rerio* (Cyprinidae, Teleostei). Brain Res. 2011;1381:95–105.
69. Parker MO, Brennan CH. Zebrafish (*Danio rerio*) models of substance abuse: harnessing the capabilities. Behaviour. 2012;149:1037–62.

Zebrafish Models of Attention-Deficit/ Hyperactivity Disorder (ADHD)

William Norton, Merlin Lange, Laure Bally-Cuif, and Klaus-Peter Lesch

Abstract Attention-deficit/hyperactivity disorder (ADHD) is a common, early onset neuropsychiatric disorder that is characterized by developmentally inappropriate inattention, hyperactivity, increased impulsivity and motivational/emotional dysregulation. However, although there is a significant genetic component to ADHD, relatively few risk genes have been identified and characterized. Furthermore, despite the effectiveness of pharmacological therapies such as methylphenidate, the long-term treatment outcome varies considerably depending on the psychosocial environment. The development of novel drug treatments has been hampered by a lack of knowledge regarding the genetics and neurobiology of ADHD. It is therefore necessary to develop animal models of ADHD in order to better understand its etiology and to improve the treatment options that are available. The aim of this chapter is to explore how we can develop zebrafish as a translational model for ADHD. We will first discuss the genetics and neurobiology of the disease. We will then consider existing animal models of ADHD and examine how the unique attributes of zebrafish can be used to extend this research. Finally, we will propose promising avenues for future research using zebrafish as an ADHD-like model.

W. Norton (✉)
Department of Neuroscience, Psychology and Behaviour, Adrian Building,
University of Leicester, University Rd, Leicester LE1 7RH, UK
e-mail: whjn1@le.ac.uk

M. Lange
Laboratory for Developmental Gene Regulation, RIKEN Brain Science Institute,
Saitama 351-0198, Japan

L. Bally-Cuif
Paris-Saclay Institute for Neuroscience (Neuro-PSI), UMR 9197, CNRS – Université
Paris-Sud, Avenue de la Terrasse, Bldg 5, 91190 Gif-sur-Yvette, France

K.-P. Lesch
Division of Molecular Psychiatry, Laboratory of Translational Neuroscience,
Department of Psychiatry, Psychosomatics, and Psychotherapy,
University of Würzburg, Füchsleinstrasse 15, 97080 Würzburg, Germany

A.V. Kalueff (ed.), *The rights and wrongs of zebrafish: Behavioral phenotyping of zebrafish*, DOI 10.1007/978-3-319-33774-6_7

Keywords Animal model • Zebrafish • Larvae • Translational model • Psychiatric disorder • Attention-deficit/hyperactivity disorder • ADHD • Endophenotype • Hyperactivity • Impulsivity • Inattention

Abbreviations

5CSRTT	Five choice serial reaction time task
6-OHDA	6-Hydroxydopamine
ACC	Anterior cingulate cortex
ADHD	Attention-deficit/hyperactivity disorder
ATO	Atomoxetine
CDCV	Common disease-common variant hypothesis
DA	Dopamine
daMCC	Dorsal anterior midcingulate cortex
DAT-KO	*Dat* knock-out mice
DSM-III	Diagnostic and Statistical Manual of the American Psychiatric Association
fMRI	Functional magnetic resonance imaging
G×E	Gene×environment interaction
GWAS	Genome-wide association study
ITI	Inter-trial interval
LED	Light emitting diode
MPH	Methyphenidate
MRI	Magnetic resonance imaging
NA	Noradrenaline
NPD	Neuropsychiatric disorder
PCBs	Polychlorinated biphenyls
PET	Photon emission tomography
PFC	Prefrontal cortex
PT	Posterior tuberculum
SHR	Spontaneous hypertensive rat
SNP	Single nucleotide polymorphism
SPECT	Single-photon emission computed tomography
SSRI	Selective serotonin (5-HT) reuptake inhibitor
VNTR	Variable number of tandem repeats
WKHA	Wistar-Kyoto hyperactive rat
WKY	Wistar-Kyoto rat

1 Introduction to Attention-Deficit Hyperactivity Disorder

Mental illnesses, or neuropsychiatric disorders, are an extremely diverse set of diseases that affect all aspects of mental function including thinking, feeling and mood as well as the ability to relate to other people [1]. Neuropsychiatric disorders (NPD) place a massive strain on society; mental illness ranks second in the burden of diseases in established market economies [2]. Nevertheless, in spite of their prevalence, the drug therapies available to treat NPDs frequently fail to prove satisfactory long-term outcomes due to variable efficacy and intolerable side-effects. Despite the clear need for better treatments, many of the pharmacological compounds used to treat NPDs were discovered serendipitously 60 years ago and have not been significantly improved since [1]. The development of novel drugs has in part been hampered by a lack of knowledge about the underlying neurobiology of NPDs. Therefore, research into the etiopathogenesis of psychiatric disorders, led by a combination of human genetic studies and animal modeling of the identified gene variants, is mandatory in order to improve drug treatments and develop early interventions that could prevent or delay disease onset.

Attention-deficit/hyperactivity disorder (ADHD) is a common, early onset neuropsychiatric disorder that is characterised by developmentally inappropriate inattention, hyperactivity, increased impulsivity and motivational/emotional dysregulation. ADHD has similar prevalence rates across different cultural settings [3, 4], resulting in poor performance in school and impairments in multiple other domains of personal and professional life. ADHD has long been considered a childhood disorder that gradually resolves itself with maturation during adolescence. However, this view has been contested by systematic follow-up studies documenting the persistence of ADHD into adulthood [5]. The adult form of ADHD is associated with considerable risk for co-morbidity with other psychiatric disorders as well as failure of psychosocial adaptation [5]. ADHD can be further divided into different subtypes in the clinic: a predominantly inattentive subtype, a predominantly hyperactive-impulsive subtype and a combined form of both, which is the most common form of the disease [6]. The behavioral symptoms of ADHD may result from alterations to underlying cognitive and motivational/emotional processes such as behavioral/response inhibition, delayed gratification (choosing a smaller earlier reward rather than a larger later one) and executive function (higher order integrated cognitive processes that allow selective attention and the use of information in problem-solving tasks) [7, 8]. The symptoms of ADHD are thought to be—at least partially—controlled by neuronal networks in the prefrontal cortex, anterior cingulate cortex and parietal cortex, the dorsal and ventral striatum and cerebellum.

Although generally accepted as being a neurodevelopmental disorder with a risk of life-long impairments and disability, the variable combination of syndromal dimensions and the diversity of comorbid disorders render diagnosis of ADHD difficult and sometimes even contentious. At one extreme the energy, exuberance and demanding behavior of ADHD patients is said to be part of the normal spectrum and psychiatrists are accused of needlessly using medication to pacify children. The evidence for this point of view is based in part upon the increasing number of people diagnosed with ADHD each year, and the difference in prescription policies between countries despite similar prevalence rates across cultural settings. At the other extreme, ADHD is presented as a purely biological construct that is caused by the interaction of genes and the environment and is treatable with medication (reviewed in [9]). The symptoms of ADHD were first described more than 160 years ago. An early description of ADHD-like symptoms can be found in a children's book written by the pediatrician and psychiatrist Heinrich Hoffmann. In "Die Geschichte vom Zappel-Philipp" (the story of fidgety Phil), Hoffmann described a boy who "won't sit still; he wriggles and giggles and then, I declare, swings backwards and forwards and tilts up his chair" [10]. In 1902 George Frederick Still wrote an account of 43 children with poor "moral control" who were aggressive, defiant, resistant to discipline and excessively emotional [11]. By the beginning of the twentieth century, diseases with similar behavioral phenotypes were described as minimal brain damage and then minimal brain dysfunction, even though there was no evidence of brain damage in the patients studied. By the 1970s, the symptoms of ADHD were first recognized as including attention deficits [12]. The symptoms of ADHD were then formalized in the third edition of the Diagnostic and Statistical Manual of the American Psychiatric Association.

ADHD is one of the most common neurodevelopmental disorders and affects around 3–5 % of children worldwide regardless of ethnicity or cultural setting [13, 14]. Although it is predominantly considered a juvenile disease, the symptoms of ADHD also persist into adulthood in about 30–50 % of cases [15, 16]. However, it is not yet clear to what extent the genes and symptoms linked to ADHD are similar in children and adults [5]. ADHD patients generally experience significant impairment of academic, behavioral and social performance [17, 18]. ADHD can also lead to life-threatening conditions. For example, children with ADHD show an increased risk of injury in traffic accidents [19]. ADHD patients are more likely to suffer from other NPDs, including depression, anxiety and substance use disorder [20–22]. ADHD is also the most common NPD to develop following brain injury [23], giving credence to idea that ADHD can be linked to damage or dysfunction of the brain. However, in common with all NPDs, it is difficult to untangle the neurobiological root cause of the multiple symptoms that are presented by patients. For example, the comorbid symptoms of ADHD could be caused secondarily to disease pathology; poor performance at school due to inattention could in turn lead to increased anxiety and depression [24]. Finally, although ADHD is often perceived as a negative attribute it may also have a positive impact on a person's life. ADHD patients are often very creative, good at problem solving and able to focus selectively on certain captivating tasks. If channeled in the correct way, the increased drive and energy shown by ADHD patients may even be enviable.

1.1 Areas of the Brain that Are Altered in ADHD

One of the difficulties facing research into the pathophysiology of ADHD is that the neural networks controlling cognitive and emotional processes in the non-ADHD brain (including executive function, working memory and attention) are poorly understood. Nevertheless, some of the brain areas that control the symptoms of the disease have already been identified. This research has combined imaging studies with knowledge of neural network architecture such as the catecholaminergic systems in the brain [24]. Photon emission tomography (PET), single-photon emission computed tomography (SPECT), magnetic resonance imaging (MRI) and functional MRI (fMRI) have all been used to examine ADHD patients [24]. MRI is particularly suited to studies of young ADHD patients since it is non-invasive and does not require the injection of a radioactive tracer. However, there are also limitations to MRI studies. ADHD patients tend to move a lot during the scan (due to increased motor activity), the disease is etiopathogenetically very heterogeneous making standardization problematic, and it is difficult to find subjects who have not already been exposed to drug treatments that could potentially skew the results [8]. Combined data from a number of neuroimaging studies have reported a 3 % reduction in white matter volume in the brain of ADHD patients [7, 25, 26]. This reduction of brain volume does not progress during adolescence suggesting that it occurs early during embryonic development [8]. The reduction of brain volume may reflect a developmental delay in the maturation of cortical circuits in ADHD patients. In agreement with this, longitudinal studies of cortical thickness have reported that the initial reduction normalizes over time in ADHD children [27]. The thickening of the cortex to normal levels with age is paralleled by a reduction of symptoms. ADHD may thus be caused by a delay rather than a disruption of brain development [27]. Studies of potential asymmetries in the brain of ADHD patients have reported differences between the left and right caudate nucleus, globus pallidus and putamen [7, 8, 28]. However, the data from these studies are inconsistent making it difficult to draw any firm conclusion about the role of brain asymmetries in ADHD.

Structural MRI studies have also identified localized reductions in the size of some ADHD brain areas. Localized reductions in the size of the prefrontal cortex (PFC) [25, 26], corpus callosum [29] and cerebellum have already been reported. These areas all contain high densities of DA neuron terminals, a neurotransmitter which is intricately linked to the symptoms of the disease. So far, the majority of studies have focused on the frontal-striatal network, made up of the PFC (including the dorsolateral- and ventrolateral prefrontal cortex) and the striatum (caudate nucleus and putamen; reviewed in [30]). The PFC is known to be important for the control of attention, working memory and executive function [31, 32], key mental processes that underlie the behavioral symptoms of ADHD. Therefore, a reduction of PFC function is in accordance with the symptoms of the disease. In agreement with this, fMRI studies have already demonstrated a hypofunction of DA in the PFC of ADHD patients. There are also some behavioral similarities between ADHD and patients with frontal lobe injuries [33]. Finally, both methylphenidate and atomoxetine (drugs which are given to ADHD patients) ameliorate the symptoms of ADHD by increasing catecholamine levels in the PFC [34, 35]. The striatum (caudate/putamen) has an important role in controlling executive function and motor output, and is connected to the dorsal anterior midcingulate cortex (daMCC, commonly known as the anterior cingulate cortex or ACC) and dorsolateral PFC. A decrease in striatum volume is also seen in ADHD patients, a phenotype which recovers by around 16 years of age [7]. Other areas of the brain which have been linked to the symptoms of ADHD include the parietal cortex and the cerebellum (reviewed in [24]). Cerebellar areas including the vermis and inferior lobes may affect both the control of motor movements and attention via connections to the PFC [36, 37].

1.2 Dopamine and Noradrenaline Control the Symptoms of ADHD

Data from drug treatments and genetic analyses have suggested that alterations in dopamine (DA) and noradrenaline (NA) (and to a lesser extent 5-HT) signaling most likely underlie the symptoms of ADHD. The combination of synthesis enzymes, transporters and metabolizing enzymes that mediate neurotransmitter signalling provide many mutable targets that can potentially lead to expression of the disease. Furthermore, DA and NA act via a large number of receptors which transduce neurotransmitter signaling in discrete neural circuits. The role of DA and NA in ADHD can be understood by re-examining the function of the PFC. DA and NA neurons have a dual function in the brain, acting both tonically to maintain a basal arousal state and phasically in response to external stimuli [37]. The prefrontal cortex is very sensitive to the levels of catecholamines in the brain which act via the D1 receptor (DA) and Alpha 2 adrenoceptor (NA) [38]. Moderate levels of NA act on the PFC to increase the "signal" or response to stimulation, whereas levels of DA act to decrease background "noise" [38]. Thus, small fluctuations in DA and NA can dramatically affect the neural circuits that control attention, arousal and executive function. Furthermore, both increases and decreases of DA and NA signaling have been found to modify PFC function in an inverted U-shaped dose response [39, 40]. In summary, the symptoms of ADHD can be ultimately thought of as being triggered by fluctuations in DAergic and NAergic tone at the level of the PFC—an area of the brain that has been suggested to underlie the symptoms of many NPDs [41].

1.3 Pharmacological Therapies for ADHD

The majority of compounds used to treat ADHD to date interact with monoaminergic neurotransmitter signaling pathways. Indeed, it was the observation that methylphenidate (MPH), a dopamine (DA) pathway drug, could be used to treat ADHD that orientated research towards monoaminergic signaling. MPH is now the most frequently prescribed ADHD treatment, and under its trade name Ritalin has even achieved a certain dubious celebrity, having been mentioned in several American cartoon series including South Park. MPH is a synthetic amphetamine derivative that causes several behavioral improvements including sustained attention, impulse control, reduction of task-irrelevant behavior and decreased disruptive behavior in ADHD patients [42–44]. MPH treatment amplifies the duration and tone of DA signaling in multiple ways, including blocking the DA transporter Slc6a3/Dat, disinhibiting DA D2 autoreceptors and activating D1 receptors on postsynaptic neurons [44]. However, MPH treatment also increases DA signaling in the nucleus accumbens, an area of the brain associated with reward behavior; thus in common with other psychostimulants MPH has the potential to be highly addictive. Drugs which target the noradrenergic (NA) system have also been successfully used to treat ADHD patients, including bupoprion and atomoxetine [42]. Atomoxetine (ATO) [45] is a selective inhibitor of the presynaptic NA transporter with minimal affinity for other neurotransmitter transporters and receptors [35]. ATO treatment increases the level of both NA and DA in the PFC (due to increased firing of anterior projections from the locus coeruleus to the PFC DA neurons) thereby improving attention and cognition [35, 46]. Although ATO does not work in all ADHD patients, it appears to be a safe and efficacious treatment for children, adolescents, and adults [42]. However, around 30 % of ADHD patients do not respond to MPH treatment, and 40 % to ATO [5], highlighting the need to improve drug therapies. Other non-stimulant medication has also been successfully used to treat ADHD patients. For example, guanfacine is a selective alpha 2A adrenoceptor agonist that stimulates prefrontal cortical networks, reducing the symptoms of the disease [47, 48]. Conversely, selective 5-HT reuptake inhibitors (SSRIs) do not reduce ADHD symptoms, questioning the importance of 5-HT in this disorder [49]. Although pharmacological treatment of ADHD (often used in combination with behavioral therapy) has been highly successful in controlling disease symptoms, the drugs which are available are only palliative and the long-term effects of stimulant medication on adolescent development are not known. There is thus a clear need to improve the drug treatments available for ADHD at every possible level, including the efficacy, number of side effects and potential for abuse.

1.4 The Genetic Basis of ADHD

The symptoms of ADHD are highly heritable suggesting that there is a genetic basis of the disease [5]. This substantial heritability, with estimates of up to 80 %, has been documented in numerous family, twin and adoption studies. However, despite this strong genetic basis, relatively few ADHD-risk genes have been identified and characterized following genome-wide approaches or candidate gene studies [20, 50–52]. ADHD has been shown to be caused by both a combination of multiple common mutations as well as polymorphisms in single neurodevelopmental genes [53–55].

The majority of ADHD-susceptibility genes examined to date have been linked to monoaminergic signaling [56–60]. For example, multiple DA signaling-related genes have been linked to ADHD. In particular, association with polymorphisms in the gene encoding the DA D4 receptor (*DRD4* [57, 59]), the DA D5 receptor (*DRD5*, [61]) and the DA transporter gene (*DAT/SLC6A3* [62]) have been reported. DAT terminates synaptic activity by reducing DA to a level at which it can no longer activate receptors. Thus, efficient DAT activity is needed in order control the strength and duration of DA neurotransmission. Most studies of *DAT* have focussed on a 40 base-pair variable number tandem repeat (VNTR) found in the 3′ untranslated region of the gene [63–65]. There is also some evidence associating the DA synthesis enzyme *Dopamine-beta hydroxylase* (*DBH*) with ADHD [66]. In the serotonin (5-HT) pathway, the 5-HT synthesis enzymes *TPH1* and *TPH2* [67, 68] and the 5-HT transporter gene (*SERT/SLC6A4* [69]) have all been linked to the disease. There is also conflicting evidence linking the receptors *HTR1B* and *HTR2A* to ADHD [70]. Recent GWAS and candidate gene studies have also identified polymorphisms in genes that are involved in cell adhesion (including *ASTN2* and *CDH13*) and synaptogenesis (S*NAP25*, *CTNNA2*, and *KLRN* [51, 55]). As well as being caused by direct modification of neurotransmitter signaling, ADHD may be triggered by more general alterations in brain formation, including cell signaling, morphogenesis and migration during development.

Environmental factors also play a significant role in the risk of suffering from ADHD. Environmental factors that have been linked to expression of the disease include exposure to nicotine, alcohol or psychosocial adversity (including child abuse, single-parenthood, marital discord or parental psychiatric disorders) [71–73]. For example, interaction of *DAT* and nicotine [73–75], alcohol [71] and psychosocial adversity [76] have been found to increase ADHD susceptibility. The gene-by-environment (G x E) interactions that increase ADHD susceptibility are still relatively poorly understudied, perhaps reflecting the difficulty inherent in conducting this research. It is also important to remember that the environmental factors that increase the risk of ADHD may also be under genetic control; the propensity of a mother to drink or smoke could be caused by genetic influences related to maternal ADHD [7, 76]. Thus environmental influences could ultimately reflect the interaction of the parent's and child's genome in controlling the expression of mental illness.

2 Translational Models of Human Diseases

Although recent studies have uncovered some of the genes linked to PDs, only few of these have been validated experimentally. It is therefore essential to use animal models in order to investigate whether a loss- or gain-of-function leads to disease pathology in each case. The complicated genetic basis of NPD makes it difficult to fully recreate them in animal models. One way to simplify this problem is to measure endophenotypes, neuropsychological or biological markers that correlate to a disease-gene's activity [77, 78]. An ideal endophenotype should be controlled by a single gene, be associated with expression of the disease in the population and be both heritable and state independent (meaning that it is expressed even when the illness is not active; [79]). Although endophenotypes for NPDs have rarely fulfilled all of these criteria, successful markers have been developed for mood disorders [80, 81], Alzheimer's disease [82, 83] and ADHD [7, 84]. Endophenotypes may help to translate information from animal models to human patients. Furthermore, the division of NPDs according to endophenotypes may help redefine psychiatric diseases. Diseases could thus be reclassified on the basis of their molecular pathology instead of behavioral symptoms, both simplifying their diagnosis and providing an explanation for co-morbidity with other diseases [77].

Despite the difficulty of modeling NPDs, studies in animals still have the potential to give insights into the etiology of mental illness making them critically important. The first animal models were established on the basis of behavioral phenotyping. Behavioral changes that appeared to mimic some aspect of the human disease were rescued with specific treatment drugs, thus validating the model. However, the advent of tools to modify genes or their expression has now allowed the creation of animal models that are firmly based upon the genetic pathways underlying a disease. A perfect animal model should have three main attributes: construct validity (meaning that it conforms to the underlying rationale of the disease), face validity (mimicking some of the characteristics of the disease) and predictive validity (allowing the prediction of novel disease symptoms, or identification of disease treatments [85–87]). The animal model should also combine genetic tractability, tools to visualize and manipulate neurons in vivo, and the ability to translate findings to patients based upon conserved neurobiology.

2.1 *Modeling Psychiatric Disorders in Zebrafish*

Although rats, mice and fruit flies have been extensively used to model human diseases, recent work has demonstrated that zebrafish are poised to become a valuable translational model [88]. Zebrafish have already been established as one of the premiere organisms to study vertebrate development. In parallel, a battery of tests for behavioral analysis of both larval and adult zebrafish has also been developed [89, 90]. Zebrafish develop rapidly outside of the mother, making it easy to collect and manipulate embryos. By 6 days, larval fish swim continuously, search for food and are able to escape from predators thus demonstrating a range of innate behaviors. Zebrafish are optically transparent until larval stages allowing the study and manipulation of neural circuits at the cellular level in the intact brain [91]. The large number of identified mutant lines, genetic tools (such as TALENs and Zinc-Finger nucleases to knock-out genes [92–94], genetic ablation [95] and optogenetics [96, 97]) and techniques to monitor neural activity (including calcium indicators and electrophysiology [98]) make zebrafish an ideal model for neuroscience. Although the formation, position and function of neurotransmitter signaling pathways sometimes differ between zebrafish and other vertebrates, comparative studies are beginning to precisely map these differences, allowing the transfer of information gained in zebrafish to other species [99]. These attributes have already been used to investigate the genetic basis of complex behaviors including reward and prepulse inhibition as an endophenotype for schizophrenia [100–102].

In spite of the experimental advantages of zebrafish it does not yet rival rat or mouse as a translational model for human disease. Thus, the challenge faced by zebrafish researchers is to design studies that harness the strengths of fish as a model system including live imaging, optogenetic interrogation of neural circuits and high-throughput screening of novel compounds. Indeed, although zebrafish are often touted as an excellent high-throughput system, this potential has been relatively under used (but see [103, 104]). Another challenge facing the field of translational research (in zebrafish as well as other animals) is to develop models for other, less-well characterized, diseases. In the rest of this chapter we will concentrate on one such NPD—attention-deficit/hyperactivity disorder.

3 Modeling ADHD in Zebrafish

Although the existing animal models for ADHD have provided novel insights into the genetics and neurobiology of the disease, there is clearly still room for development of new models. Despite the numerous advantages of zebrafish for developmental biology and neuroscience, there are currently very few studies that have reported an ADHD-like model in zebrafish [105, 106].

Since we do not know a priori which of the innate behaviors shown by fish could constitute an ADHD endophenotype we have to start by using a candidate gene approach. Starting with data from genome-wide screening approaches of ADHD patients we can identify and clone the homologous gene in zebrafish. We can characterize the expression of the ADHD-linked gene during neural development and then abrogate gene activity by either injecting morpholino oligonucleotides [107] or creating a novel mutant line. We can then assay the behavioral changes that are manifested by morphant (gene-specific morpholino injected) or mutant zebrafish in an attempt to identify novel ADHD-linked endophenotypes. The ensuing behavioral changes can be measured with- and without application of an ADHD treatment drug, thus providing face validity for the zebrafish model. Finally, the morphants can be used to investigate alterations to neurotransmitter signaling triggered by loss of gene function, as well as to screen for novel therapeutic compounds. In this approach we use endophenotype in the loosest sense of the word—a measurable behavioral phenotype that corresponds to the activity of a disease gene and represents a subset of the symptoms of the disease.

Morpholinos are an excellent tool to transiently knock-down gene activity, but they cannot be used to mimic gain-of-function mutations or to assay the impact of SNPs on gene function. One method that could be used to address this issue (but which has not yet been used in zebrafish ADHD studies) is to combine morpholino knock-down with co-injection of mRNA encoding either a "humanized" form of the zebrafish ortholog or the human disease gene itself (reviewed in [108]). Thus the human gene can replace the function of the zebrafish paralog during development and alterations to neuroanatomy and behavior can be assessed. In this way, the importance of SNP polymorphisms that have been identified in human genes to disease progression can be studied. This method constitutes a promising avenue for linking novel SNP polymorphisms to the formation of ADHD and should be further explored in the future.

Of the three major symptoms of ADHD—inattention, hyperactivity and increased impulsivity—it is easiest to devise tests to measure hyperactivity in zebrafish. Although hyperactivity can be readily measured in both adult [109, 110] and larval fish [90, 111–113], for the purposes of this chapter we will concentrate on larvae since they are more amenable to high-throughput analysis and live-imaging. Conversely, it is perhaps better to measure impulsivity and inattention in adult fish since we do not know at which age the larval brain is mature enough to mediate these behaviors. The small number of protocols available to measure impulsivity and inattention perhaps reflects the difficulty of designing tests to measure them (see [114]), as well as the general under-appreciation of adult zebrafish as a behavioral model [115]. Behavioral analysis of adult zebrafish may require the use of stable zebrafish mutant lines, since morpholino knock-down is transient and gene activity will recover by 3–4 days of development [107]. Thus morpholino-mediated knock-down may not be suitable for studies of impulsivity or inattention. Regardless of this drawback however, injection of morpholinos can still alter the expression of ADHD-linked behaviors in adult fish. For example, reduction of *nr4a2* (an ADHD-linked dopaminergic orphan nuclear receptor; [116]) activity during development leads to permanent hyperactivity, indicating that a critical developmental process was affected that does not appear to recover over time [117]. Correlated permanent changes to the neuroanatomy of adult *nr4a2* morphants have not yet been studied.

3.1 Zebrafish ADHD-Like Endophenotypes: Hyperactivity and Motor Impulsivity

Zebrafish larvae hatch from their chorion at around 4–5 days post fertilization at which point frequent bouts of swimming occur [118–120]. Changes in the speed of locomotion are easy to measure and can even be quantified without using sophisticated equipment. For example, the number of times that larvae cross gridlines drawn on a Petri dish within a defined time-window could be counted. Nevertheless, we prefer to use videotracking software to measure the locomotory behavior of larval fish (such as Zebralab from Viewpoint Life Sciences, or Daniovision from Noldus) [90] (Fig. 1). Videotracking allows the automated tracking of multiple animals at the same time, reduces both observer bias and inter-observer variability and permits the simultaneous measurement of multiple parameters (including speed and distance swum, turning angle, time spent at the side or middle of an arena and the total

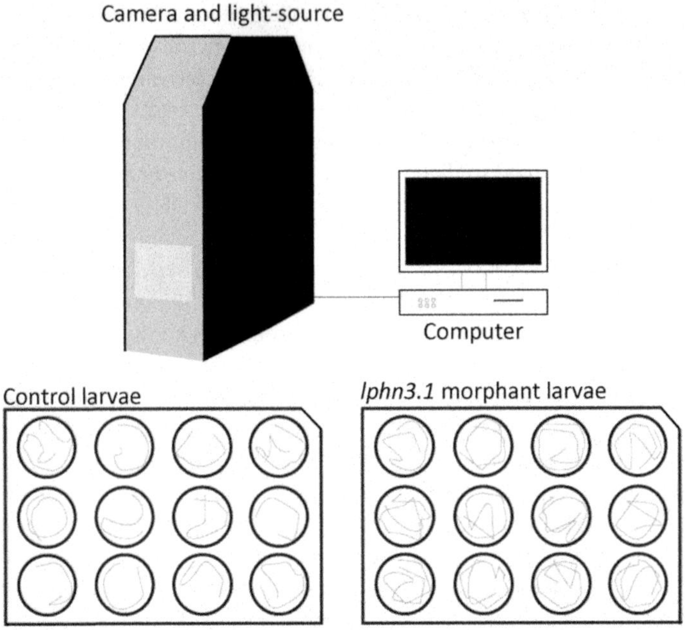

Fig. 1 Equipment used to measure larval locomotion. Cartoon representation of setup used to measure locomotion in *lphn3.1* and control morphants. Larvae are placed into twelve-well plates which are then mounted in a box containing a light-source and camera. A computer running specialized software tracks the position of the larvae during a 5-min experiment, allowing the distance swum and speed of locomotion to be calculated

time spent resting during the experiment). The hyperactivity shown by ADHD patients can be remarkably stable over time and may also be maintained at night [121–123]. Computer-automated setups use infra-red light to detect motion, thus allowing the activity of larvae to be recorded in the dark. As well as the total increase in the distance swum in a defined time-window, the pattern of larval swimming can be measured by looking at the bursts of acceleration that 6-day old larvae use to propel themselves. ADHD-associated increases in impulsivity can be subdivided into both motor and cognitive components [124, 125]. The locomotion curves of hyperactive larvae may show sharper peaks of acceleration than animals with normal activity levels, a pattern interpreted as motor impulsivity.

3.2 *Zebrafish ADHD-Like Endophenotypes: Inattention*

The ability to pay attention is a complex behavior that includes a number of hypothetical cognitive processes. Indeed, the measurement of attention in animals remains a controversial subject among neuroscientists [126]. However, despite the difficulty of measuring attention in animals, its importance in a number of NPDs makes it necessary to design experiments to probe this issue. In a recent review of attention studies in animals, Bushnell proposed that attention can be divided into five types of cognitive process: orienting, expectancy, stimulus differentiation (selecting between two stimuli), sustained attention, and parallel processing [126]. Therefore, each of these processes could be used as a basis to develop endophenotypes for attention in animals. To date there is no single behavioral test that can directly measure attention (or inattention) in zebrafish. However, it might be possible to infer information about attention from the results of other behavioral tests [127]. For example, behavioral paradigms include visual discrimination in a T-maze or plus maze [128, 129], or appetitive instrumental conditioning in a choice assay [130, 131] could be used. Indeed, the appetitive conditioning tests are fairly similar to the 5 choice serial reaction time task (5CSRTT) described below, without a variation in the inter-trial interval [185,186]. Nevertheless, whilst these tests demonstrate a certain amount of cognitive ability in zebrafish, it is still not clear to what extent attention is being measured, or whether there is any link to ADHD. It is clear that inattention tasks for adult zebrafish require more development before they can be proposed as endophenotypes for ADHD.

3.3 *Zebrafish ADHD-Like Endophenotypes: Cognitive Impulsivity*

There are very few studies that have reported measurements of cognitive impulsivity in zebrafish. In rodents a five choice serial reaction time task (5CSRTT) has been established, in which impulsivity is defined as a premature response during an inter-trial interval (ITI)—the animal is unable to wait for a stimulus presentation before performing a behavioral response (usually a nose-poke [132]). Brennan and colleagues have developed a 5 choice serial reaction time task (5CSRTT) that can be used to measure impulsivity [114, 133, 134]. The 5CSRTT is measured in a tank that has a green LED on one side and five yellow LEDs in separate compartments on the other. Following illumination of the green LED, adult zebrafish are taught to only enter the compartment where the yellow LED is switched on. The correct execution of this behavior is reinforced with a food reward. Following a training period, in which the fish learns to associate the yellow light with a reward, the 5CSRTT can begin. The green stimulus LED is first activated and is then followed by a 10 s ITI. Following this pause, one of the yellow LEDs is lit, and the fish is rewarded with food upon entering the correct compartment. However, entry into any compartment before the end of the ITI, perhaps indicative of impulsivity, will result in a punishment (a 10 s time-out with no food). Entry into an incorrect compartment on the other side (i.e. one in which the yellow LED is not illuminated) will also trigger the punishment. Interestingly, ATO treatment reduces- and MPH increases anticipatory responses in this test [134], providing a possible link to ADHD-like behavior. The 5CSRTT is an impressive test for adult zebrafish, and appears to be a promising paradigm to measure ADHD-linked cognitive impulsivity. Nevertheless, it has only been tested on wild-type animals and so needs to be applied to zebrafish lacking the function of an ADHD-linked gene before we can decide whether or not it constitutes an ADHD-related endophenotype.

3.4 Reduction of Lphn3.1 Activity During Development Triggers ADHD-Linked Alterations in Larval Zebrafish Behavior

As an example of how zebrafish can be used to analyze the function of ADHD-linked genes, we have recently conducted an analysis of *latrophilin 3.1* (*lphn3.1*) during zebrafish development [105]. *LPHN3* is an orphan adhesion-G protein-coupled receptor whose gene contains a variation that conveys a risk haplotype for ADHD. *LPHN3* was identified by linkage analysis of a genetically isolated European population in Columbia (that originated from Spain), followed by fine-mapping in several North American and European populations [135]. Replication of the finding in a cohort of Spanish ADHD patients suggests a role for LPHN3 in the adult form of the disease [136]. *LPHN3* was also identified as one of 86 risk genes in a genome-wide association study of patients with substance abuse disorders, suggesting that ADHD and substance dependence share a high degree of comorbidity [137]. LPHN3 has the capacity to moderate cell-cell interactions. It can act as one of the receptors for α-latrotoxin, a component of black widow spider venom, causing exocytosis of neurotransmitter-containing presynaptic vesicles. The connection between Latrophilin activity and synaptic signaling has been strengthened by the recent identification of two families of endogenous ligands for Latrophilins, the Teneurins and the FLRTs (Fibronectin leucine-rich repeat transmembrane proteins; [138, 139]). For example, FLRT3 appears to specifically interact with LPHN3, is expressed in restricted areas of the developing mouse brain and may control the number of Glutamatergic synapses which are formed [138]. Thus, although the normal physiological function of LPHN3 is not well understood, its function in relation to the formation of synapses during brain development is a particularly promising area for future research.

latrophilin3.1 is one of two zebrafish homologues of human *LPHN3*, both of which are expressed in differentiated neurons throughout the brain up to 6 days post fertilization. We reduced *lphn3.1* function during zebrafish development by injecting one of two gene-specific morpholinos. We then measured larval behavior at 6 days and found an increase in the distance swum by the morphants, a hyperactive phenotype (Fig. 2). The hyperactivity of *lphn3.1* was maintained during the night, suggesting a permanent increase in locomotion compared to control-injected animals (data not shown). *lphn3.1* morphants also show an increase in the number of bursts of acceleration while swimming, indicative of motor impulsivity (Fig. 3). In order to probe the link between changes in *lphn3.1* morphant behavior and ADHD, we rescued the hyperactivity and motor impulsivity by applying the ADHD treat-

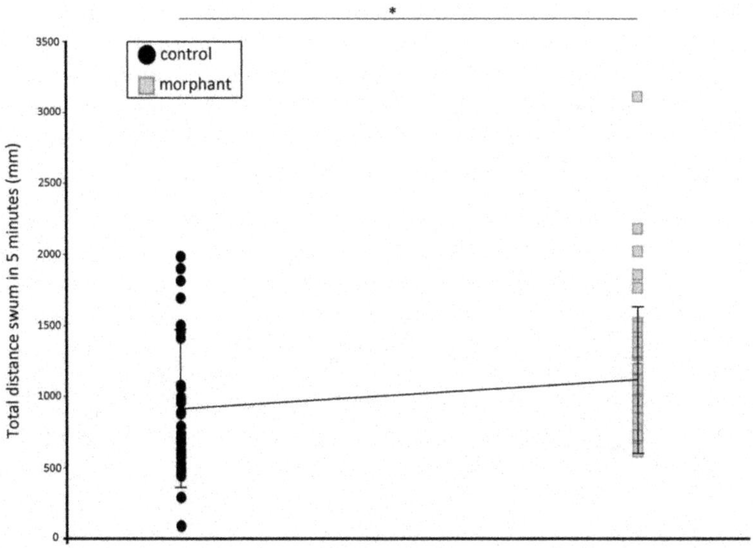

Fig. 2 *lphn3.1* morphant larvae are hyperactive. Mean distance swum in a 5-min time interval by 6 dpf larvae injected with either a control morpholino or *lphn3.1*-specific morpholino. Control larvae n = 39 and *lphn3.1* morphant larvae n = 44. *t*-test reveals a significant difference between the two groups, $*p<0.03$

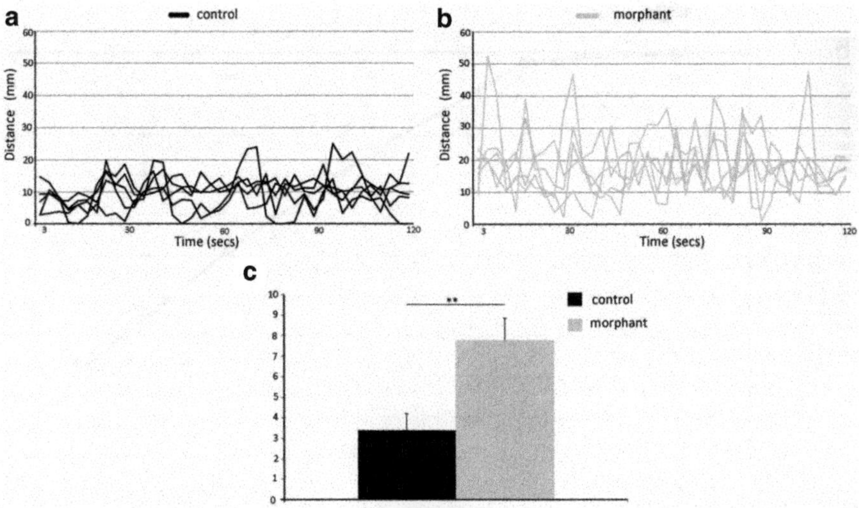

Fig. 3 *lphn3.1* morphant larvae show motor impulsivity. (**a**, **b**) Hyperactive *lphn3.1* larvae display motor impulsivity, revealed by the sharp peaks of locomotion in each of separate morphant locomotion curves (**b**) compared to those of control animals (**a**). (**c**) Lphn3-MO1 morphant larvae exhibit more activity peaks compared to the Lphn3-CO. The number of activity peaks for the two populations (control larvae and *lphn3.1* morpholino-injected larvae) is significantly different during a 120-s experiment. A peak is defined by 5 mm acceleration in at least 12 s. n = 6 for each group. *t*-test, $**p<0.01$ for number of peaks

ment drugs MPH and ATO. Acute treatment of either drug had no effect on control-injected larval behavior at the doses used (10 μM MPH or 1 μM ATO for 1 h), but rescued morphant behavior, bringing locomotion back to control levels (Fig. 4). *lphn3.1* morphants also display a parallel reduction of dopaminergic cells in the posterior tuberculum (PT), a prominent group of dopaminergic neurons in the ventral diencephalon. The PT acts as a locomotory centre in the larval brain [117, 140,

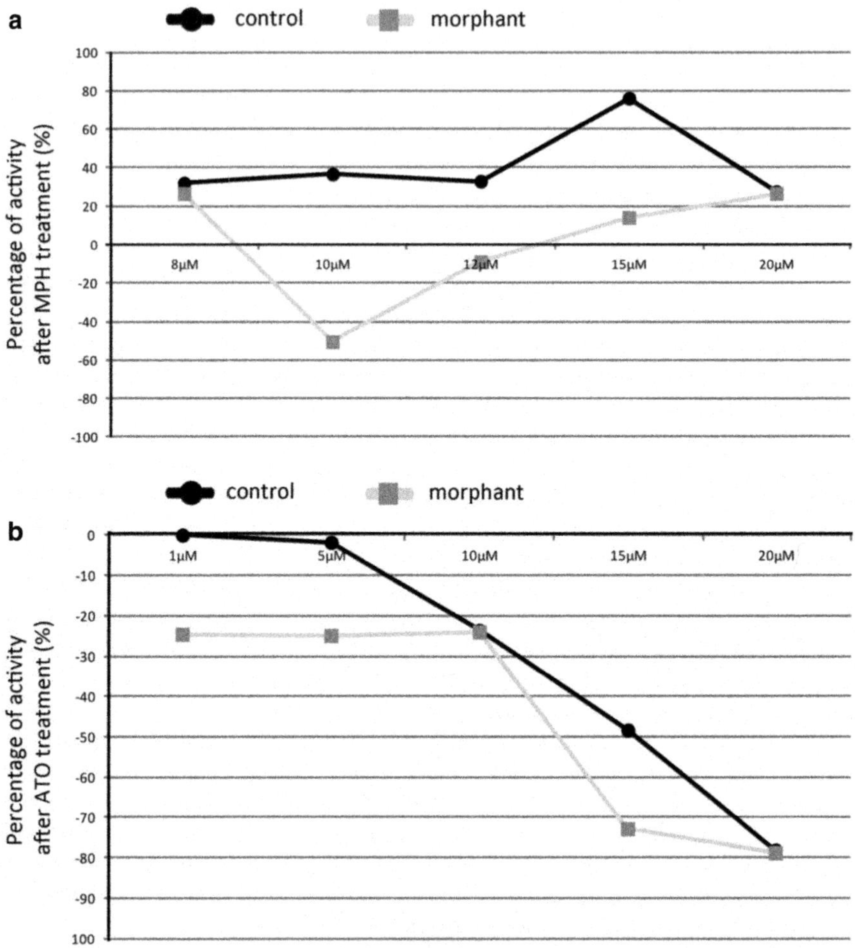

Fig. 4 Application of ADHD treatment drugs rescues *lphn3.1* morphant hyperactivity. (**a**) Dose response curve showing locomotion following methylphenidate (MPH) treatment. Values depict the percentage of change in the distance swum following a 1-h MPH treatment (8, 10, 12, 15 or 20 μM). Control larvae n = 12 and *lphn3.1* morphant larvae n = 12, for each drug concentration. (**b**) Dose response curve showing locomotion following atomoxetine (ATO) treatment. Values depict the percentage of change in the distance swum following a 1-h ATO treatment (1, 5, 10, 15, or 20 μM). Control larvae n = 12 and *lphn3.1* morphant larvae n = 12, for each drug concentration

141] and sends projections both anteriorly to the telencephalon and posteriorly to motorneurons of the spinal cord [142]. In parallel, immunohistochemical, in situ hybridisation and high pressure liquid chromatography (HPLC) studies of other neurotransmitter systems suggest that NA, 5-HT, GABA and glutamate are not affected by loss of *lphn3.1* function [105].

Our study of *lphn3.1* morphant larvae provides several pieces of information regarding the use of zebrafish as an ADHD-like model. Firstly, we have identified ADHD-like endophenotypes in larval zebrafish including hyperactivity (both during the day and night) and motor impulsivity. Secondly, we have provided novel insights into the expression of *lphn3.1* during embryonic development and identified a critical role in controlling the development of dopaminergic neurons. Finally we have provided some of the first concrete evidence that zebrafish may constitute a valid model organism to study ADHD. *lphn3.1* morphant larvae are an excellent tool to begin tease apart the genetics and neurobiology of ADHD. Nevertheless, future work will be required in order to understand how a gene that is expressed in a seemingly wide-spread pattern can lead to such a restricted loss of a few dopaminergic neurons. The possible maintenance of the phenotype into adulthood also needs to be analysed, since morpholino knock-down is only transient. *lphn3.1* morphants provide the ideal tool to search for novel ADHD-like endophenotypes in zebrafish. If the hyperactivity is maintained into adulthood (meaning that *lphn3.1*-mediated changes to embryonic development are sufficient to trigger permanent alterations to behavior) then it would be fascinating to use the 5CSRTT to measure impulsivity.

3.5 *period1b* Mutant Zebrafish

In an interesting recent study, Huang and colleagues have studied *period1b* (*per1b*) mutant zebrafish in connection with ADHD [106]. A key symptom of ADHD is hyperactivity that can result in sleep deprivation [143]. Furthermore, GWAS studies of ADHD patients have identified circadian clock genes [144], and mice with lacking *Clock* gene function exhibit hyperactivity and reduced sleep as well as other behavioural changes [145]. Zebrafish *per1b* mutants are hyperactive at both larval and adult stages and spend more time attacking a mirror, behaviors that can be rescued with the ADHD drugs MPH and deprenyl [106]. They also need more time to learn in an active avoidance test and are more impulsive in a reaction-time task similar to the 5CSRTT. These behavioral phenotypes are correlated with a reduction- and misplacement of DA neurons in the posterior tuberculum (similar to *lphn3.1* morphant animals) and global alterations to DA and NA turnover. Importantly, *Per1b* knock-out mice show a similar phenotype (hyperactivity, learning impairment and reduced DA levels in the brain) demonstrating a conserved function for this gene across species [106]. The possible connection between ADHD and circadian biology is fascinating and should form the basis for further research in the future.

3.6 Future Directions in Zebrafish ADHD Research

The ease of generating zebrafish morphants in large numbers makes our *lphn3.1* larvae an ideal platform with which to identify novel potential ADHD treatment drugs. Zebrafish are the perfect model system for pharmacological studies, since compounds can be directly diluted in small volumes of embryo medium and embryos, thus reducing the amount (and cost) of the compounds used. Therefore, an automated screening setup could be developed that would allow the comparison of hundreds of chemical compounds under standardized conditions. One area of research that has been explored by several groups is the use of zebrafish to look at the effect of ADHD-linked environmental toxins on development. For instance, both lead and bisphenol exposure during embryonic development have been linked to increased susceptibility for ADHD [146], and these compounds have already been applied to zebrafish during development [111–113]. The behavioral effect of MPH during zebrafish development has also been reported by Levin and colleagues [147]. Acute MPH application during the first 5 days of development leads to an increase in DA, NA and 5-HT in the 6-day old larval brain, as well as behavioral changes in adult fish. Drug-treated zebrafish show a reduction of anxiety (measured by a tank-diving assay) and decreased learning in a choice assay compared to mock-treated controls [147].

However, despite the promise shown by zebrafish as an ADHD-like model, there is still a clear need to expand the number of endophenotypes that can be measured, in particular to include those that quantify impulsivity and attention. Such work will be mandatory in order to demonstrate that we are specifically modeling ADHD rather than general NPD-related changes in behavior. The large number of groups that are now beginning to develop and validate protocols to measure adult zebrafish behavior suggest that the search for endophenotypes of NPDs may well be fruitful. The future of zebrafish as a translational model for NPDs looks bright.

Acknowledgements We are grateful to all members of the Bally-Cuif laboratory for their helpful discussions related to this work.

References

1. Mitchell K. The miswired brain: making connections from neurodevelopment to psychopathology. BMC Biol. 2011;9:23.
2. WHO (World Health Organisation). The global burden of disease. 2004 update. Geneva: WHO; 2008.
3. Faraone SV, Sergeant J, Gillberg C, Biederman J. The worldwide prevalence of ADHD: is it an American condition? World Psychiatry. 2003;2:104–13.
4. Polanczyk G, de Lima MS, Horta BL, Biederman J, Rohde LA. The worldwide prevalence of ADHD: a systematic review and metaregression analysis. Am J Psychiatry. 2007;164:942–8. doi:10.1176/appi.ajp.164.6.942.
5. Franke B, Faraone SV, Asherson P, Buitelaar J, Bau CH, Ramos-Quiroga JA, et al. The genetics of attention deficit/hyperactivity disorder in adults, a review. Mol Psychiatry. 2011;17: 960–87.

6. APA (American Psychiatric Association). Diagnostic and statistical manual of mental disorders, fourth edition (DSM-IV). Washington, DC: American Psychiatric Association; 1994.
7. Castellanos FX, Tannock R. Neuroscience of attention-deficit/hyperactivity disorder: the search for endophenotypes. Nat Rev Neurosci. 2002;3:617–28.
8. Krain AL, Castellanos FX. Brain development and ADHD. Clin Psychol Rev. 2006;26: 433–44.
9. Taylor E. Antecedents of ADHD: a historical account of diagnostic concepts. Atten Defic Hyperact Disord. 2011;3:69–75.
10. Hoffman K, Webster TF, Weisskopf MG, Weinberg J, Vieira VM. Exposure to polyfluoroalkyl chemicals and attention deficit/hyperactivity disorder in U.S. children 12–15 years of age. Environ Health Perspect. 2010;118:1762–7. doi:10.1289/ehp.1001898.
11. Still GF. The Goulstonian lectures on some abnormal psychical conditions in children. Lancet. 1902;1:1008–12.
12. Dykman RA, Ackerman PT, Clements SD, Peters JE. Specific learning disabilities: an attentional deficit syndrome. In: Progress in learning disabilities, vol. 2. New York: Grune and Stratton; 1971.
13. Polanczyk G, Rohde LA. Epidemiology of attention-deficit/hyperactivity disorder across the lifespan. Curr Opin Psychiatry. 2007;20:386–92.
14. Swanson J, Castellanos FX, Murias M, LaHoste G, Kennedy J. Cognitive neuroscience of attention deficit hyperactivity disorder and hyperkinetic disorder. Curr Opin Neurobiol. 1998;8:263–71.
15. Biederman J, Faraone SV. Attention-deficit hyperactivity disorder. Lancet. 2005;366: 237–48.
16. Jacob CP, Romanos J, Dempfle A, Heine M, Windemuth-Kieselbach C, Kruse A, et al. Co-morbidity of adult attention-deficit/hyperactivity disorder with focus on personality traits and related disorders in a tertiary referral center. Eur Arch Psychiatry Clin Neurosci. 2007;257:309–17.
17. Barkley RA, Fischer M, Smallish L, Fletcher K. Young adult outcome of hyperactive children: adaptive functioning in major life activities. J Am Acad Child Adolesc Psychiatry. 2006;45:192–202.
18. Schmidt S, Petermann F. Developmental psychopathology: attention deficit hyperactivity disorder (ADHD). BMC Psychiatry. 2009;9:58.
19. Stavrinos D, Biasini FJ, Fine PR, Hodgens JB, Khatri S, Mrug S, et al. Mediating factors associated with pedestrian injury in children with attention-deficit/hyperactivity disorder. Pediatrics. 2011;128:296–302.
20. Lesch KP, Timmesfeld N, Renner TJ, Halperin R, Roser C, Nguyen TT, et al. Molecular genetics of adult ADHD: converging evidence from genome-wide association and extended pedigree linkage studies. J Neural Transm. 2008;115:1573–85.
21. Molina BS, Pelham Jr WE. Childhood predictors of adolescent substance use in a longitudinal study of children with ADHD. J Abnorm Psychol. 2003;112:497–507.
22. Sharp SI, McQuillin A, Gurling HM. Genetics of attention-deficit hyperactivity disorder (ADHD). Neuropharmacology. 2009;57:590–600.
23. Max JE, Sharma A, Qurashi MI. Traumatic brain injury in a child psychiatry inpatient population: a controlled study. J Am Acad Child Adolesc Psychiatry. 1997;36:1595–601.
24. Bush G. Attention-deficit/hyperactivity disorder and attention networks. Neuropsy chopharmacology. 2010;35:278–300.
25. Seidman LJ, Valera EM, Bush G. Brain function and structure in adults with attention-deficit/ hyperactivity disorder. Psychiatr Clin North Am. 2004;27:323–47.
26. Valera EM, Faraone SV, Murray KE, Seidman LJ. Meta-analysis of structural imaging findings in attention-deficit/hyperactivity disorder. Biol Psychiatry. 2007;61:1361–9.
27. Shaw P, Rabin C. New insights into attention-deficit/hyperactivity disorder using structural neuroimaging. Curr Psychiatry Rep. 2009;11:393–8.
28. Liston C, Malter Cohen M, Teslovich T, Levenson D, Casey BJ. Atypical prefrontal connectivity in attention-deficit/hyperactivity disorder: pathway to disease or pathological end point? Biol Psychiatry. 2011;69:1168–77.

29. Makris N, Buka SL, Biederman J, Papadimitriou GM, Hodge SM, Valera EM, et al. Attention and executive systems abnormalities in adults with childhood ADHD: a DT-MRI study of connections. Cereb Cortex. 2008;18:1210–20.
30. Rubia K. "Cool" inferior frontostriatal dysfunction in attention-deficit/hyperactivity disorder versus "hot" ventromedial orbitofrontal-limbic dysfunction in conduct disorder: a review. Biol Psychiatry. 2011;69:e69–87.
31. Duncan J, Owen AM. Common regions of the human frontal lobe recruited by diverse cognitive demands. Trends Neurosci. 2000;23:475–83.
32. Sawaguchi T, Iba M. Prefrontal cortical representation of visuospatial working memory in monkeys examined by local inactivation with muscimol. J Neurophysiol. 2001;86:2041–53.
33. Barkley RA. Behavioral inhibition, sustained attention, and executive functions: constructing a unifying theory of ADHD. Psychol Bull. 1997;121:65–94.
34. Berridge CW, Devilbiss DM, Andrzejewski ME, Arnsten AF, Kelley AE, Schmeichel B, et al. Methylphenidate preferentially increases catecholamine neurotransmission within the prefrontal cortex at low doses that enhance cognitive function. Biol Psychiatry. 2006;60: 1111–20.
35. Bymaster FP, Katner JS, Nelson DL, Hemrick-Luecke SK, Threlkeld PG, Heiligenstein JH, et al. Atomoxetine increases extracellular levels of norepinephrine and dopamine in prefrontal cortex of rat: a potential mechanism for efficacy in attention deficit/hyperactivity disorder. Neuropsychopharmacology. 2002;27:699–711.
36. Berquin PC, Giedd JN, Jacobsen LK, Hamburger SD, Krain AL, Rapoport JL, et al. Cerebellum in attention-deficit hyperactivity disorder: a morphometric MRI study. Neurology. 1998;50:1087–93.
37. Arnsten AF. Catecholamine and second messenger influences on prefrontal cortical networks of "representational knowledge": a rational bridge between genetics and the symptoms of mental illness. Cereb Cortex. 2007;17 Suppl 1:i6–15.
38. Arnsten AF, Pliszka SR. Catecholamine influences on prefrontal cortical function: relevance to treatment of attention deficit/hyperactivity disorder and related disorders. Pharmacol Biochem Behav. 2011;99:211–6.
39. Arnsten AF, Goldman-Rakic PS. Analysis of alpha-2 adrenergic agonist effects on the delayed nonmatch-to-sample performance of aged rhesus monkeys. Neurobiol Aging. 1990;11:583–90.
40. Arnsten AF, Goldman-Rakic PS. Noise stress impairs prefrontal cortical cognitive function in monkeys: evidence for a hyperdopaminergic mechanism. Arch Gen Psychiatry. 1998;55: 362–8.
41. Goldman-Rakic PS. Circuitry of the frontal association cortex and its relevance to dementia. Arch Gerontol Geriatr. 1987;6:299–309.
42. Barkley RA. Adolescents with attention-deficit/hyperactivity disorder: an overview of empirically based treatments. J Psychiatr Pract. 2004;10:39–56.
43. Buitelaar J, Medori R. Treating attention-deficit/hyperactivity disorder beyond symptom control alone in children and adolescents: a review of the potential benefits of long-acting stimulants. Eur Child Adolesc Psychiatry. 2010;19:325–40.
44. Wilens TE. Effects of methylphenidate on the catecholaminergic system in attention-deficit/hyperactivity disorder. J Clin Psychopharmacol. 2008;28:S46–53.
45. Newcorn JH. New treatments and approaches for attention deficit hyperactivity disorder. Curr Psychiatry Rep. 2001;3:87–91.
46. Pliszka SR. Comorbidity of attention-deficit/hyperactivity disorder with psychiatric disorder: an overview. J Clin Psychiatry. 1998;59 Suppl 7:50–8.
47. Sallee FR, Lyne A, Wigal T, McGough JJ. Long-term safety and efficacy of guanfacine extended release in children and adolescents with attention-deficit/hyperactivity disorder. J Child Adolesc Psychopharmacol. 2009;19:215–26.
48. Sallee FR, McGough J, Wigal T, Donahue J, Lyne A, Biederman J. Guanfacine extended release in children and adolescents with attention-deficit/hyperactivity disorder: a placebo-controlled trial. J Am Acad Child Adolesc Psychiatry. 2009;48:155–65.

49. Popper CW. Pharmacologic alternatives to psychostimulants for the treatment of attention-deficit/hyperactivity disorder. Child Adolesc Psychiatr Clin N Am. 2000;9:605–46. viii.
50. Congdon E, Poldrack RA, Freimer NB. Neurocognitive phenotypes and genetic dissection of disorders of brain and behavior. Neuron. 2010;68:218–30.
51. Faraone SV, Perlis RH, Doyle AE, Smoller JW, Goralnick JJ, Holmgren MA, et al. Molecular genetics of attention-deficit/hyperactivity disorder. Biol Psychiatry. 2005;57:1313–23.
52. Zhou K, Dempfle A, Arcos-Burgos M, Bakker SC, Banaschewski T, Biederman J, et al. Meta-analysis of genome-wide linkage scans of attention deficit hyperactivity disorder. Am J Med Genet B Neuropsychiatr Genet. 2008;147B:1392–8.
53. Mitchell K, Porteus DJ. Rethinking the genetic architecture of schizophrenia. Psychol Med. 2011;41:19–32.
54. Elia J, Gai X, Xie HM, Perin JC, Geiger E, Glessner JT, et al. Rare structural variants found in attention-deficit hyperactivity disorder are preferentially associated with neurodevelopmental genes. Mol Psychiatry. 2009;15:637–46.
55. Lesch KP, Selch S, Renner TJ, Jacob C, Nguyen TT, Hahn T, et al. Genome-wide copy number variation analysis in attention-deficit/hyperactivity disorder: association with neuropeptide Y gene dosage in an extended pedigree. Mol Psychiatry. 2010;16:491–503.
56. Cook EH, Scherer SW. Copy-number variations associated with neuropsychiatric conditions. Nature. 2008;455:919–23. doi:10.1038/nature07458.
57. Ebstein RP, Novick O, Umansky R, Priel B, Osher Y, Blaine D, et al. Dopamine D4 receptor (D4DR) exon III polymorphism associated with the human personality trait of novelty seeking. Nat Genet. 1996;12:78–80.
58. Faraone SV, Mick E. Molecular genetics of attention deficit hyperactivity disorder. Psychiatr Clin North Am. 2010;33:159–80.
59. LaHoste GJ, Swanson JM, Wigal SB, Glabe C, Wigal T, King N, et al. Dopamine D4 receptor gene polymorphism is associated with attention deficit hyperactivity disorder. Mol Psychiatry. 1996;1:121–4.
60. Neale BM, Medland SE, Ripke S, Asherson P, Franke B, Lesch KP, et al. Meta-analysis of genome-wide association studies of attention-deficit/hyperactivity disorder. J Am Acad Child Adolesc Psychiatry. 2010;49:884–97.
61. Hawi Z, Lowe N, Kirley A, Gruenhage F, Nothen M, Greenwood T, et al. Linkage disequilibrium mapping at DAT1, DRD5 and DBH narrows the search for ADHD susceptibility alleles at these loci. Mol Psychiatry. 2003;8:299–308.
62. Cook EH, Stein MA, Krasowski MD, Cox NJ, Olkon DM, Kieffer JE, et al. Association of attention-deficit disorder and the dopamine transporter gene. Am J Hum Genet. 1995;56:993–8.
63. Curran S, Mill J, Tahir E, Kent L, Richards S, Gould A, et al. Association study of a dopamine transporter polymorphism and attention deficit hyperactivity disorder in UK and Turkish samples. Mol Psychiatry. 2001;6:425–8.
64. Purper-Ouakil D, Wohl M, Mouren MC, Verpillat P, Ades J, Gorwood P. Meta-analysis of family-based association studies between the dopamine transporter gene and attention deficit hyperactivity disorder. Psychiatr Genet. 2005;15:53–9.
65. Vandenbergh DJ, Persico AM, Hawkins AL, Griffin CA, Li X, Jabs EW, et al. Human dopamine transporter gene (DAT1) maps to chromosome 5p15.3 and displays a VNTR. Genomics. 1992;14:1104–6.
66. Comings DE, Wu S, Chiu C, Ring RH, Gade R, Ahn C, et al. Polygenic inheritance of Tourette syndrome, stuttering, attention deficit hyperactivity, conduct, and oppositional defiant disorder: the additive and subtractive effect of the three dopaminergic genes--DRD2, D beta H, and DAT1. Am J Med Genet. 1996;67:264–88.
67. Li J, Wang Y, Zhou R, Zhang H, Yang L, Wang B, et al. Association between polymorphisms in serotonin transporter gene and attention deficit hyperactivity disorder in Chinese Han subjects. Am J Med Genet B Neuropsychiatr Genet. 2007;144B:14–9.
68. Walitza S, Renner TJ, Dempfle A, Konrad K, Wewetzer C, Halbach A, et al. Transmission disequilibrium of polymorphic variants in the tryptophan hydroxylase-2 gene in attention-deficit/hyperactivity disorder. Mol Psychiatry. 2005;10:1126–32.

69. Gizer IR, Ficks C, Waldman ID. Candidate gene studies of ADHD: a meta-analytic review. Hum Genet. 2009;126:51–90.
70. Hawi Z, Dring M, Kirley A, Foley D, Kent L, Craddock N, et al. Serotonergic system and attention deficit hyperactivity disorder (ADHD): a potential susceptibility locus at the 5-HT(1B) receptor gene in 273 nuclear families from a multi-centre sample. Mol Psychiatry. 2002;7:718–25.
71. Brookes KJ, Mill J, Guindalini C, Curran S, Xu X, Knight J, et al. A common haplotype of the dopamine transporter gene associated with attention-deficit/hyperactivity disorder and interacting with maternal use of alcohol during pregnancy. Arch Gen Psychiatry. 2006;63: 74–81.
72. Deault LC. A systematic review of parenting in relation to the development of comorbidities and functional impairments in children with attention-deficit/hyperactivity disorder (ADHD). Child Psychiatry Hum Dev. 2010;41:168–92.
73. Kahn RS, Khoury J, Nichols WC, Lanphear BP. Role of dopamine transporter genotype and maternal prenatal smoking in childhood hyperactive-impulsive, inattentive, and oppositional behaviors. J Pediatr. 2003;143:104–10.
74. Becker K, El-Faddagh M, Schmidt MH, Esser G, Laucht M. Interaction of dopamine transporter genotype with prenatal smoke exposure on ADHD symptoms. J Pediatr. 2008;152: 263–9.
75. Neuman RJ, Lobos E, Reich W, Henderson CA, Sun LW, Todd RD. Prenatal smoking exposure and dopaminergic genotypes interact to cause a severe ADHD subtype. Biol Psychiatry. 2007;61:1320–8.
76. Laucht M, Skowronek MH, Becker K, Schmidt MH, Esser G, Schulze TG, et al. Interacting effects of the dopamine transporter gene and psychosocial adversity on attention-deficit/hyperactivity disorder symptoms among 15-year-olds from a high-risk community sample. Arch Gen Psychiatry. 2007;64:585–90.
77. Gottesman I, Gould TD. The endophenotype concept in psychiatry: etymology and strategic intentions. Am J Psychiatry. 2003;160:636–45.
78. Kendler KS, Neale MC. Endophenotype: a conceptual analysis. Mol Psychiatry. 2010;15: 789–97.
79. Rommelse NN. Endophenotypes in the genetic research of ADHD over the last decade: have they lived up to their expectations? Expert Rev Neurother. 2008;8:1425–9.
80. Lenox RH, Gould TD, Manji HK. Endophenotypes in bipolar disorder. Am J Med Genet. 2002;114:391–406.
81. Niculescu AB, Akiskal HS. Proposed endophenotypes of dysthymia: evolutionary, clinical and pharmacogenomic considerations. Mol Psychiatry. 2001;6:363–6.
82. Kurz A, Riemenschneider M, Drzezga A, Lautenschlager N. The role of biological markers in the early and differential diagnosis of Alzheimer's disease. J Neural Transm Suppl. 2002; 62:127–33.
83. Neugroschl J, Davis KL. Biological markers in Alzheimer disease. Am J Geriatr Psychiatry. 2002;10:660–77.
84. Gould TD, Bastain TM, Israel ME, Hommer DW, Castellanos FX. Altered performance on an ocular fixation task in attention-deficit/hyperactivity disorder. Biol Psychiatry. 2001;50: 633–5.
85. Arime Y, Kubo Y, Sora I. Animal models of attention-deficit/hyperactivity disorder. Biol Pharm Bull. 2011;34:1373–6.
86. Einat H, Manji HK, Belmaker RH. New approaches to modeling bipolar disorder. Psychopharmacol Bull. 2003;37:47–63.
87. Sarter M, Hagan J, Dudchenko P. Behavioral screening for cognition enhancers: from indiscriminate to valid testing: part I. Psychopharmacology (Berl). 1992;107:144–59.
88. Ingham PW. The power of the zebrafish for disease analysis. Hum Mol Genet. 2009;18: R107–12.
89. Fero K, Yokogawa T, Burgess HA. The behavioral repertoire of larval zebrafish. In: Zebrafish models in neurobehavioral research. New York: Humana Press; 2010.

90. Norton W. Measuring larval zebrafish behavior: locomotion, thigmotaxis and startle. In Zebrafish Protocols for Neurobehaviorual Research. Humana Press New Yorl. Kalueff AV (ed). 2. 2011.
91. Fetcho JR, Liu KS. Zebrafish as a model system for studying neuronal circuits and behavior. Ann N Y Acad Sci. 1998;860:333–45.
92. Amacher SL. Emerging gene knockout technology in zebrafish: zinc-finger nucleases. Brief Funct Genomic Proteomic. 2008;7:460–4.
93. Huang P, Xiao A, Zhou M, Zhu Z, Lin S, Zhang B. Heritable gene targeting in zebrafish using customized TALENs. Nat Biotechnol. 2011;29:699–700.
94. Sander JD, Cade L, Khayter C, Reyon D, Peterson RT, Joung JK, et al. Targeted gene disruption in somatic zebrafish cells using engineered TALENs. Nat Biotechnol. 2011;29:697–8.
95. Curado S, Anderson RM, Jungblut B, Mumm J, Schroeter E, Stainier DY. Conditional targeted cell ablation in zebrafish: a new tool for regeneration studies. Dev Dyn. 2007;236: 1025–35.
96. Nagel G, Szellas T, Huhn W, Kateriya S, Adeishvili N, Berthold P, et al. Channelrhodopsin-2, a directly light-gated cation-selective membrane channel. Proc Natl Acad Sci U S A. 2003;100:13940–5.
97. Zhang F, Wang LP, Brauner M, Liewald JF, Kay K, Watzke N, et al. Multimodal fast optical interrogation of neural circuitry. Nature. 2007;446:633–9.
98. Higashijima S. Transgenic zebrafish expressing fluorescent proteins in central nervous system neurons. Dev Growth Differ. 2008;50:407–13.
99. Tropepe V, Sive HL. Can zebrafish be used as a model to study the neurodevelopmental causes of autism? Genes Brain Behav. 2003;2:268–81.
100. Webb KJ, Norton WH, Trümbach D, Meijer AH, Ninkovic J, Topp S, et al. Zebrafish reward mutants reveal novel transcripts mediating the behavioral effects of amphetamine. Genome Biol. 2009;10:R81. doi:10.1186/gb-2009-10-7-r81.
101. Wood JD, Bonath F, Kumar S, Ross CA, Cunliffe VT. Disrupted-in-schizophrenia 1 and neuregulin 1 are required for the specification of oligodendrocytes and neurones in the zebrafish brain. Hum Mol Genet. 2009;18:391–404.
102. Norton WH. Toward developmental models of psychiatric disorders in zebrafish. Front Neural Circuits. 2013;7:79. doi:10.3389/fncir.2013.00079.
103. Kokel D, Bryan J, Laggner C, White R, Cheung CY, Mateus R, et al. Rapid behavior-based identification of neuroactive small molecules in the zebrafish. Nat Chem Biol. 2010;6:231–7.
104. Rihel J, Prober DA, Arvanites A, Lam K, Zimmerman S, Jang S, et al. Zebrafish behavioral profiling links drugs to biological targets and rest/wake regulation. Science. 2010;327:348–51.
105. Lange M, Norton W, Coolen M, Chaminade M, Merker S, Proft F, et al. The ADHD-susceptibility gene lphn3.1 modulates dopaminergic neuron formation and locomotor activity during zebrafish development. Mol Psychiatry. 2012. doi:10.1038/mp.2012.29.
106. Huang J, Zhong Z, Wang M, Chen X, Tan Y, Zhang S, et al. Circadian modulation of dopamine levels and dopaminergic neuron development contributes to attention deficiency and hyperactive behavior. J Neurosci. 2015;35:2572–87. doi:10.1523/JNEUROSCI.2551-14.2015.
107. Draper BW, Morcos PA, Kimmel CB. Inhibition of zebrafish fgf8 pre-mRNA splicing with morpholino oligos: a quantifiable method for gene knockdown. Genesis. 2001;30:154–6.
108. Kabashi E, Brustein E, Champagne N, Drapeau P. Zebrafish models for the functional genomics of neurogenetic disorders. Biochim Biophys Acta. 2011;1812:335–45.
109. Blaser RE, Chadwick L, McGinnis GC. Behavioral measures of anxiety in zebrafish (*Danio rerio*). Behav Brain Res. 2010;208:56–62.
110. Lopez-Patino MA, Yu L, Cabral H, Zhdanova IV. Anxiogenic effects of cocaine withdrawal in zebrafish. Physiol Behav. 2008;93:160–71.
111. Chen TH, Wang YH, Wu YH. Developmental exposures to ethanol or dimethylsulfoxide at low concentrations alter locomotor activity in larval zebrafish: implications for behavioral toxicity bioassays. Aquat Toxicol. 2011;102:162–6.
112. Saili KS, Corvi MM, Weber DN, Patel AU, Das SR, Przybyla J, et al. Neurodevelopmental low-dose bisphenol A exposure leads to early life-stage hyperactivity and learning deficits in adult zebrafish. Toxicology. 2012;291:83–92.

113. Seibt KJ, Oliveira Rda L, Zimmermann FF, Capiotti KM, Bogo MR, Ghisleni G, et al. Antipsychotic drugs prevent the motor hyperactivity induced by psychotomimetic MK-801 in zebrafish (*Danio rerio*). Behav Brain Res. 2010;214:417–22. doi:10.1016/j.bbr.2010.06.014.
114. Parker MO, Millington ME, Combe FJ, Brennan CH. Development and implementation of a three-choice serial reaction time task for zebrafish (*Danio rerio*). Behav Brain Res. 2012;227:73–80. doi:10.1016/j.bbr.2011.10.037.
115. Norton W, Bally-Cuif L. Adult zebrafish as a model organism for behavioural genetics. BMC Neurosci. 2010;11:90. doi:10.1186/1471-2202-11-90.
116. Smith KM, Bauer L, Fischer M, Barkley R, Navia BA. Identification and characterization of human NR4A2 polymorphisms in attention deficit hyperactivity disorder. Am J Med Genet B Neuropsychiatr Genet. 2005;133B:57–63.
117. Blin M, Norton W, Bally-Cuif L, Vernier P. NR4A2 controls the differentiation of selective dopaminergic nuclei in the zebrafish brain. Mol Cell Neurosci. 2008;39:592–604. doi:10.1016/j.mcn.2008.08.006.
118. Budick SA, O'Malley DM. Locomotor repertoire of the larval zebrafish: swimming, turning and prey capture. J Exp Biol. 2000;203:2565–79.
119. McLean DL, Fan J, Higashijima S, Hale ME, Fetcho JR. A topographic map of recruitment in spinal cord. Nature. 2007;446:71–5.
120. Saint-Amant L, Drapeau P. Time course of the development of motor behaviors in the zebrafish embryo. J Neurobiol. 1998;37:622–32.
121. Porrino LJ, Rapoport JL, Behar D, Sceery W, Ismond DR, Bunney WE. A naturalistic assessment of the motor activity of hyperactive boys. I. Comparison with normal controls. Arch Gen Psychiatry. 1983;40:681–7.
122. Gruber R. Sleep characteristics of children and adolescents with attention deficit-hyperactivity disorder. Child Adolesc Psychiatr Clin N Am. 2009;18:863–76.
123. Taylor E. Clinical foundations of hyperactivity research. Behav Brain Res. 1998;94:11–24.
124. Patton JH, Stanford MS, Barratt ES. Factor structure of the Barratt impulsiveness scale. J Clin Psychol. 1995;51:768–74.
125. Sagvolden T, Johansen EB, Aase H, Russell VA. A dynamic developmental theory of attention-deficit/hyperactivity disorder (ADHD) predominantly hyperactive/impulsive and combined subtypes. Behav Brain Sci. 2005;28:397–419. discussion 419–68.
126. Bushnell PJ. Behavioral approaches to the assessment of attention in animals. Psychopharmacology (Berl). 1998;138:231–59.
127. Echevarria DJ, Jouandot DJ, Toms CN. Assessing attention in the zebrafish: are we there yet? Prog Neuropsychopharmacol Biol Psychiatry. 2011;35:1416–20.
128. Colwill RM, Raymond MP, Ferreira L, Escudero H. Visual discrimination learning in zebrafish (*Danio rerio*). Behav Processes. 2005;70:19–31.
129. Sison M, Gerlai R. Associative learning in zebrafish (*Danio rerio*) in the plus maze. Behav Brain Res. 2010;207:99–104.
130. Bilotta J, Risner ML, Davis EC, Haggbloom SJ. Assessing appetitive choice discrimination learning in zebrafish. Zebrafish. 2005;2:259–68.
131. Risner ML, Lemerise E, Vukmanic EV, Moore A. Behavioral spectral sensitivity of the zebrafish (*Danio rerio*). Vision Res. 2006;46:2625–35.
132. Robbins TW. The 5-choice serial reaction time task: behavioural pharmacology and functional neurochemistry. Psychopharmacology (Berl). 2002;163:362–80. doi:10.1007/s00213-002-1154-7.
133. Parker MO, Ife D, Ma J, Pancholi M, Smeraldi F, Straw C, et al. Development and automation of a test of impulse control in zebrafish. Front Syst Neurosci. 2013;7:65. doi:10.3389/fnsys.2013.00065.
134. Parker MO, Brock AJ, Sudwarts A, Brennan CH. Atomoxetine reduces anticipatory responding in a 5-choice serial reaction time task for adult zebrafish. Psychopharmacology (Berl). 2014;231:2671–9. doi:10.1007/s00213-014-3439-z.
135. Arcos-Burgos M, Jain M, Acosta MT, Shively S, Stanescu H, Wallis D, et al. A common variant of the latrophilin 3 gene, LPHN3, confers susceptibility to ADHD and predicts effectiveness of stimulant medication. Mol Psychiatry. 2010;15:1053–66.

136. Ribases M, Bosch R, Hervas A, Ramos-Quiroga JA, Sanchez-Mora C, Bielsa A, et al. Case–control study of six genes asymmetrically expressed in the two cerebral hemispheres: association of BAIAP2 with attention-deficit/hyperactivity disorder. Biol Psychiatry. 2009;66:926–34.
137. Liu QR, Drgon T, Johnson C, Walther D, Hess J, Uhl GR. Addiction molecular genetics: 639,401 SNP whole genome association identifies many "cell adhesion" genes. Am J Med Genet B Neuropsychiatr Genet. 2006;141B:918–25.
138. O'Sullivan ML, de Wit J, Savas JN, Comoletti D, Otto-Hitt S, Yates JR, et al. FLRT proteins are endogenous latrophilin ligands and regulate excitatory synapse development. Neuron. 2012;73:903–10.
139. Silva JP, Lelianova VG, Ermolyuk YS, Vysokov N, Hitchen PG, Berninghausen O, et al. Latrophilin 1 and its endogenous ligand Lasso/teneurin-2 form a high-affinity transsynaptic receptor pair with signaling capabilities. Proc Natl Acad Sci U S A. 2011;108:12113–8.
140. Bretaud S, Lee S, Guo S. Sensitivity of zebrafish to environmental toxins implicated in Parkinson's disease. Neurotoxicol Teratol. 2004;26:857–64.
141. Sallinen V, Torkko V, Sundvik M, Reenila I, Khrustalyov D, Kaslin J, et al. MPTP and MPP+ target specific aminergic cell populations in larval zebrafish. J Neurochem. 2009;108: 719–31.
142. Tay TL, Ronneberger O, Ryu S, Nitschke R, Driever W. Comprehensive catecholaminergic projectome analysis reveals single-neuron integration of zebrafish ascending and descending dopaminergic systems. Nat Commun. 2011;2:171.
143. Philipsen A, Hornyak M, Riemann D. Sleep and sleep disorders in adults with attention deficit/hyperactivity disorder. Sleep Med Rev. 2006;10:399–405. doi:10.1016/j.smrv.2006.05.002.
144. Lasky-Su J, Anney RJL, Neale BM, Franke B, Zhou K, Maller JB, et al. Genome-wide association scan of the time to onset of attention deficit hyperactivity disorder. Am J Med Genet B Neuropsychiatr Genet. 2008;147B:1355–8. doi:10.1002/ajmg.b.30869.
145. Roybal K, Theobold D, Graham A, DiNieri JA, Russo SJ, Krishnan V, et al. Mania-like behavior induced by disruption of CLOCK. Proc Natl Acad Sci U S A. 2007;104:6406–11. doi:10.1073/pnas.0609625104.
146. Sagvolden T. Behavioral validation of the spontaneously hypertensive rat (SHR) as an animal model of attention-deficit/hyperactivity disorder (AD/HD). Neurosci Biobehav Rev. 2000;24: 31–9.
147. Levin ED, Sledge D, Roach S, Petro A, Donerly S, Linney E. Persistent behavioral impairment caused by embryonic methylphenidate exposure in zebrafish. Neurotoxicol Teratol. 2011;33:668–73.

Zebrafish Neurobehavioral Assays for Drug Addiction Research

Henning Schneider

Abstract Adult and larval zebrafish have been employed for studies of drugs of addiction in numerous excellent studies. This chapter outlines zebrafish neurobehavioral assays such as unconditioned (acute) locomotor responses, sensitization test, tolerance test, conditioned place preference, and self-administration for drugs including ethanol, cocaine, morphine, nicotine, D-amphetamine, heroin/THC and the opiate salvinorin A. Methodological details of assays including environmental parameters, design of testing chambers, treatment regimens and analysis are provided. Test for adult and larval zebrafish are addressed in separate sections. Established robust neurobehavioral assays in combination with powerful molecular genetic tools manifest strong prospects for discovery of underlying mechanism of drug use-behavior and development of pharmacotherapeutics for treatment of drug dependence using zebrafish.

Keywords Behavioral screening • Zebrafish • Sensitization • Conditioned place preference • Self-administration • Ethanol • Cocaine • Morphine • Nicotine • D-amphetamine • Heroin • Opiate • Drug-use • Drug dependence • Drug-discovery • Pharmacotherapeutics

1 The Zebrafish Model for Studies of Drugs of Addiction

The ultimate goal of studies on drugs of addiction is to develop treatments that help people to overcome drug dependence. To reach this goal and develop new drug cessation treatments studies aim for the development of new pharmacotherapeutics and the discovery of neural and molecular mechanisms of drug-use behavior, which are still poorly understood [1–4]. Experimental strategies developed for rodent models were enormously successful for advancing the understanding of mechanisms of addiction and drug-use behavior but challenges remain [5–10].

Great progress has been made for utilizing zebrafish in studies of drugs of addiction [11–18]. The behavioral screening for mutants with abnormal behavioral responses to cocaine in a conditioned place preference (CPP) paradigm represents

H. Schneider (✉)
Department of Biology, DePauw University, 1 East Hanna St., Greencastle, IN 46135, USA
e-mail: hschneider@depauw.edu

A.V. Kalueff (ed.), *The rights and wrongs of zebrafish: Behavioral phenotyping of zebrafish*, DOI 10.1007/978-3-319-33774-6_8

one of the first behavioral studies on drugs of addiction in zebrafish [12]. Modified neurobehavioral responses to morphine have been identified in too few mutants that have a reduced number of dopamine and serotonin neurons [15]. Random screening for mutants with modified neurobehavioral responses to nicotine led to the identification of two genes with potential roles in nicotine-response behavior, metabotropic GABA receptors (gababr1.2; hbog) and the cct8 tgene (chaperone containing protein 8 gene; bdav) [17]. Overall, these examples indicate that both larval and adult zebrafish are excellent model systems for studying the function of drugs of addiction.

Since the zebrafish genome contains human gene orthologs involved in the action of drugs of addiction such as nicotinic acetylcholine receptors, opioid peptides, opioid receptors, alcohol dehydrogenase genes, glutamate receptors, and downstream neurotransmitter receptors for dopamine, serotonin, noradrenaline, glutamate and GABA findings made in zebrafish could translate easily into mammalian and human studies [19, 20]. The fast development from fertilization to swimming larval stages takes 5 days and makes larval zebrafish attractive for screening of chemical libraries and drug discovery [14, 17, 19, 21–25].

This chapter provides an overview of zebrafish behavioral assays and methodological details used for the study of drugs of addiction. Details concerning the design of behavioral chambers/tanks, treatment regimens and range of concentrations used for different compounds are provided. Behavioral assays for adult and larval zebrafish are presented in separate sections. The information provided gives a comprehensive overview of current zebrafish behavioral assays and may be helpful when getting started with utilizing zebrafish for studies of drugs of addiction.

2 Overview of Zebrafish Neurobehavioral Assays

The acute response of animals to chemical compounds represents the simplest neurobehavioral assay and provides insight into the biological activity (or sensitivity) of the tested chemical compound. Various doses applied over different exposure times are commonly used to obtain a dose–response relationship to a drug of addiction. Zebrafish, have a rich behavioral repertoire of at least 190 neurobehavioral patterns that range from simple patterns, such as swimming distance, swimming speed, thigmotaxis, turning behavior, to complex patters, such as anxiety, reward, learning and memory, and shoaling [20, 26–30]. Since many drugs of addiction cause a biphasic repose following an inverted U-shaped dose–response curve, a dose–response relationship determines whether a substance has similar response kinetics in zebrafish compared to mammals. The dosing is dependent on chemical properties of a chemical as well as the application method. An optimal concentration of a chemical or drug of addiction produces the strongest significant behavioral response without adverse effects. Once the optimal concentration of a drug of addiction has been determined, an acute response test can then be employed to screen for chemicals that reduce the drug-induced behavioral response.

Chronic treatments of animals have been used to measure neuroadaptations induced by a drug of addiction. In contrast to acute responses, chronic treatments may include exposure to a drug for several days either continuously, intermittently or repeatedly for short time periods each day [26, 31–34]. Repeated exposure can cause sensitization or tolerance to a drug [26, 29, 35, 36]. Whereas low to moderate concentrations appear to generate sensitization [17], high concentrations and longer exposure over many days or weeks appear to generate tolerance [18, 26, 36–38]. However, the neurobiological basis of sensitization and tolerance in zebrafish is unknown. In mammals, changes in dopamine function are involved but underlying neural and molecular mechanisms of sensitization and tolerance are mostly still unclear [5, 8, 10, 39–42].

Reinforcing effects of a drug of addiction that cause dependence in humans have been studied in many animal species using conditioned place preference training (CPP) and self-administration paradigms [7]. In CPP training, zebrafish are exposed to a drug of addiction (unconditioned stimulus) in the presence of cues (conditioned stimulus), which are mostly visual [12, 15, 43–46]. Following repeated conditioning trials, zebrafish will chose an environment with conditioned cues in the absence of the drug of addiction. Thus in contrast to sensitization or tolerance, conditioned place preference includes elements of associative learning that could be enforced by a drug of addiction [11]. Neurobehavioral states that closely represent a state of dependence include active avoidance of withdrawal or active seeking a dug (craving). Both drug seeking and avoiding withdrawal has be assessed using a self-administration paradigm. In self-administration test using mammals, animals have free access to a drug of addiction through water or pressing of a lever [10]. While testing paradigms for CPP, sensitization and tolerance that have been carried out successfully in mammals have been adopted successfully for zebrafish, establishing a self-administration test for zebrafish, for example, has been challenging [47, 48].

The rich behavioral repertoire of different developmental stages of zebrafish represents opportunities and challenges for studies on addiction. Which developmental stage to select depends on the primary goal of a proposed study. For example, larval zebrafish up to 14 days post fertilization (dpf) do not show shoaling behavior [49, 50]. Thus behavioral experiments on larval zebrafish can be carried out using groups of animals facilitating high throughput screening. In contrast, neurobehavioral experiments on adult zebrafish have been carried out by testing fish individually to eliminate shoaling behavior, which could complicate the interpretation of behavioral results unless the effects of drugs on shoaling behavior have been studied [49, 51].

Gender differences in the biology of cocaine addiction are known in humans and mammals [52] and have been described also for zebrafish [53]. Female zebrafish develop anxiety-like withdrawal symptoms to cocaine earlier than male zebrafish. Hormones appear to influence the neural circuits of addiction. Potentially, studies of addiction in female and male zebrafish could facilitate the understanding of underlying mechanisms of gender difference of addiction biology through the use of genetic methods.

Behavioral responses to drugs of addiction can vary with the strain of zebrafish. Adult zebrafish of the AB and SF strain showed different responses to acute and chronic ethanol [54]. In the SF strain, higher levels of expression have been reported for the dopamine receptor D1 (drd1) while lower levels of expression have been measured for the serotonin transporter gene slc64a and the GABA B receptor 1 (gababr1). Moreover, the amount of biogenic amines dopamine, serotonin, GABA, glutamate and glycine were lower in brain tissue of the adult SF strain [55]. Thus the selection of the zebrafish strain for studies on addiction could affect the outcome of experiments.

While the neural circuitry of addiction in zebrafish is unknown the neuroanatomy of dopamine-neurons, which are central for reward behavior in mammals, has been described in detail using immunohistochemical staining for tyrosine hydroxylase and genetic methods [56–59]. Moreover, dopamine receptor and dopamine transporter genes have been sequenced and their expression has been mapped in larval zebrafish [60–62]. More information about neurotransmitter systems in the zebrafish central nervous system can be found in the following publications: [56–58, 63–73].

3 General Methodology for Studies of Drugs of Addiction in Zebrafish

3.1 Environmental Conditions

Zebrafish are tropical poikilothermic animals and are raised and kept at 28 °C. Locomotor activity decreases with temperature in both adults and larvae. Thus constant temperature conditions have been used when measuring behavioral responses to drugs or during conditioning trials. Only few studies indicate if behavioral testing has been carried out at room temperature or 28 °C. Light and time of the day were additional factors that affected a behavioral response. Most studies indicate that experiments were carried out at the same time of the day to control for circadian variations of locomotor activity [74]. Locomotor activity has been reported to be higher in the morning and lower in the afternoon in zebrafish larvae [75]. Light conditions also affected behavioral response and outcome. Larval zebrafish prefer a lighter environment whereas adult zebrafish prefer darker over lighter environments [75, 76]. Thus, the duration of acclimatization periods depends on the testing environment, developmental stage of zebrafish and the type of experiment carried out.

3.2 General Procedures for Application of Chemicals

In zebrafish the most widely used method of drug administration is the immersion method in which substances were applied directly to the tank water. Optimal doses of chemical compounds have been established in dose–response experiments and were mainly depend on chemical properties of compounds, uptake mechanisms and developmental stage of animals. However, even water soluble chemicals may not be able to pass through the blood–brain barrier and enter the brain [22]. Water-insoluble chemicals have been dissolved first in DMSO and then diluted in water. Alternatively in adult zebrafish, chemicals can also be administered by intraperitoneal or intramuscular injection [77, 78]. In studies on zebrafish larvae, the immersion method was the method of choice. When substances needed to be dissolved in DMSO, the final DMSO concentration in the experiment can reach 1 %, which was tolerated by larvae and had no known adverse effects [79, 80]. Since application and tissue penetration of drugs differ, new drugs would need to be tested for the optimal application procedure as well as duration of drug application for larval and adult zebrafish, such as ethanol [32]. Concentrations used for ethanol, nicotine, D-amphetamine, cocaine and morphine in different studies indicate a somewhat standard range for each drug that could be adopted (ethanol: [26, 35, 44, 81]; morphine: [47]; cocaine: [12, 82, 83]; D-amphetamine: [46, 82]; nicotine: [17, 44, 84, 85]). To confirm that drugs entered the brain, concentrations of applied chemicals have been measured in brain tissue over an incubation period. Brain tissue was removed quickly from adult zebrafish, weighed and frozen or homogenized and processed for the appropriate chemical assay (ethanol: [32]; amphetamine: [77]; cocaine: [83]). While the application of chemicals directly into the water is simple, behavioral changes could be due to effects of chemicals on sensory systems or the peripheral nervous system and increases the potential of off-target effects and require control experiments [17]. Higher concentrations of a substance have been used for acute treatment (minutes) while lower concentrations were used for chronic treatment (days, weeks) to avoid toxicity. Since several factors determine the final concentration of a compound in brain tissue, optimal working concentrations of substances and uptake kinetics need to be determined carefully.

3.3 Analysis of Behavioral Experiments

The analysis of behavioral data and comparison of groups of untreated and treated animals involves mostly an ANOVA test (1- or 2-way) and depends on the behavioral test paradigm. For example, a one-way ANOVA was used to test for the significant differences between movement activity before and after withdrawal of ethanol [37]. A 2-way ANOVA was used to compare groups of fish that have been treated differently. In addition, a 2-way ANOVA was coupled to a post test such as a Bonferroni or Tukey post hoc comparison test when performing CPP experiments [12, 44]. The software used for statistical analysis frequently includes SPSS (IBM) or PRIZM, a statistics and graphic software from Graphpad (La Jolla, CA).

4 Neurobehavioral Assays for Adult Zebrafish

Studies on adult zebrafish regarding drugs of addiction outnumber the studies on larval zebrafish (Tables 1 and 2). Age of adults used in these studies is usually between 3 and 15 months. Juvenile zebrafish emerge from metamorphosis after 25–35 days after fertilization [24] and reach the adult stage with the onset of reproductive behavior and breeding 2–3 months after fertilization [24]. Adults have been tested mostly individually unless the effect of a drug on social behavior has been investigated [50, 51, 85, 96].

Table 1 Behavioral assays for adult zebrafish

Behavioral test	Tested drug	Authors	Result
Unconditioned locomotor response/ behavior to acute treatment	Ethanol	Gerlai et al. (2000) [86]	Demonstrates that intermediate doses of alcohol increase locomotor activity, location of swimming, enhance aggression, and preference for conspecifics and reduces preference for a school at high concentrations. Shows enhanced antipredatory behavior at low doses and decreased antipredatory behavior at high doses
		Dlugos and Rabin (2003) [32]	Demonstrates strain-dependent difference of behavioral responses to acute and chronic ethanol treatment
		Mathur and Guo (2011) [34]	Demonstrates that acute ethanol treatment causes anxiolytic-like behavior in the Novel Tank Diving Test and a Light/Dark Choice Assay
		Tran et al. (2015a) [37]	Demonstrates association of changed levels of anxiety-like behavior and dopamine and serotonin concentrations in brain tissue in response to 30-min ethanol exposure
	Cocaine	Lopez-Patino et al. (2008a) [83]	Demonstrates that non-anesthetic cocaine doses do not change locomotor behavior and that cocaine withdrawal causes hyperlocomotion indicating an anxiogenic effect
		Lopez-Patino et al. (2008b) [53]	Demonstrates gender difference to cocaine withdrawal
	Nicotine	Levin and Chen (2004) [87]	Demonstrates that acute nicotine at low concentrations enhances memory but decreases memory at high concentrations as tested in a 3-compartment zebrafish maze
		Levin et al. (2007) [84]	Demonstrates anxiolytic effects of acute nicotine using a novel tank diving test
		Bencan and Levin (2008) [88]	Indicates that nicotinic α_7 and $\alpha_4\beta_2$ receptors are involved in the nicotinic-induced anxiolytic effects
	Heroin, THC	Stewart and Kalueff (2014) [89]	Measures response to a 20 min THC or heroin exposure in novel tank diving test; heroin induces hyperlocomotion but no change in anxiety-like behavior; THC induces anxiogenic-like effects

Unconditioned locomotor response/behavior to chronic treatment	Nicotine	Stewart et al. (2015a) [85]	A chronic 4-day exposure to 1–2 mg/L nicotine slightly increases shoaling behavior and induces anxiogenic behavior as tested in the novel tank diving test
	Ethanol	Dlugos and Rabin (2003) [32]	Shows development of behavioral tolerance to 1-week continuous ethanol treatment
		Gerlai et al. (2006) [26]	Shows that chronic ethanol treatment affects responses to acute ethanol treatment and potential anxiolytic effects
		Egan et al. (2009) [33]	Demonstrates anxiolytic effects of 1-week ethanol treatment (increased exploration and reduced erratic swimming) and shows high-anxiety behavior in leopard and albino strains but low-anxiety behavior in wild type strains
		Mathur and Guo (2011) [34]	Demonstrates that chronic ethanol induces anxiogenic-like behavior in contrast to acute ethanol treatment, which causes anxiolytic-like behavior using the Novel Tank Diving Test and a Light/Dark Choice Assay
	Morphine	Cachat et al. (2010a) [31]	Demonstrates anxiogenic effects of repeated daily withdrawal from chronic 1-week morphine treatment
Tolerance	Ethanol	Gerlai et al. (2006) [26]	Two week continuous exposure to 0.25 % ethanol causes behavioral adaptation demonstrated by reduced locomotor response to acute ethanol
		Stewart et al. (2011) [18]	Reviews development of tolerance
		Tran and Gerlai (2013) [36]	Chronic exposure to ethanol changes the behavioral response measured over a 1-h period during which zebrafish are exposed to ethanol and confirms biphasic nature of ethanol actions
		Tran and Gerlai (2014) [38]	Reviews development of tolerance to ethanol
		Tran et al. (2015a) [37]	Demonstrates that chronic ethanol induces tolerance at the behavioral level that correlates with changed ADH activity
Sensitization	Ethanol	Blaser et al. (2010b) [35]	Demonstrates ethanol-induced locomotor sensitization in the presence of paired contextual stimuli
		Tran and Gerlai (2014) [38]	Reviews sensitization to ethanol

(continued)

Table 1 (continued)

Behavioral test	Tested drug	Authors	Result
CPP	Cocaine	Darland and Dowling (2001) [12]	Demonstrates cocaine-CPP in wild-type zebrafish in a two chamber test and identifies mutants with low morphine-CPP
		Darland et al. (2012) [90]	Demonstrates cocaine-CPP in a three-chamber test and in conjunction with an increase in expression of tyrosine hydroxylase (TH) and a significant decrease in the expression of elongation factor 1alpha
	Morphine	Lau et al. (2006) [15]	Demonstrates morphine-CPP that can be abolished by pretreatment with dopamine receptor antagonists and identifies a mutant line with no morphine-CPP
		Mathur et al. (2011b) [45]	Describes detailed protocol for single drug exposure induced CPP
	D-amphetamine	Ninkovic et al. (2006) [46]	Demonstrates D-amphetamine-CPP in wild-type and AchE mutants
		Ninkovic and Bally-Cuif (2006) [77]	Describes detailed protocol of D-amphetamine-CPP
	Nicotine	Kily et al. (2008) [44]	Demonstrates nicotine-CPP and used micro-arrays to identify genes associated with nicotine- CPP
		Kedikian et al. (2013) [91]	Demonstrates nicotine-CPP in a two-chambered test and shows a correlated increase of α_7 and $\alpha_4\beta_2$ nicotinic receptor expression using real-time qPCR and an increase in phosphorylated CREB
		Ponzoni et al. (2014) [92]	Demonstrates that nicotine-CPP can be blocked by CC4 and CC26 cytisine derivatives indicating that α_7 and $\alpha_4\beta_2$ nicotinic receptors are involved in nicotine-CPP
	Ethanol	Kily et al. (2008) [44]	Demonstrates that repeated exposure to ethanol causes CPP in a two-chamber test and used micro-arrays to identify genes associated with nicotine- CPP
		Mathur et al. (2011a) [93]	Demonstrates ethanol—CPP following a single exposure to ethanol in a three-chamber test
	Salvinorin A	Braida et al. (2007) [94]	Demonstrates salvinorin A—CPP that can be blocked with pretreatment of the opioid antagonist, nor-binaltorphimine
CPP, extinction & reinstatement	Ethanol	Parmar et al. (2011) [95]	Describes detailed protocol of ethanol-CPP in combination with extinction training and reinstatement methods

CPA	Salvinorin A	Braida et al. (2007) [94]	Demonstrates that high concentrations of salvinorin A produce conditioned place aversion whereas low concentrations produce CPP
Self-administration-like	Ethanol	Sterling et al. (2015) [48]	Uses ethanol containing gelatin food for quantification of ethanol consumption that causes increased locomotor behavior, reduced anxiety, stimulated aggressive behavior, and increased expression of galanin and orexin in the hypothalamus
Withdrawal	Ethanol	Cachat et al. (2010a) [31]	Demonstrates that discontinuation of chronic 1-week ethanol treatment increases anxiogenic-like behavior
		Mathur and Guo (2011) [34]	Demonstrates elevated anxiety-like behavior for up to at least 7 days following interruption of chronic intermittent ethanol treatment using a novel tank diving test and a light/dark choice assay
		Tran et al. (2015a) [37]	Demonstrates changes in motor patterns, cortisol levels, and neurochemical alterations associated with withdrawal from chronic 0.5 % (v/v) ethanol within 60 min
	Cocaine	Lopez-Patino et al. (2008a) [83]	Demonstrates increased hyperactivity that remains elevated for at least 5 days following withdrawal from chronic cocaine treatment
		Lopez-Patino et al. (2008b) [53]	Describes that anxiety-like behavior following withdrawal form chronic cocaine develops earlier in female zebrafish but is stronger and last longer in males
	Morphine	Cachat et al. (2010a) [31]	Demonstrates that discontinuation of 1-week chronic morphine treatment increases anxiogenic-like behavior
	Diazepam	Cachat et al. (2010a) [31]	Demonstrates that discontinuation of 1-week chronic diazepam treatment increases anxiogenic-like behavior

Table 2 Behavioral assays for larval zebrafish

Behavioral Test	Drug	Authors	Result
Unconditioned acute locomotor response	Ethanol	Lockwood et al. (2004) [28]	Demonstrates a dose-dependent locomotor response to ethanol. Intermediate doses cause hyperactivity. High doses cause hypoactivity and sedation. Ethanol causes thigmotaxis
		MacPhail et al. (2009) [74]	Shows that 1 and 2 % ethanol increases locomotor activity and that 4 % ethanol dramatically decreases locomotor activity in 96-well plates
		Irons et al. (2010) [82]	Demonstrates that ethanol at low concentrations increases locomotor activity during light periods
	Nicotine	Petzold et al. (2009) [17]	Demonstrates nicotine-induced increase in locomotor activity and identifies mutants with reduced response to acute nicotine
		Schneider et al. (2012b) [25]	Detailed description of neurobehavioral assay for acute nicotine-induced locomotor response for identifying chemicals that reduce the nicotine response
		Cousin et al. (2014) [21]	Shows screening of FDA approved chemicals that reduce the nicotine-induced locomotor response
	D-amphet-amine	Irons et al. (2010) [82]	Demonstrates that D-amphetamine increases locomotor activity in dark periods
	Cocaine	Irons et al. (2010) [82]	Demonstrates that cocaine at high concentration decreases locomotor activity during dark periods
Sensitization	Nicotine	Petzold et al. (2009) [17]	Demonstrates that nicotine pretreatment increases acute-nicotine induced locomotor response
Self-administration-like	Morphine	Bretaud et al. (2007) [47]	Introduces a self-immersion test apparatus and shows that morphine pretreated larvae prefer immersion in morphine containing water. This behavior is reduced by pretreatment with naloxone or dopamine receptor antagonists and in too few mutants

4.1 Design of Testing Environment (and Image Analysis) for Adult Zebrafish

Studies on drugs of addiction in adults are mostly carried out using a rectangular tank that is viewed from above or a trapezoid tank (1.5 l; AquaticHabitats/Pentair) that is view from the side and/or above [33, 84, 97]. Ambient lighting conditions and a room temperature of 28 °C have been recommended. The analysis of swimming behavior in neurobehavioral zebrafish assays in adult zebrafish is carried out either manually or automatically using either commercially available software (Ethovision Noldus; DanioVision, Viewpoint; LocoScan, CleverSys.) or systems developed by individual laboratories for specific behavioral assays using web-cams or consumer video cameras (e.g. [16, 97, 98]). Commercial systems and software can be used for a real-time analysis or a post-trial analysis. Ethovision and DanioVision have the capability to track swimming behavior of individually kept fish in an array of 20 tanks (one fish per tank) simultaneously [83].

4.2 *Acute and Chronic Treatment of Adult Zebrafish*

Locomotor Behavior Acute and chronic exposure of zebrafish to drugs of addiction can cause a change in a number of different behavioral patterns such as locomotor activity, swimming speed, distance from bottom of a tank (novel tank test, see below IV.F.), swimming pattern (zig-zags; turn angle), freezing, floating, aggression, group preference, antipredatory response, shoaling behavior, pigment response, and light/dark preference [27]. Locomotor activity such as distance travelled over a certain time interval has been commonly used in zebrafish studies on addictive substance as well as in rodents [99]. Position in the test tank relative to the bottom is a variable for testing the sensitivity of zebrafish to chemicals and has been often used in conjunction with studies of anxiety [84, 85, 88, 97]. The following section focuses on variables such as distance travelled and/or swimming speed for measuring locomotor activity and refer the reader to other chapters in this book that address certain behaviors such anxiety and social behavior specifically.

Experimental Design and Analysis To measure distance travelled in an observation tank in response to chemical compounds, a camera can be positioned above or besides an observation tank. Observation tanks have been placed into a separate room or space free of activity that could distract the zebrafish. Animals were allowed to adapt to the tank environment repeatedly (up to 5 times—once per day) in a pre-test phase. The duration that a fish is allowed to adapt each time matched the duration of the behavioral test when swimming activity was recorded [83]. Observation tanks or acute and chronic drug testing were either rectangular (1–2 l, ~$20 \times 10 \times 10$ cm^3), round (20 cm diameter), or trapezoid [15, 26, 32, 84, 85]. Water levels were usually kept between 5 and 12 cm so that the fish stay in focus.

Drugs of addiction have been administered either directly into the tank water or injected into individual fish intraperitoneally or intramuscularly. Used concentrations and incubations times vary between drugs. If ethanol was applied directly to tank water, brain ethanol levels peaked after the initial 30 min of ethanol exposure (or introduction of fish into the test tank) and remained constant [32, 37]. The uptake of ethanol into the brain was dose-dependent. The concentration of ethanol (v/v) in the brain of adult fish was roughly 10–20 % of ethanol applied to the water or 0.05–

0.1 % for 0.5 % ethanol and 0.1–0.2 % for 1.00 % ethanol after 20–60 min [32, 37]. In contrast, cocaine levels increased steadily over a 60 min period when they reached a maximum and remained high with cocaine present in the tank water [83]. A cocaine concentration in tank water of 1.5 μM resulted in a peak brain concentration of adult male zebrafish of ~8–12 pg/μg protein after 60 min of incubation [83]. Much shorter application times have been used for measuring the acute response to nicotine. For acute tests in adult zebrafish, application of nicotine has been performed in a dosing beaker that contains 50 ml of nicotine water at concentrations of 50–800 mg/l. Following a 3 min immersion in nicotine water, fish were transferred to the observation tank in which their behavior was recorded [87]. These treatments have not caused any adverse effects. In a different study [84], adult zebrafish were immersed as a group into a nicotine-containing water (50, 100, or 200 mg/l) for specific time periods (5 min). Then individual fish were transferred into the 1.5 l trapezoid-shaped viewing tank (22.9 cm bottom length, 27.9 cm top length, 15.2 cm height and 15.9 cm length of diagonal side) for a 5-min video recording. A camera was placed on the side of the tank so that horizontal swimming activity (swim path length and position) could be recorded (for measurement of anxiety). In this type of experimental set-up, swimming activity increased over a 5-min period in control animals and fish treated with 50 mg/l nicotine. Zebrafish treated with 100 mg/l showed the same elevated swimming activity from the 1st to the 5th minute [84]. Higher doses of 200 mg/l have been reported to cause diminished activity and more than 200 mg/l nicotine in tank water impairs activity [84]. At a water concentration of 100 mg/l nicotine almost doubled the swimming speed in acute treatments [88]. These effects could be inhibited by nicotinic acetylcholine receptor antagonists mecamylamine, methyllycaconitine and dihydro-β-erythroidine [84, 88].

For chronic treatment, zebrafish have been kept either continuously or intermittently for days or weeks in tanks containing the drug of addiction. For chronic treatment, adult zebrafish were exposed to concentrations that are lower than those used for acute treatment to avoid toxicity. For example, concentrations of 1–2 mg/l (2–4 μM) nicotine have been used for adult zebrafish and a treatment period of 4 days [85] compared to up to 800 mg/l nicotine for 3 min [87]. At lower nicotine concentrations, no adverse effects or changes of locomotor activity have been detected following a 4-day exposure phase (2 days at 1 mg/l followed by 2 days on 2 mg/l). Variables of locomotor activity such as distance travelled, average velocity, period of immobility remained unchanged. In addition, concentrations of 1–2 mg/l nicotine and chronic treatment did not increase whole-body cortisol. However, an anxiogenic-like effect has been recorded as it takes longer for zebrafish to transition to the upper compartment of a tank [85]. Chronic ethanol treatment has been used in studies on anxiety and demonstrated anxiolytic effects as fish spent significantly more time in the upper half of the test tank during a 6 min recording period [33]. Adult zebrafish were kept in a tank containing 0.3 % ethanol for 7 days. The tank water and ethanol was changed every 2 days. In another study of chronic ethanol treatment, adult zebrafish were treated with 0.25 % ethanol in tank water for 2 weeks with daily changes of tank water containing ethanol [26]. Then the response of fish to acute was measured by recording swimming activity for 10 min and analyzed in

1 min bouts. Chronic ethanol treatment appeared to involve an adaptation to ethanol and led to reduced anxiolytic effects of acute ethanol. In a third ethanol study [32], fish were treated with 0.5 % ethanol for 2 weeks. Shoaling was measured over a 30 min test period. Testing and recording following chronic treatment was not carried out in the home tank to exclude any interference caused by associative learning [32]. To better match chronic alcohol use of humans in zebrafish, zebrafish were exposed to alcohol intermittently over 8 days, each day to 1 % ethanol for 20 min with the goal to study withdrawal [34]. Chronic treatment of adult zebrafish was often carried out to test for the development of tolerance to a drug of addiction. Tolerance to ethanol could be established when zebrafish were exposed chronically to higher doses of ethanol [18, 26, 36–38]. Ethanol concentrations were raised stepwise in the tank over 11 days (day 1–4: 0.125 %; day 5–8: 0.25 %; day 9–11: 0.375 %) followed by a 10 day 0.5 % chronic incubation period. Then an ethanol challenge was provided by immersing fish in 0.5 % or 1.0 % ethanol for 60 min during which total swimming distance was recorded in 1 min interval over 60 min recording [36]. The chronic ethanol treatment reduced the total distance traveled in the acute challenge and modified other behaviors [29, 36]. Chronic morphine increased cortisol levels but no changes of an acute morphine challenge after chronic morphine exposure [31]. Overall, zebrafish are suitable for studying chronic effects of addiction using a variety of testing paradigms and drugs.

4.3 Sensitization and Tolerance

A single or repeated exposure to a drug of addiction can cause sensitization or tolerance to the drug in zebrafish [37]. Ethanol, for example, has a sensitizing effect at low concentrations but causes the development of tolerance at high concentrations (for review: [38]). Sensitization of zebrafish to drugs of abuse includes two phases: induction phase and expression phase. In the induction phase animals were exposed repeatedly to a substance for short periods. In the expression phase individuals were challenged with the same concentration of alcohol [38].

Experimental Design and Analysis The behavioral set-up and protocol for establishing sensitization or tolerance to drugs of addiction in zebrafish is simple. A rectangular tank 2-l has been used with a water level of about 12 cm. Fish were first transferred to a holding tank that has the same dimension and water levels as sensitization and test tanks. The novelty effect of this treatment was the same for treated and untreated (control) fish as they were handled in the same way whether they were exposed to substances (experimental) or not (controls) [35]. Before and during experiments tanks have been shielded so that movements in the test room did not distract fish. Experimental tanks were placed into a box that is not covered at the top. In general, adult zebrafish were exposed to a drug (ethanol at 0.5 % and 1 % ethanol v/v;) in the fish tank for one trial or for several trials (up to eight trials have been reported) [34, 35]. The exposure time varied between studies from 10 min

[86], 20 min [34] and 1 h [35]. In 1 h long recordings locomotor activity (swimming distance in cm/s) increased in sensitized animals compared to water treated controls [35]. Since brain ethanol levels in adults increased over the first 30 min of exposure and then remained constant [29], it seemed reasonable to employ 1 h long exposure trials and 1 h test trials. One exposure trial (induction phase) per animal was conducted each day during an 8-day trial period. To test if sensitization had occurred, locomotor activity (distance traveled) was determined over the entire 1-h test period on the following 2 days (once per day for 1 h; expression phase).

One study clearly demonstrated that ethanol-induced sensitization of adult zebrafish is not dependent on environmental cues but appears to be modulated by environmental cues [35]. Fish were individually placed in a standard 2 l tank (half all-back or half all-white) with or without 0.5 % or 1.0 % ethanol (v/v) for a 1-h period which was video recorded with a camera placed above the observation tank. The distance moved in 5 min intervals was determined to receive 12 data points over a 1-h observational period. Sensitization was concentration and context dependent and was recorded only in the same environment in which alcohol was administered [35, 38]. A 1.0 % but not 0.5 % concentration of ethanol caused sensitization independent of tank design (black or white) whether ethanol was paired or unpaired with a black or white design. In a control group, fish were repeatedly exposed to water instead of ethanol. Testing paradigms used for sensitization to ethanol could also be used for measuring sensitization to other drugs of addiction.

4.4 Conditioned Place Preference (with Extinction and Reinstatement)

Conditioned place preference (CPP) is the most widely used behavioral test for assessing reinforcing properties of a drug of addiction in adult zebrafish [11, 100, 101]. However, CPP is based on an associative learning model and does not measure the potential of a chemical as a drug of abuse per se [11]. Briefly, CPP can be established in three steps. First, a baseline preference for one compartment is established during a preconditioning phase whereby compartments provide different visual cues. Then, in the second step or conditioning phase, the zebrafish are restricted to the least preferred compartment and exposed to a drug (unconditioned stimulus). And finally in the third step, zebrafish have free access to all compartments and preference for individual compartments is recorded. A significant difference between baseline and conditioned preference indicates that the drug can modify the behavior of the animal, which associates a cue (conditioned stimulus) with the drug (unconditioned stimulus). Extinction of drug-induced CPP has been achieved by allowing zebrafish free access to all compartments in the absence of the drug during extinction training sessions. The preference for the conditioned compartment diminishes over several extinction training sessions and can be eventually abolished. Reinstatement of place preference has been achieved after extinction by exposing zebrafish in a separate compartment such as a beaker to the drug that was used for inducing CPP. When a fish has been returned to the CPP apparatus and was allowed free access to all chambers the fish can show a significant preference for the conditioned chamber.

Experimental Design and Analysis Most choice tank systems have been build in individual laboratories following a common design but with modifications. Visual cues have been used successfully for conditioning when testing cocaine, ethanol, nicotine, D-amphetamine, morphine, and the opiate salvinorin A [12, 15, 44, 77, 90–94]. Preconditioning and conditioning paradigms are very similar among experimental approaches, which differ mainly in the way drugs are applied, duration of habituation periods in the preconditioning phase and the number of repeated conditioning trials.

Design of Conditioning Tank (Fig. 1) Adult zebrafish naturally prefer a dark over a light environment [76]. However, a reduced contrast brown/white environment turned out to produce more robust conditioning than a high-contrast black/white-conditioning environment [46, 77]. Similar two chamber tank designs have been used for cocaine, D-amphetamine and morphine conditioning [12, 44, 77, 91]. The bottom and sides of the preferred (half) side of a rectangular tank (about 26 cm long, 20 cm wide, 20 cm deep; 2 l volume; water level 10–12 cm) have been covered with brown paper in two studies [77, 91]. The bottom and the sides of the avoided half of the tank were covered with white paper. In addition, two black discs were placed at the bottom of the avoided side. For cocaine conditioning, the two different compartments of the conditioning tank have been separated by a perforated divider that somewhat impedes the movement from one side of the tank to the other and the tank was placed into a white rectangular container [12]. The two sides of the divider used to restrict movements of the fish to one side had matching patterns such as brown color and white color or white with black dots or white with black stripes [44]. However, also a clear plastic divider has been used so that fish still have visual contact with the preferred compartment while being kept in the avoided compartment [77]. In modified designs of conditioning tanks, the sides of one half of a 2 l rectangular tank were covered with uniformly distributed 1.5 cm black spots whereas the sides of the other half of the tank were covered with alternating 0.5 cm wide black and white stripes [44] or a glass tank with transparent walls, one half of the bottom with a solid white and the other half with a dotted pattern (blue or black dots on white background) separated by a central grey or brown alley [15, 45, 93]. All described tank designs produced robust CPP. The design of conditioning tank systems can cause an initial preference (biased design) or no initial preference (unbiased design) and may be adjusted until desired experimental conditions have been achieved [77].

Authors	Design of test chamber	Schematics of test chambers
Darland and Dowling 2001	2 compartment; white rectangular container holding a greenish breeding tank insert paper on bottom and sides; perforated divider (see Fig. 1a in Darland and Dowling, 2001)	
Lau et al. 2006	3 compartments (bottom: one white, one white with black dots separated by gray central alley); clear glass walls; glass tank paced into rectangular box (Fig. 1a in Lau et al. 2006)	
Ninkovic and Bally-Cuif 2006	2 liter rectangular clear plastic tank placed in isolated white cabined; top illumination; water level 10 cm; 2 compartments with solid divider: one brown bottom and walls; one white bottom with two black spots on bottom (Fig. 2a in Ninkovic and Bally-Cuif 2006)	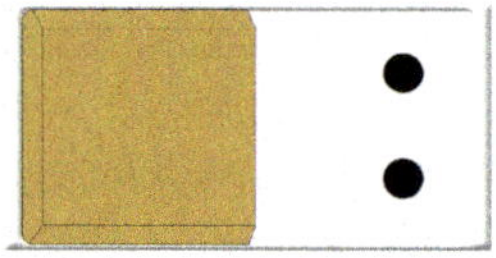
Braida et al. 2007	similar to Darland & Dowling (2001) with slight modifications (no Fig. in Braida et al. 2007)	
Kily et al. 2008	2 liter rectangular plastic tank; 2 compartments separated by solid divider; one white with black dots (1.5 cm in diameter) on bottom and walls; one with white and black vertical stripes (0.5 cm width) (see Fig. 3a in Parker and Brennan 2012)	
Mathur et al. 2011a,b	3 compartments (bottom: one white, one white with bluedots separated by browncentral alley); clear glass walls; glass tank paced into rectangular box (Fig. 1a in Lau et al. 2006)	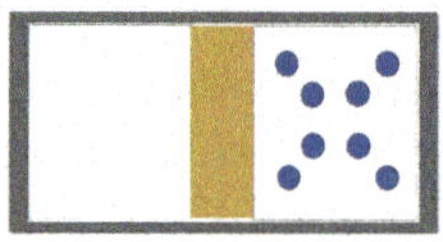

Fig. 1 Design of test chambers for conditioned place preference (CPP) in adult zebrafish

Authors	Design of test chamber	Schematics of test chambers
Darland et al. 2012	3 compartment (rear, middle, front); 2 same size compartments (rear, front) separated by larger (double-length) middle section/compartment; clear wall in front compartment; grey (duct-tape) walls in rear compartment and bottom; different lighting conditions in front and rear	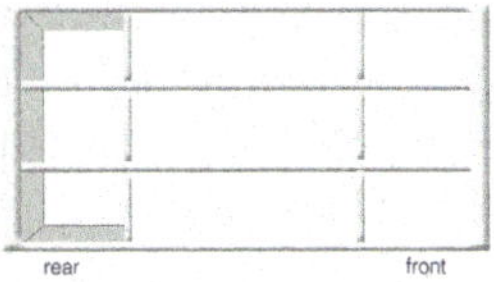
Kedikian et al. 2013	2 compartment; similar to Ninkovic & Bally-Cuif 2006 with modification; brown paper on bottom and sides & white paper on bottom and sides with black spots on bottom (see Fig. 2a Ninkovic & Bally-Cuif, 2006)	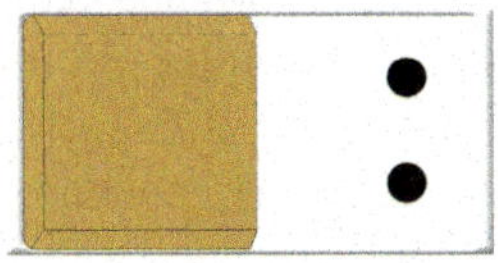
Ponzoni et al. 2014	same as Braida et al. 2007	

Fig. 1 (continued)

Habituation An accommodation phase of 2 days has been suggested. During this phase fish can adapt to the new tank environment before the baseline preference is determined [44, 77]. Since behavioral adaptation to a conditioning tank system may vary among studies, habituation to the novel environment could be assessed over a 5-day period to determine the optimal duration of preconditioning phase.

Baseline Place Preference Following the introduction of an adult zebrafish into the conditioning 2-l tank, animals have been allowed to adapt to the tank environment for at least 5 min in most studies. After this adaptation period, the baseline preference has been determined over a 2-min period during which the time spent in each compartment has been recorded. A change in preference has been presented either in absolute time (seconds) or as percentage relative to the baseline [12, 15, 44]. However, a minimum time of 15 min for measuring place preference has been suggested [77]. In this alternative approach, a fish was restricted first to the preferred side of a conditioning tank for about 20 min and then to the least-preferred side for 20 min. At this point, fish had been excluded from conditioning if they showed a high preference (>70 %) for one compartment or if fish showed excessive or deficient swimming activity. Between 10 and 25 % of zebrafish qualified for exclusion from conditioning in the precondition phase of the experiment [12, 77].

The design of chambers of the conditioning tank system had been modified if a high, unconditioned preference for one compartment persisted.

Conditioning To condition zebrafish, they were placed and restricted for 20–30 min into the least preferred compartment of the conditioning tank and the drug. Drugs that dissolve in water were added directly into the water of the compartment using different methods. For example, in one study they were either added using a "saturated wick" or by pipetting a small (10 ml) volume of a high concentrated stock solution to achieve a desired final concentration following dilution in the compartment [44]. For adult zebrafish the following final concentrations were used in conditioning trials: 0–1.5 % ethanol [44], 0–300 μM nicotine [44], 0–10 mg/l cocaine [12], 3 μM morphine sulfate [15] and 0.5 μg/kg (injected intramuscular) Salvinorin A [94]. Drugs that do not dissolve in water such as D-amphetamine have been applied by intraperitoneal injection of 40 μg D-Amphetamine per gram body weight and 3 μg methylene-blue per gram of body weight in 10 μl 110 mM NaCl [46, 77]. If fish were injected with a drug solution, they were anesthetized first for 1 min, then injected and then placed into the non-preferred compartment of the conditioning tank to recover for 45 min. For each drug and concentration 10–25 zebrafish have been conditioned across studies. Before animals were returned into their home tank after each conditioning trial, they were placed into a separate tank for rinsing off a drug if it was applied directly into the water. In most studies, one conditioning trial has been carried out per day.

Final place preference test trial: Following a single or more conditioning trials the place preference has been tested in the test trial during which zebrafish can access each compartment of the conditioning tank system freely in the same way as they could during baseline and baseline preference trials. The duration of the test trials depended on testing environments and conditioning paradigms and ranged from 2 to 15 min [15, 44, 77, 94]. A detailed method to determine the optimal duration of the preference test trial has been described [77]. If the drug induced a potential rewarding behavioral state, zebrafish spent more time in the compartment that was paired with the drug of addiction.

Control Groups When chemicals are administered into the water, a standard control group was given the same volume of water instead of a drug [12, 44]. In one study on nicotine conditioning, a counter-balanced nicotine group served as additional control and involved alternating nicotine-exposure to the white and light-brown compartment of the tank (trial 1 and 3 in light-brown compartment, trial 2 in white compartment; [91]). When chemicals were injected intraperitoneally, control animals were handled in the same way as drug injected fish and a 110 mM salt solution was injected intraperitoneally as control solution [46, 77]. For intramuscular injections into the caudal axial muscles animals received 0.0001–0.1 microgram nicotine per kilogram body weight in a volume of 2 μl per gram body weight [92] or 0.1–80 micrograms Salvinorin A per kilogram body weight in in a volume of 2 μl per gram body weight [94]. Under described experimental conditions, preference for the non-preferred or least-preferred compartment was established after one trial for nicotine or ethanol. Nicotine at a final concentration of around 30 μM was optimal for generating a preference for the treatment side after one single trial [44]. A single conditioning trial for ethanol (175 mM) was optimal for conditioning [44]. However, a significant change in preference for the treatment side was established after three conditioning trials. As mentioned above, no place preference data were recorded for up to 5 min immediately following placement of fish into the conditioning compartment or test tank on each conditioning trial and the test trial during which place preference is measured.

Modifications A drug-induced change in preference depends on several factors, which include variables of the testing environment, zebrafish used and properties of the drug of addiction, for example. Variations in conditioning paradigms between studies involved mainly the number of repeats of conditioning trials or the length of the conditioning phase. For example, nicotine and ethanol conditioning phases have included 1 trial [44], 3 trials over 3 days [44, 91], or 28 trials over a 4-week period (one trial/day) [44]. In one study, each conditioning trial included restriction of zebrafish to the preferred compartment before fish were conditioned to the drug in the non-preferred compartment [91].

A three-compartment chamber was used for CPP to cocaine [12]. The two test chambers or section (of same dimensions) flanked a longer section (same width, double length as test chambers/sections; Fig. 1). The volume of the entire three-chamber compartment was 1 l. Walls of one test chamber (rear) were wrapped with (grey) duct tape, whereas walls of the other chamber remained clear. For conditioning, adult zebrafish (6–8 months old) were treated with 10 mg/l cocaine for 45 min in the grey compartment. In this study no fish were excluded. In test trials, the time spent in the grey compartment after conditioning relative to the preference during baseline measurement was calculated. CPP could be established reliably in this set-up. One advantage of the three-chamber arrangement was that the number of entries and the time spend in the front or rear compartment could easily be recorded manually without the need of a tracking software. A three choice test has been described for amphetamine that was subsequently modified into an automated 5-choice test for impulse control and could be implemented for studying drugs of addiction [98, 102].

Dosing Dose–response curves have been generated for determining the optimal dose for the CPP in most studies. The shape of dose–response relationships of drugs of addiction is typically an inverted U-shape (or Gaussian or Bell-shape): behavioral responses to drugs of addiction are weak or not existent at low concentrations, increase with concentration until a maximum is reached and decrease at high concentrations that exceed the optimal concentration [77]. The dosing is critical as it has been shown in mice where lower doses produced CPP, higher doses produced conditioned place aversion (CPA) and doses in-between had no effect (for review [7]). Toxic or diminishing effects occur at high concentrations. In adult zebrafish concentrations that have caused a maximum response were 30 μM (5 mg/l; applied to water) for nicotine, 175 mM (1 %; applied to water) for ethanol, and 60 μg/g D-amphetamine (injected) [44, 77].

Validation Validation of a CPP in zebrafish has been approached in different ways including application of aversive stimuli to test if conditioning persists (punishment—[44]; aversion—[77]) and extinction and reinstatement experiments (for review: [11]). An extinction-reinstatement test could be associated with quitting-relapse behavior in humans. Using adult zebrafish in extinction trials, the conditioned stimulus (color/pattern of non-preferred compartment) was presented in the absence of the unconditioned stimulus (drug). Fish were placed for 20 min into the conditioning compartment to provide visual cues, but no drug was applied. After repeated extinction trials (once per day for 1–2 weeks), fish eventually did not associate the visual cue (color/pattern) with the drug and the CPP was extinct [95]. For testing reinstatement of CPP, zebrafish were exposed to the drug (unconditioned stimulus; for example, 0.5 or 1.0 % ethanol for 10 min) and then transferred to a tank without ethanol but with conditioned visual cues. If fish preferred the compartment with visual cues, CPP had been reinstated. One study reported conditioning of CPP with 1 % EtOH and extinction of CPP after 2 weeks of extinction training [95].

Conditioned place aversion (CPA) to a drug can develop at higher doses in mice whereas a lower concentration of the same drug produces CPP (for review: [7]). CPA has not been intensively studied in zebrafish. An exception is a study on salvinorin A. At concentrations of 0.2–0.5 μg/kg salvinorin A, CPP was established whereas a concentration of 80 μg/kg produced CPA [94]. Tricaine methanesulfonate (TMS) has been reported to produce CPA but not clove oil in a light/dark preference test [103]. CPP for nicotine and ethanol could be established despite an adverse stimulus in form of restriction to the "non-drug-treatment" compartment if zebrafish entered the conditioned "drug-treatment" compartment [44].

In addition, behavioral tests for vision, learning and memory and locomotion have been carried out to probe for potential adverse effects caused by drug treatments or for behavioral deficiencies of tested mutants that could potentially interfere with conditioning [12, 15, 77].

4.5 Self-Administration and Consumption of Drugs of Addiction

A good measure of reward in addiction research is the consumption of a drug [9]. One study has used gelatin-based food to measure the effect of voluntary ethanol intake on ethanol-use associated behavior [48]. First, adult zebrafish were fed with gelatin-containing brine shrimp. For ethanol consumption, ethanol (10 or 20 % w/v) was added to gelatin containing brine shrimp. The consumption of gelatin was measured by comparing total weight of gelatin-drops added to the tank and removed from tank after a 5- min feeding period. The gelatin contained 20 % ethanol. Using this method, blood alcohol levels raised to ~140 and ~160 mg/dl at 5 min and 15 min after the end of the 5 min feeding period, respectively. Thirty minutes after the feeding period, blood alcohol levels had decreased to ~60 mg/dl. Gelatin containing 10 % ethanol increased blood ethanol levels by about half the level that was measured when gelatin containing 20 % ethanol was fed. The 20 % ethanol-containing gelatin significantly increased locomotion by about 10–20 % and decreased the latency to reach the top in a novel tank diving test. Overall, the study demonstrated that voluntary ethanol intake could increase locomotion, reduce anxiety and increase aggression. The ethanol consumption method could potentially be used for studies of ethanol dependence in adult zebrafish.

Self-administration via intra-venous injection or injection into the brain in rodents reflects drug consumption in human closely and has been used widely for studies on addiction in mammals [9, 104]. However, no such self-administration protocol or method has been developed for adult zebrafish.

4.6 *Withdrawal*

Behavioral assays for withdrawal, which is a characteristic condition in humans caused by drugs of addiction, have been developed for rodents [7, 18, 99]. In adult zebrafish, withdrawal has been addressed in three ways: (1) removal of a substance after a chronic exposure period, (2) removal of a substance after repeated exposure [34] and (3) removal of a substance after CPP has been established [31]. Measurement of withdrawal overlapped somewhat with behavioral assays for anxiety, which represents a characteristic withdrawal symptom in mammals and humans [18, 31, 105]. In zebrafish, the novel tank test has been used for the measuring anxiety that occurs during withdrawal from drugs of addiction (ethanol, morphine, diazepam, caffeine: [31]; cocaine: [83]; ethanol: [34, 106]). In a novel tanks diving test, the following parameters have been measured: latency to reach the top half of the tank, time spent in the top, number of transitions to the top, number of erratic movements, and number and duration of freezing bouts [31]. Withdrawal symptoms are indicated if a fish spends more time at or close to the bottom of the observation tank. Also, an increased number of freezing bouts and erratic movements compared to controls have been interpreted as withdrawal symptoms in adult zebrafish [31, 97]. Withdrawal from chronic ethanol treatment initiated a number of behavioral changes such as shoaling behavior and swimming pattern and correlated with an increase in cortisol levels [18, 31, 81, 107].

4.7 *Drug Discovery Studies Using Adult Zebrafish*

To discover potential pharmacotherapeutic agents for the treatment of drug addiction, animals have been exposed to candidate chemical compounds such as neurotransmitter agonists or antagonist at some point of behavioral testing. In adult zebrafish, acute neurobehavioral effects of nicotine have been studied by co-administering antagonists for α_7 (Methyllycaconitine, 200 mg/l) or $\alpha_4\beta_2$ (Dihydro-β-erythroidine; 200 mg/l) nicotinic acetylcholine receptors together with nicotine (100 mg/l) for 3 min by immersion (followed by a 5 min water immersion without chemicals) before carrying out a novel tank diving test [88]. CPP could be prevented when sulpiride (10 μM) was co-administered with cocaine (10 mg/l) in conditioning trials [90]. A study on the function of the hallucinogen salvinorin A, antagonists of the CB1 receptor (rimonabant, SR 141716A) or antagonists of the k-opioid receptor (nor-BNI) were injected into the caudal musculature of zebrafish 15 min (rimonabant) or 2 h (nor-BNI) before salvinorin A was injected into the caudal musculature to show that antagonists block development of salvinorin A induced CPP [94]. The general opioid receptor antagonists naloxone was used to block morphine induced CPP [15]. After the initial baseline preference had been determined, adult zebrafish were exposed to 2.7 μM naloxone for 1 h outside the CPP chamber, before morphine-CPP was carried out.

4.8 Combination with Other Behavioral Tests

Described behavioral treatments and experiments (acute and chronic exposure, sensitization and conditioned place preference) have been combined with other behavioral assays to better characterize the activity of a drug of addiction. As mentioned above, the described novel tank diving test has been used as a measure of anxiety induced by drugs of addiction [33, 84, 97, 108–110]. Visual acuity and T-maze tests have been combined with conditioned place preference experiments for cocaine or morphine to control for memory functions or potential effects on the visual system [12, 15].

In adults the analysis of social behavior such as shoaling or group preferences, aggression, and antipredatory behavior has been used to find neurobehavioral changes caused by the treatment with drugs of addiction [50, 86]. Most of these studies measure behavioral changes to acute and chronic drug treatment.

Overall, adult zebrafish are excellent model organisms for the study of drugs of addiction as different actions of drugs can be probed with different behavioral paradigms. However, using behavioral assays for larval zebrafish could facilitate the discovery of pharmacotherapeutics.

5 Neurobehavioral Assays for Larval Zebrafish

Exploring reinforcing and addictive properties of substances using zebrafish larvae has certain advantages over working with adult zebrafish. Handling caused less stress in larval zebrafish than it does in adults [14, 28] and thus reduced the time for adaptation to new environments. Using early developmental stages (5–7 days post fertilization—5–7 dpf) facilitated the screening for gene mutations [17]. Moreover, the yolk sac is still providing nutrients until 7 dpf, which abolishes the need for feeding. Most IACUC protocols do not require specific handling until 8 dpf, when animals require feeding. Larvae younger than 14 dpf do not show shoaling behavior, which allows the observation and analysis of many larvae in a group instead of individual observations as in adults [49, 50]. Thus, screening larger number of fish as in high throughput screens is facilitated. Chemicals have been directly delivered into the water for uptake through the skin. The smaller size of larvae (2–10 mm) also facilitates the uptake of chemicals into the brain. Chemical compounds that are not soluble in water can be first dissolved in DMSO and applied to embryo water containing up to 1 % DMSO without causing any adverse effects. Oxygen is taken up across the skin until 12–14 dpf [111, 112]. The blood-barrier in larval zebrafish does not develop fully until 10 dpf thus facilitating the penetration of brain tissue [113]. The 5 dpf larval stage has been the earliest larval stage used in most neurobehavioral assays since the swim bladder is fully inflated and larvae show normal swimming patterns and a rich repertoire of other behavioral patterns [22, 27, 114, 115]. However, swimming behavior in zebrafish larvae is episodic and not as continuous as in adult zebrafish [116]. Moreover, neurotransmitter systems have matured at 5 dpf [56, 57, 59, 68, 69, 73, 117–119]. Most dopamine, serotonin, noradrenaline neurons have developed and most of their neurotransmitter receptor and transporter genes are expressed [19, 56, 57, 59–61, 67, 69, 120–123]. Since a single female can produce 200–800 eggs/week large numbers of progeny can be obtained without difficulty. Larvae are optically transparent which facilitates the use of optogenetic and calcium-imaging methods for neural network studies in intact animals [124–129]. Methods for genome modification such as TALENs and CRISPR based techniques have been established for zebrafish and produced new zebrafish lines in shorter time periods and at lower cost than in rodents [130–136]. However, compared to adult zebrafish, fewer studies have employed larvae for studies of drugs of addiction (Table 2).

5.1 Design of Testing Environments and Analysis

Behavioral experiments on zebrafish larvae have been carried out in Petri dishes or multi-well plates placed onto a light box or light panel. A camera has been used to take images of the top view. Behavioral testing environments such as the size of Petri dishes (35, 60, 100 mm in diameter), shape of plastic dishes or containers (round vs. rectangular), and 6–96 multi-well plates varied among studies. Locomotor activity of larvae does not appear to depend on the diameter of wells in 6, 12, 24 and 48 well pates but on the depth of wells which have been recommended to be at least 5.5 mm deep [137]. Less variation in experimental outcomes could be achieved by using older larvae (7 dpf vs. 4 dpf) and deeper wells. Space constrains in wells of 96-well plates affected locomotor activity of larvae [138–140].

With the commercial availability of automated recording systems such as ZebraBox (Viewpoint) and DanioVision (Noldus) that are designed for multi-well plates, behavioral assays and analyses are user friendly and somewhat standardized. For example, the Noldus Ethovision/DanioVision and Viewpoint ZebraLab software allow the analysis of variables such as distance traveled, and swimming speed from video recordings in real-time, but these systems are expensive. Systems designed by individual laboratories are less expensive and provide some flexibility of data analyses but may not have real-time data analysis capability, For example, LSRtrack is a MATLAB based program that allows the use of less expensive video camera systems [141, 142]. Additional custom-made viewing and analysis systems for zebrafish larvae have been developed [137, 143, 144]. Video analysis has also been developed for high-throughput chemical screens using larval zebrafish and embryos [23, 145] but are not addressed in this chapter because they are not employed specifically in addiction studies (for review: [146]).

5.2 *Acute and Chronic Responses to Drugs of Addiction in Larvae (Unconditioned Locomotor Response)*

Experimental Design and Analysis To measure locomotor activity in response to chemical compounds and drugs of addiction, chemical stock solutions were directly added to Petri dishes or wells in multi-well plates. Ethanol has been easily taken up through the skin of larval zebrafish if added to embryo water [28]. To assess the effects of ethanol, 10 larvae were placed into a rectangular viewing chamber and allowed to habituate for 5 min. Ethanol was added to the viewing chamber (containing larvae) to obtain a final concentration of 0.5, 1, 1.5, 2, or 4 %. Larvae could withstand higher concentrations of ethanol than adult zebrafish [28]. Then, larvae were exposed to nicotine for 20 min. During this time locomotor activity was video recorded. Swimming speed and location of larvae was analyzed at multiple time points (0, 1, 3, 5, 7, 10, 13, 16, 20 min) using a Dynamic Image Analysis System and plotted using Excel. Ethanol at concentrations of 1, 1.5 and 2 % caused a significant increase in locomotor activity (hyperactivity) in the acute test (20 min). However, no change in locomotor activity was recorded in response to 0.5 % ethanol. As in adult zebrafish, the dose–response relationship of ethanol in larvae had an inverted-U shape with the maximum locomotor activity at 1.5 % ethanol. A 4 % ethanol concentration caused hypoactivity and sedation. Methanol (1.5 %) was used as a control for ethanol induced-hyperactivity and did not show a significant change of locomotor activity. Described changes in locomotor activity defined as swimming speed were not immediate and occurred about 7 min after ethanol had been added to the viewing chamber. The delay in the response suggests that behavioral changes are not based on an effect of ethanol on the sensory system [28].

Nicotine elicited robust behavioral increase in locomotor activity in 5–7 dpf larvae and has been applied directly into the embryo water [17, 21, 25]. In these studies, 15–20 larvae were added to 60 mm Petri dishes that were placed onto a LED light panel. To increase the number of larvae screened three dishes with 15–20 larvae are routinely used in a single experiment for the measurement of the locomotor response to acute nicotine which is video captured [21]. “Alternatively, photo shots were taken at different time points (90 and 30 s before application of nicotine and 15, 30, 60, 90, 120, 270 and 300 s after addition of nicotine to Petri dishes) [25]". A single photo shot has been taken after 30 s of nicotine application for faster screening [17]. At each time point, two photo shots were taken in a 1.6 or 2.0 s interval. Locomotor behavior was then analyzed as percentage of larvae that had moved at each time point during the 1.6 or 2 s time interval [25]. This manual method could be performed quickly and required fewer resources than commercially available systems. Nicotine working concentrations in these studies ranged between 10 and 50 μM. For acute nicotine responses higher concentrations (50 μM) caused about 80 % of larvae to move within 1–2 min. For a 2-day treatment period (5–7 dpf), 10 μM nicotine has been used and did not cause adverse or toxic effects. Using the acute nicotine response, three zebrafish mutants with significantly reduced responses to nicotine have been identified and characterized [17]. Mutated genes were identified as a GABA B receptor gene (gababr1.2; hbog) and a chaperone containing protein 8 gene (cct8; bdav). Thus, the use of zebrafish larvae in behavior demonstrates the potential of this model organism for the isolation of new genes with functions in addiction biology.

A third example of measuring acute locomotor responses of zebrafish larvae to drug exposure used 96-well plates and light/dark challenges [82]. The assay took advantage of reduced locomotor activities of larvae in bright light compared to the dark. When light was turned off, zebrafish larvae began to move more (measured by distance traveled) until the light had been turned on again. Thus, alternating light and dark conditions were used to stimulate locomotor activity of treated and untreated zebrafish larvae. In this robust assay, solutions with chemicals were added to well plates, which were subsequently placed onto a light box for both infrared and visible light (Noldus). After a 10 min adaptation and incubation time, video recording of locomotor activity was started with only the infrared (dark) light on. Then light was turned on for 10 min. The dark–light challenge (10 min dark followed by 10 min light) was repeated two more times followed by a 10-min dark period at the end of the experiment. The entire recording phase was 70 min long. In this light–dark challenge approach, acute ethanol, D-amphetamine and cocaine produced robust dose-dependent locomotor responses. Both ethanol and D-amphetamine showed a typical inverted U-shaped dose response with larvae showing the largest increase in locomotor activity at 1 % ethanol and 0.7 μM D-amphetamine both in the dark and light phase of recording phases. However, inverted U-shaped dose response characteristics for a ethanol and D-amphetamine differed in the dark and light phase. D-amphetamine had inverted U-shaped dose–response kinetics in the dark but not in the light, whereas ethanol showed inverted U-shaped dose response kinetics in the light but not the dark phase. Acute cocaine application (0.2–50 μM) in this experimental set-up resulted in progressively decreasing locomotor activity (increasing hypo-locomotion) relative to controls (water) in both dark and light phases.

Pre-treatment for Drug Discovery Using Zebrafish Larvae Pretreatment of larvae with potential pharmacotherapeutic chemical compounds have been conducted on a larger number of larvae in a short time, thus facilitating high throughput screening. Moreover, less chemicals were needed because of the smaller volume in Petri dishes. Because of the smaller size of larvae and faster penetration of brain tissue, pretreatment of larvae has been short (about 1 h) but could also be conducted over a 24-h period [17, 21, 25]. Physicochemical properties of individual chemical compounds were used to determine the timing as well as dosing and required careful optimization of incubation times and chemical concentrations in the range of 10^{-6}–10^{-4} M. Concentrations of 10^{-6} M rarely caused any change in the acute response to a drug of addiction in larvae if applied directly to the water, while concentrations of 10^{-4} M were often toxic or diminished mobility if larvae were exposed for 24 h [25].

5.3 Sensitization

Sensitization is defined as an increased response to a substance following repeated application. Sensitization has been described for nicotine in 5–7 dpf zebrafish larvae, but not in earlier developmental stages (2–4 dpf) [17]. In the sensitization phase of the experiment, larvae were exposed to a chemical once or repeatedly for a short time period. In the test phase, the locomotor response to acutely applied drugs was measured. Control animals received water during the sensitization phase. For detecting sensitization, locomotor activity was measured as described in the previous section (acute locomotor responses). For the sensitization, 20 larvae were kept in a single 60 mm plastic Petri dish, exposed to 10 μM nicotine for 1 min and then returned to a clean 60 mm Petri dish with embryo water for about 8 h until the test phase. During the 8 h phase following a sensitization trial and before testing, larvae were kept in a 28 °C bottom-lit incubator. Petri dishes were kept at room temperature in the behavioral recording chamber during both the sensitization and test phase. Using this sensitization paradigm, the study showed that sensitization to nicotine occurred in 5, 6 and 7 dpf larvae but not in earlier developmental stages on days 2, 3 and 4 post fertilization [17]. About 10–30 % more larvae moved in response to nicotine compared to controls. The nicotinic acetylcholine receptor antagonist mecamylamine (10 μM for 4 h) significantly reduced the locomotor response to acute nicotine by 10–40 % in naive and 40–50 % in sensitized larvae [17]. The experimental approach was used for the identification of zebrafish mutants (bette davis (bdav), humphrey bogard (hbog) and yul brunner (yulb)) that develop only a reduced sensitization to nicotine [17].

5.4 Conditioned Place Preference (CPP)

A CPP test for drugs of abuse has not been established successfully in larval zebrafish. In a study concerning ethanol [28], thigmotaxis, or the preference for the outer edges of a viewing chamber [27] has been reported to change following acute ethanol exposure. However, thigmotaxis is not a measure of CPP. A significant difference between changes in thigmotaxis of treated and untreated larvae could be established. It is unknown if certain developmental factors could prevent CPP for drugs of addiction in early larvae. An associative place-conditioning test for 6–8 dpf larvae indicated that certain learning and memory functions occur at these early larval stages [147]. In the absence of larval CPPs for drugs of abuse, no measurements on extinction and reinstatement of CPP have been conducted either.

5.5 Self-Administration-Like and Choice

Although a classical self-administration assay as routinely used in rodents has not been developed for neither adults nor larvae, one study used a choice test in which zebrafish larvae could chose to immerse themselves in morphine-containing water or just water [47]. A flow chamber was developed for larval zebrafish that maintains one aquatic test tank compartment with morphine and one without morphine (water only) in this self-administration-like test. Both compartments were not separated by a barrier, which allowed free movement of larval zebrafish between compartments similar to a light/dark environment in the light/dark test tanks. Larvae were pretreated (or pre-exposed) by submerging them in a morphine sulfate solution (0.4–1.5 μM) for 1 h. On the next day, 10 pretreated larvae were transferred as a group to the choice tank without any morphine present in both chambers and allowed to adapt to the test tank during a habituation period of 5 min. Then a morphine sulfate solution (15 μM) was applied to one compartment of the test chamber and water flow was turned on in both chambers. The number of larvae in the morphine-containing compartment was counted every 10 s over a 10 min observational period resulting in 60 time points at which the number of larvae in the morphine containing section of the test tank was counted and added. If all larvae would be on the morphine side, the number counted would be 600 equaling 100 %. The percent preference was then calculated by subtracting the percentage of larvae on the water-only containing compartment from the morphine-containing compartment. For testing if chemicals can change the preference for morphine, larvae were first exposed to a chemical (such as an agonist or antagonist for dopamine receptors) for 1 h before the morphine solution was added to the same dish. The same test paradigm was also used for the application of food instead of morphine. The concentration of morphine was measured in the test compartment and an homogenate of about 400 morphine-exposed 2-week old larvae after a 1-h exposure period using HPLC. Morphine concentrations varied substantially between 0.02 and 0.8 μM in four quadrants of the test chamber into which morphine was added. Spontaneous swimming activity was not effected by 0.8 μM morphine. As control exposure and to rule out that chemical exposure alone generated a change in behavior, experiments were also carried out with morpholine (570 μM) instead of morphine and showed that morpholine does not cause reward-like effects. Two-week old larvae but not 1-week old larvae showed a trend towards a preference for the morphine-containing compartment. The self-immersion method has great potential for the development of a more robust assay in which concentrations of drugs are stable.

6 Conclusion and Outlook

The short history of zebrafish neurobehavioral studies on drugs of addiction has generated an impressive list of robust behavioral assays similar to those used in rodents. Studies on cocaine, D-amphetamine, ethanol, morphine and nicotine in adult zebrafish have demonstrated that conditioned place preference tests, for example, are effective in measuring drug induced neurobehavioral changes. Adult and larval zebrafish have been used successfully to screen for chemicals that change the neurobehavioral response of zebrafish to drugs of addiction.

However, the list of zebrafish neurobehavioral assays could be expanded to include a robust larval CPP, improved tests for withdrawal and relapse, self-administration-like tests for both larval and adult zebrafish, and other assays that are routinely used in mice [7]. Regardless of pharmacological studies on larval and adult zebrafish that indicate drugs are acting on similar molecular targets as in mice and humans [148] and that are supported by genomic data [19], pharmacological characterizations of most zebrafish neurotransmitter receptors and transporters still have to be carried out to advance translating results from zebrafish to mammals and humans. Moreover, drug assays for larval zebrafish can be easily scaled for screening of chemical libraries [23, 145, 149]. Straightforward genome modification techniques for the design of disease models represent a major advantage of the zebrafish model organism [131–133].

Thus, existing and future robust zebrafish neurobehavioral assays provide a strong foundation for studying molecular mechanisms of drugs of addiction and discovering new pharmacotherapeutics that will ultimately advance treatment of drug-use behavior.

References

1. Gipson CD, Kupchik YM, Kalivas PW. Rapid, transient synaptic plasticity in addiction. Neuropharmacology. 2014;76(Pt B):276–86.
2. Koob GF. Theoretical frameworks and mechanistic aspects of alcohol addiction: alcohol addiction as a reward deficit disorder. Curr Top Behav Neurosci. 2013;13:3–30.
3. Koob GF, Volkow ND. Neurocircuitry of addiction. Neuropsychopharmacology. 2010;35:217–38.
4. Muller CP, Homberg JR. The role of serotonin in drug use and addiction. Behav Brain Res. 2015;277:146–92.
5. Dalia A, Uretsky NJ, Wallace LJ. Dopaminergic agonists administered into the nucleus accumbens: effects on extracellular glutamate and on locomotor activity. Brain Res. 1998;788:111–7.
6. Gardner EL. Use of animal models to develop antiaddiction medications. Curr Psychiatry Rep. 2008;10:377–84.
7. Hall FS, Markou A, Levin ED, Uhl GR. Mouse models for studying genetic influences on factors determining smoking cessation success in humans. Ann N Y Acad Sci. 2012;1248:39–70.

8. Robinson TE, Berridge KC. Review. The incentive sensitization theory of addiction: some current issues. Philos Trans R Soc Lond B Biol Sci. 2008;363:3137–46.
9. Stephens DN, Duka T, Crombag HS, Cunningham CL, Heilig M, Crabbe JC. Reward sensitivity: issues of measurement, and achieving consilience between human and animal phenotypes. Addict Biol. 2010;15:145–68.
10. Tzschentke TM. Measuring reward with the conditioned place preference (CPP) paradigm: update of the last decade. Addict Biol. 2007;12:227–462.
11. Collier AD, Echevarria DJ. The utility of the zebrafish model in conditioned place preference to assess the rewarding effects of drugs. Behav Pharmacol. 2013;24:375–83.
12. Darland T, Dowling JE. Behavioral screening for cocaine sensitivity in mutagenized zebrafish. Proc Natl Acad Sci U S A. 2001;98:11691–6.
13. Gerlai R. A small fish with a big future: zebrafish in behavioral neuroscience. Rev Neurosci. 2011;22:3–4.
14. Guo S. Linking genes to brain, behavior and neurological diseases: what can we learn from zebrafish? Genes Brain Behav. 2004;3:63–74.
15. Lau B, Bretaud S, Huang Y, Lin E, Guo S. Dissociation of food and opiate preference by a genetic mutation in zebrafish. Genes Brain Behav. 2006;5:497–505.
16. Parker MO, Brennan CH. Zebrafish (*Danio rerio*) model of substance abuse: harnessing the capabilities. Behaviour. 2012;149:1037–62.
17. Petzold AM, Balciunas D, Sivasubbu S, Clark KJ, Bedell VM, Westcot SE, Myers SR, Moulder GL, Thomas MJ, Ekker SC. Nicotine response genetics in the zebrafish. Proc Natl Acad Sci U S A. 2009;106:18662–7.
18. Stewart A, Wong K, Cachat J, Gaikwad S, Kyzar E, Wu N, Hart P, Piet V, Utterback E, Elegante M, Tien D, Kalueff AV. Zebrafish models to study drug abuse-related phenotypes. Rev Neurosci. 2011;22:95–105.
19. Klee EW, Schneider H, Clark KJ, Cousin MA, Ebbert JO, Hooten WM, Karpyak VM, Warner DO, Ekker SC. Zebrafish: a model for the study of addiction genetics. Hum Genet. 2012;131:977–1008.
20. Stewart AM, Ullmann JF, Norton WH, Parker MO, Brennan CH, Gerlai R, Kalueff AV. Molecular psychiatry of zebrafish. Mol Psychiatry. 2015;20:2–17.
21. Cousin MA, Ebbert JO, Wiinamaki AR, Urban MD, Argue DP, Ekker SC, Klee EW. Larval zebrafish model for FDA-approved drug repositioning for tobacco dependence treatment. PLoS One. 2014;9:e90467.
22. Drapeau P, Saint-Amant L, Buss RR, Chong M, McDearmid JR, Brustein E. Development of the locomotor network in zebrafish. Prog Neurobiol. 2002;68:85–111.
23. Kokel D, Bryan J, Laggner C, White R, Cheung CY, Mateus R, Healey D, Kim S, Werdich AA, Haggarty SJ, Macrae CA, Shoichet B, Peterson RT. Rapid behavior-based identification of neuroactive small molecules in the zebrafish. Nat Chem Biol. 2010;6:231–7.
24. Schilling TF. The morphology of larval and adult zebrafish. In: Nüsslein-Volhard C, Dahm R, editors. Zebrafish. Oxford: Oxford University Press; 2002. p. 59–94.
25. Schneider H, Fritzky L, Williams J, Heumann C, Yochum M, Pattar K, Noppert G, Mock V, Hawley E. Cloning and expression of a zebrafish 5-HT(2C) receptor gene. Gene. 2012;502:108–17.
26. Gerlai R, Lee V, Blaser R. Effects of acute and chronic ethanol exposure on the behavior of adult zebrafish (*Danio rerio*). Pharmacol Biochem Behav. 2006;85:752–61.
27. Kalueff AV, Gebhardt M, Stewart AM, Cachat JM, Brimmer M, Chawla JS, Craddock C, Kyzar EJ, Roth A, Landsman S, Gaikwad S, Robinson K, Baatrup E, Tierney K, Shamchuk A, Norton W, Miller N, Nicolson T, Braubach O, Gilman CP, Pittman J, Rosemberg DB, Gerlai R, Echevarria D, Lamb E, Neuhauss SC, Weng W, Bally-Cuif L, Schneider H. Towards a comprehensive catalog of zebrafish behavior 1.0 and beyond. Zebrafish. 2013;10:70–86.
28. Lockwood B, Bjerke S, Kobayashi K, Guo S. Acute effects of alcohol on larval zebrafish: a genetic system for large-scale screening. Pharmacol Biochem Behav. 2004;77:647–54.

29. Tran S, Nowicki M, Chatterjee D, Gerlai R. Acute and chronic ethanol exposure differentially alters alcohol dehydrogenase and aldehyde dehydrogenase activity in the zebrafish liver. Prog Neuropsychopharmacol Biol Psychiatry. 2015;56:221–6.
30. Tran S, Nowicki M, Muraleetharan A, Chatterjee D, Gerlai R. Differential effects of acute administration of SCH-23390, a D1 receptor antagonist, and of ethanol on swimming activity, anxiety-related responses, and neurochemistry of zebrafish. Psychopharmacology (Berl). 2015;232:3709–18.
31. Cachat J, Canavello P, Elegante M, Bartels B, Hart P, Bergner C, Egan R, Duncan A, Tien D, Chung A, Wong K, Goodspeed J, Tan J, Grimes C, Elkhayat S, Suciu C, Rosenberg M, Chung KM, Kadri F, Roy S, Gaikwad S, Stewart A, Zapolsky I, Gilder T, Mohnot S, Beeson E, Amri H, Zukowska Z, Soignier RD, Kalueff AV. Modeling withdrawal syndrome in zebrafish. Behav Brain Res. 2010;208:371–6.
32. Dlugos CA, Rabin RA. Ethanol effects on three strains of zebrafish: model system for genetic investigations. Pharmacol Biochem Behav. 2003;74:471–80.
33. Egan RJ, Bergner CL, Hart PC, Cachat JM, Canavello PR, Elegante MF, Elkhayat SI, Bartels BK, Tien AK, Tien DH, Mohnot S, Beeson E, Glasgow E, Amri H, Zukowska Z, Kalueff AV. Understanding behavioral and physiological phenotypes of stress and anxiety in zebrafish. Behav Brain Res. 2009;205:38–44.
34. Mathur P, Guo S. Differences of acute versus chronic ethanol exposure on anxiety-like behavioral responses in zebrafish. Behav Brain Res. 2011;219:234–9.
35. Blaser RE, Koid A, Poliner RM. Context-dependent sensitization to ethanol in zebrafish (*Danio rerio*). Pharmacol Biochem Behav. 2010;95:278–84.
36. Tran S, Gerlai R. Time-course of behavioural changes induced by ethanol in zebrafish (*Danio rerio*). Behav Brain Res. 2013;252:204–13.
37. Tran S, Chatterjee D, Gerlai R. An integrative analysis of ethanol tolerance and withdrawal in zebrafish (*Danio rerio*). Behav Brain Res. 2015;276:161–70.
38. Tran S, Gerlai R. Recent advances with a novel model organism: alcohol tolerance and sensitization in zebrafish (*Danio rerio*). Prog Neuropsychopharmacol Biol Psychiatry. 2014;55:87–93.
39. Pignatelli M, Bonci A. Role of dopamine neurons in reward and aversion: a synaptic plasticity perspective. Neuron. 2015;86:1145–57.
40. Schultz W. Behavioral theories and the neurophysiology of reward. Annu Rev Psychol. 2006;57:87–115.
41. Steketee JD, Kalivas PW. Drug wanting: behavioral sensitization and relapse to drug-seeking behavior. Pharmacol Rev. 2011;63:348–65.
42. Tzschentke TM. Measuring reward with the conditioned place preference paradigm: a comprehensive review of drug effects, recent progress and new issues. Prog Neurobiol. 1998;56:613–72.
43. Braida D, Limonta V, Capurro V, Fadda P, Rubino T, Mascia P, Zani A, Gori E, Fratta W, Parolaro D, Sala M. Involvement of kappa-opioid and endocannabinoid system on salvinorin A-induced reward. Biol Psychiatry. 2008;63:286–92.
44. Kily LJ, Cowe YC, Hussain O, Patel S, McElwaine S, Cotter FE, Brennan CH. Gene expression changes in a zebrafish model of drug dependency suggest conservation of neuro-adaptation pathways. J Exp Biol. 2008;211:1623–34.
45. Mathur P, Lau B, Guo S. Conditioned place preference behavior in zebrafish. Nat Protoc. 2011;6:338–45.
46. Ninkovic J, Folchert A, Makhankov YV, Neuhauss SC, Sillaber I, Straehle U, Bally-Cuif L. Genetic identification of AChE as a positive modulator of addiction to the psychostimulant D-amphetamine in zebrafish. J Neurobiol. 2006;66:463–75.
47. Bretaud S, Li Q, Lockwood BL, Kobayashi K, Lin E, Guo S. A choice behavior for morphine reveals experience-dependent drug preference and underlying neural substrates in developing larval zebrafish. Neuroscience. 2007;146:1109–16.
48. Sterling ME, Karatayev O, Chang GQ, Algava DB, Leibowitz SF. Model of voluntary ethanol intake in zebrafish: effect on behavior and hypothalamic orexigenic peptides. Behav Brain Res. 2015;278:29–39.

49. Buske C, Gerlai R. Shoaling develops with age in zebrafish (*Danio rerio*). Prog Neuropsychopharmacol Biol Psychiatry. 2011;35:1409–15.
50. Gerlai R. Social behavior of zebrafish: from synthetic images to biological mechanisms of shoaling. J Neurosci Methods. 2014;234:59–65.
51. Miller N, Greene K, Dydinski A, Gerlai R. Effects of nicotine and alcohol on zebrafish (*Danio rerio*) shoaling. Behav Brain Res. 2013;240:192–6.
52. Bobzean SA, Dennis TS, Perrotti LI. Acute estradiol treatment affects the expression of cocaine-induced conditioned place preference in ovariectomized female rats. Brain Res Bull. 2014;103:49–53.
53. Lopez-Patino MA, Yu L, Yamamoto BK, Zhdanova IV. Gender differences in zebrafish responses to cocaine withdrawal. Physiol Behav. 2008;95:36–47.
54. Gerlai R, Chatterjee D, Pereira T, Sawashima T, Krishnannair R. Acute and chronic alcohol dose: population differences in behavior and neurochemistry of zebrafish. Genes Brain Behav. 2009;8:586–99.
55. Pan Y, Chatterjee D, Gerlai R. Strain dependent gene expression and neurochemical levels in the brain of zebrafish: focus on a few alcohol related targets. Physiol Behav. 2012;107:773–80.
56. Filippi A, Mahler J, Schweitzer J, Driever W. Expression of the paralogous tyrosine hydroxylase encoding genes th1 and th2 reveals the full complement of dopaminergic and noradrenergic neurons in zebrafish larval and juvenile brain. J Comp Neurol. 2010;518:423–38.
57. Kastenhuber E, Kratochwil CF, Ryu S, Schweitzer J, Driever W. Genetic dissection of dopaminergic and noradrenergic contributions to catecholaminergic tracts in early larval zebrafish. J Comp Neurol. 2010;518:439–58.
58. Schweitzer J, Driever W. Development of the dopamine systems in zebrafish. Adv Exp Med Biol. 2009;651:1–14.
59. Tay TL, Ronneberger O, Ryu S, Nitschke R, Driever W. Comprehensive catecholaminergic projectome analysis reveals single-neuron integration of zebrafish ascending and descending dopaminergic systems. Nat Commun. 2011;2:171.
60. Boehmler W, Carr T, Thisse C, Thisse B, Canfield VA, Levenson R. D4 Dopamine receptor genes of zebrafish and effects of the antipsychotic clozapine on larval swimming behaviour. Genes Brain Behav. 2007;6:155–66.
61. Boehmler W, Obrecht-Pflumio S, Canfield V, Thisse C, Thisse B, Levenson R. Evolution and expression of D2 and D3 dopamine receptor genes in zebrafish. Dev Dyn. 2004;230:481–93.
62. Holzschuh J, Ryu S, Aberger F, Driever W. Dopamine transporter expression distinguishes dopaminergic neurons from other catecholaminergic neurons in the developing zebrafish embryo. Mech Dev. 2001;101:237–43.
63. Higashijima S, Mandel G, Fetcho JR. Distribution of prospective glutamatergic, glycinergic, and GABAergic neurons in embryonic and larval zebrafish. J Comp Neurol. 2004;480:1–18.
64. Higashijima S, Masino MA, Mandel G, Fetcho JR. Engrailed-1 expression marks a primitive class of inhibitory spinal interneuron. J Neurosci. 2004;24:5827–39.
65. Kaslin J, Nystedt JM, Ostergard M, Peitsaro N, Panula P. The orexin/hypocretin system in zebrafish is connected to the aminergic and cholinergic systems. J Neurosci. 2004;24:2678–89.
66. Kaslin J, Panula P. Comparative anatomy of the histaminergic and other aminergic systems in zebrafish (*Danio rerio*). J Comp Neurol. 2001;440:342–77.
67. Lillesaar C, Stigloher C, Tannhauser B, Wullimann MF, Bally-Cuif L. Axonal projections originating from raphe serotonergic neurons in the developing and adult zebrafish, *Danio rerio*, using transgenics to visualize raphe-specific pet1 expression. J Comp Neurol. 2009;512:158–82.
68. McLean DL, Fetcho JR. Ontogeny and innervation patterns of dopaminergic, noradrenergic, and serotonergic neurons in larval zebrafish. J Comp Neurol. 2004;480:38–56.
69. McLean DL, Fetcho JR. Relationship of tyrosine hydroxylase and serotonin immunoreactivity to sensorimotor circuitry in larval zebrafish. J Comp Neurol. 2004;480:57–71.

70. Mueller T, Vernier P, Wullimann MF. The adult central nervous cholinergic system of a neurogenetic model animal, the zebrafish *Danio rerio*. Brain Res. 2004;1011:156–69.
71. Panula P, Chen Y-C, Priyadarshini M, Kudo H, Semenova S, Sundvik M, Sallinen V. The comparative neuroanatomy and neurochemistry of zebrafish CNS systems of relevance to human neuropsychiatric diseases. Neurobiol Dis. 2010;40:46–57.
72. Peitsaro N, Sundvik M, Anichtchik OV, Kaslin J, Panula P. Identification of zebrafish histamine H1, H2 and H3 receptors and effects of histaminergic ligands on behavior. Biochem Pharmacol. 2007;73:1205–14.
73. Rink E, Wullimann MF. Development of the catecholaminergic system in the early zebrafish brain: an immunohistochemical study. Brain Res Dev Brain Res. 2002;137:89–100.
74. MacPhail RC, Brooks J, Hunter DL, Padnos B, Irons TD, Padilla S. Locomotion in larval zebrafish: influence of time of day, lighting and ethanol. Neurotoxicology. 2009;30:52–8.
75. Burgess HA, Granato M. Modulation of locomotor activity in larval zebrafish during light adaptation. J Exp Biol. 2007;210:2526–39.
76. Serra EL, Medalha CC, Mattioli R. Natural preference of zebrafish (*Danio rerio*) for a dark environment. Braz J Med Biol Res. 1999;32:1551–3.
77. Ninkovic J, Bally-Cuif L. The zebrafish as a model system for assessing the reinforcing properties of drugs of abuse. Methods. 2006;39:262–74.
78. Stewart A, Cachat JM, Suciu C, Hart PC, Siddarth G, Utterback E, DiLeon J, Kalueff AV. Intraperitoneal injection as a method of psychotropic drug delivery in adult zebrafish. In: Kalueff AV, Cachat JM, editors. Zebrafish neurobehavioral protocols, vol. 51. New York: Humana Press/Springer; 2010. p. 169–79.
79. Murphey RD, Stern HM, Straub CT, Zon LI. A chemical genetic screen for cell cycle inhibitors in zebrafish embryos. Chem Biol Drug Des. 2006;68:213–9.
80. Usenko CY, Harper SL, Tanguay RL. In vivo evaluation of carbon fullerene toxicity using embryonic zebrafish. Carbon N Y. 2007;45:1891–8.
81. Echevarria DJ, Toms CN, Jouandot DJ. Alcohol-induced behavior change in zebrafish models. Rev Neurosci. 2011;22:85–93.
82. Irons TD, MacPhail RC, Hunter DL, Padilla S. Acute neuroactive drug exposures alter locomotor activity in larval zebrafish. Neurotoxicol Teratol. 2010;32:84–90.
83. Lopez-Patino MA, Yu L, Cabral H, Zhdanova IV. Anxiogenic effects of cocaine withdrawal in zebrafish. Physiol Behav. 2008;93:160–71.
84. Levin ED, Bencan Z, Cerutti DT. Anxiolytic effects of nicotine in zebrafish. Physiol Behav. 2007;90:54–8.
85. Stewart AM, Grossman L, Collier AD, Echevarria DJ, Kalueff AV. Anxiogenic-like effects of chronic nicotine exposure in zebrafish. Pharmacol Biochem Behav. 2015;139(Pt B):112–20.
86. Gerlai R, Lahav M, Guo S, Rosenthal A. Drinks like a fish: zebra fish (*Danio rerio*) as a behavior genetic model to study alcohol effects. Pharmacol Biochem Behav. 2000;67:773–82.
87. Levin ED, Chen E. Nicotinic involvement in memory function in zebrafish. Neurotoxicol Teratol. 2004;26:731–5.
88. Bencan Z, Levin ED. The role of alpha7 and alpha4beta2 nicotinic receptors in the nicotine-induced anxiolytic effect in zebrafish. Physiol Behav. 2008;95:408–12.
89. Stewart AM, Kalueff AV. The behavioral effects of acute Δ^9-tetrahydrocannabinol and heroin (diacetylmorphine) exposure in adult zebrafish. Brain Res. 2014;1543:109–19.
90. Darland T, Mauch JT, Meier EM, Hagan SJ, Dowling JE, Darland DC. Sulpiride, but not SCH23390, modifies cocaine-induced conditioned place preference and expression of tyrosine hydroxylase and elongation factor 1alpha in zebrafish. Pharmacol Biochem Behav. 2012;103:157–67.
91. Kedikian X, Faillace MP, Bernabeu R. Behavioral and molecular analysis of nicotine-conditioned place preference in zebrafish. PLoS One. 2013;8:e69453.
92. Ponzoni L, Braida D, Pucci L, Andrea D, Fasoli F, Manfredi I, Papke RL, Stokes C, Cannazza G, Clementi F, Gotti C, Sala M. The cytisine derivatives, CC4 and CC26, reduce nicotine-

induced conditioned place preference in zebrafish by acting on heteromeric neuronal nicotinic acetylcholine receptors. Psychopharmacology (Berl). 2014;231:4681–93.
93. Mathur P, Berberoglu MA, Guo S. Preference for ethanol in zebrafish following a single exposure. Behav Brain Res. 2011;217:128–33.
94. Braida D, Limonta V, Pegorini S, Zani A, Guerini-Rocco C, Gori E, Sala M. Hallucinatory and rewarding effect of salvinorin A in zebrafish: kappa-opioid and CB1-cannabinoid receptor involvement. Psychopharmacology (Berl). 2007;190:441–8.
95. Parmar A, Parmar M, Brennan CH. Zebrafish conditoned place preference models of drug reinforcement and relapse to drug seeking. In: Kalueff AV, Cachat JM, editors. Zebrafish neurobehavioral protocols, vol. 51. New York: Springer/Humana Press; 2011. p. 75–84.
96. Crosby EB, Bailey JM, Oliveri AN, Levin ED. Neurobehavioral impairments caused by developmental imidacloprid exposure in zebrafish. Neurotoxicol Teratol. 2015;49:81–90.
97. Cachat J, Stewart A, Grossman L, Gaikwad S, Kadri F, Chung KM, Wu N, Wong K, Roy S, Suciu C, Goodspeed J, Elegante M, Bartels B, Elkhayat S, Tien D, Tan J, Denmark A, Gilder T, Kyzar E, Dileo J, Frank K, Chang K, Utterback E, Hart P, Kalueff AV. Measuring behavioral and endocrine responses to novelty stress in adult zebrafish. Nat Protoc. 2010;5:1786–99.
98. Parker MO, Ife D, Ma J, Pancholi M, Smeraldi F, Straw C, Brennan CH. Development and automation of a test of impulse control in zebrafish. Front Syst Neurosci. 2013;7:65.
99. Brennan CH. Zebrafish behavioural assays of translational relevance for the study of psychiatric disease. Rev Neurosci. 2011;22:37–48.
100. Brennan CH, Parmar A, Kily LKM, Ananthathevan A, Doshi A, Patel S, Kaleuff AV. Conditioned place preference models of drug dependence and relapse to drug seeking: studies with nicotine and ethanol. Neuromethods. 2011;52:163–80.
101. Collier AD, Khan KM, Caramillo EM, Mohn RS, Echevarria DJ. Zebrafish and conditioned place preference: a translational model of drug reward. Prog Neuropsychopharmacol Biol Psychiatry. 2014;55:16–25.
102. Parker MO, Millington ME, Combe FJ, Brennan CH. Development and implementation of a three-choice serial reaction time task for zebrafish (*Danio rerio*). Behav Brain Res. 2012;227:73–80.
103. Wong D, von Keyserlingk MA, Richards JG, Weary DM. Conditioned place avoidance of zebrafish (*Danio rerio*) to three chemicals used for euthanasia and anaesthesia. PLoS One. 2014;9:e88030.
104. Thomsen M, Caine SB. Intravenous drug self-administration in mice: practical considerations. Behav Genet. 2007;37:101–18.
105. Blaser RE, Rosemberg DB. Measures of anxiety in zebrafish (*Danio rerio*): dissociation of black/white preference and novel tank test. PLoS One. 2012;7:e36931.
106. Tran S, Nowicki M, Muraleetharan A, Chatterjee D, Gerlai R. Neurochemical factors underlying individual differences in locomotor activity and anxiety-like behavioral responses in zebrafish. Prog Neuropsychopharmacol Biol Psychiatry. 2016;65:25–33.
107. Tran S, Chatterjee D, Gerlai R. Acute net stressor increases whole-body cortisol levels without altering whole-brain monoamines in zebrafish. Behav Neurosci. 2014;128:621–4.
108. Blaser R, Gerlai R. Behavioral phenotyping in zebrafish: comparison of three behavioral quantification methods. Behav Res Methods. 2006;38:456–69.
109. Blaser RE, Chadwick L, McGinnis GC. Behavioral measures of anxiety in zebrafish (*Danio rerio*). Behav Brain Res. 2010;208:56–62.
110. Stewart A, Gaikwad S, Kyzar E, Green J, Roth A, Kalueff AV. Modeling anxiety using adult zebrafish: a conceptual review. Neuropharmacology. 2012;62:135–43.
111. Pelster B, Sanger AM, Siegele M, Schwerte T. Influence of swim training on cardiac activity, tissue capillarization, and mitochondrial density in muscle tissue of zebrafish larvae. Am J Physiol Regul Integr Comp Physiol. 2003;285:R339–47.
112. Schwerte T. Cardio-respiratory control during early development in the model animal zebrafish. Acta Histochem. 2009;111:230–43.

113. Fleming A, Diekmann H, Goldsmith P. Functional characterisation of the maturation of the blood–brain barrier in larval zebrafish. PLoS One. 2013;8:e77548.
114. Budick SA, O'Malley DM. Locomotor repertoire of the larval zebrafish: swimming, turning and prey capture. J Exp Biol. 2000;203:2565–79.
115. Colwill RM, Creton R. Locomotor behaviors in zebrafish (*Danio rerio*) larvae. Behav Processes. 2011;86:222–9.
116. Wiggin TD, Anderson TM, Eian J, Peck JH, Masino MA. Episodic swimming in the larval zebrafish is generated by a spatially distributed spinal network with modular functional organization. J Neurophysiol. 2012;108:925–34.
117. Brustein E, Chong M, Holmqvist B, Drapeau P. Serotonin patterns locomotor network activity in the developing zebrafish by modulating quiescent periods. J Neurobiol. 2003;57:303–22.
118. Buss RR, Drapeau P. Synaptic drive to motoneurons during fictive swimming in the developing zebrafish. J Neurophysiol. 2001;86:197–210.
119. McDearmid JR, Drapeau P. Rhythmic motor activity evoked by NMDA in the spinal zebrafish larva. J Neurophysiol. 2005;95:401–17.
120. Holzschuh J, Barrallo-Gimeno A, Ettl AK, Durr K, Knapik EW, Driever W. Noradrenergic neurons in the zebrafish hindbrain are induced by retinoic acid and require tfap2a for expression of the neurotransmitter phenotype. Development. 2003;130:5741–54.
121. Lillesaar C. The serotonergic system in fish. J Chem Neuroanat. 2011;41:294–308.
122. Norton WHJ, Folchert A, Bally-Cuif L. Comparative analysis of serotonin receptor (HTR1A/HTR1B families) and transporter (slc6a4a/b) gene expression in the zebrafish brain. J Comp Neurol. 2008;511:521–42.
123. Schneider H, Klee EW, Clark KJ, Petzold AM, Mock VL, Abarr JM, Behrens JL, Edelen RE, Edwards BA, Hobgood JS, Pogue ME, Singh NK, Ekker SC. Zebrafish and drug development: a behavioral assay system for probing nicotine function in larval zebrafish. In: Kaleuff AV, Smith AJ, editors. Zebrafish neurobehavioral protocols, vol. 2. New York: Springer; 2012.
124. Ahrens MB, Li JM, Orger MB, Robson DN, Schier AF, Engert F, Portugues R. Brain-wide neuronal dynamics during motor adaptation in zebrafish. Nature. 2012;485:471–7.
125. McLean DL. Optogenetics: illuminating sources of locomotor drive. Curr Biol. 2013;23:R441–3.
126. Portugues R, Severi KE, Wyart C, Ahrens MB. Optogenetics in a transparent animal: circuit function in the larval zebrafish. Curr Opin Neurobiol. 2013;23:119–26.
127. Simmich J, Staykov E, Scott E. Zebrafish as an appealing model for optogenetic studies. Prog Brain Res. 2012;196:145–62.
128. Vladimirov N, Mu Y, Kawashima T, Bennett DV, Yang CT, Looger LL, Keller PJ, Freeman J, Ahrens MB. Light-sheet functional imaging in fictively behaving zebrafish. Nat Methods. 2014;11:883–4.
129. Wyart C, Del Bene F, Warp E, Scott EK, Trauner D, Baier H, Isacoff EY. Optogenetic dissection of a behavioural module in the vertebrate spinal cord. Nature. 2009;461:407–10.
130. Auer TO, Del Bene F. CRISPR/Cas9 and TALEN-mediated knock-in approaches in zebrafish. Methods. 2014;69:142–50.
131. Bedell VM, Ekker SC. Using engineered endonucleases to create knockout and knockin zebrafish models. Methods Mol Biol. 2015;1239:291–305.
132. Bedell VM, Wang Y, Campbell JM, Poshusta TL, Starker CG, Krug 2nd RG, Tan W, Penheiter SG, Ma AC, Leung AY, Fahrenkrug SC, Carlson DF, Voytas DF, Clark KJ, Essner JJ, Ekker SC. In vivo genome editing using a high-efficiency TALEN system. Nature. 2012;491:114–8.
133. Blackburn PR, Campbell JM, Clark KJ, Ekker SC. The CRISPR system--keeping zebrafish gene targeting fresh. Zebrafish. 2013;10:116–8.
134. Gonzales AP, Yeh JR. Cas9-based genome editing in zebrafish. Methods Enzymol. 2014;546:377–413.

135. Hwang WY, Fu Y, Reyon D, Gonzales AP, Joung JK, Yeh JR. Targeted mutagenesis in zebrafish using CRISPR RNA-guided nucleases. Methods Mol Biol. 2015;1311:317–34.
136. Neff KL, Argue DP, Ma AC, Lee HB, Clark KJ, Ekker SC. Mojo Hand, a TALEN design tool for genome editing applications. BMC Bioinf. 2013;14:1.
137. Ingebretson JJ, Masino MA. Quantification of locomotor activity in larval zebrafish: considerations for the design of high-throughput behavioral studies. Front Neural Circuits. 2013;7:109.
138. Farrell TC, Cario CL, Milanese C, Vogt A, Jeong JH, Burton EA. Evaluation of spontaneous propulsive movement as a screening tool to detect rescue of Parkinsonism phenotypes in zebrafish models. Neurobiol Dis. 2011;44:9–18.
139. Padilla S, Hunter DL, Padnos B, Frady S, MacPhail RC. Assessing locomotor activity in larval zebrafish: influence of extrinsic and intrinsic variables. Neurotoxicol Teratol. 2011;33:624–30.
140. Selderslaghs IWT, Hooyberghs J, De Coen W, Witters HE. Locomotor activity in zebrafish embryos: a new method to assess developmental neurotoxicity. Neurotoxicol Teratol. 2010;32:460–71.
141. Cario CL, Farrell TC, Milanese C, Burton EA. Automated measurement of zebrafish larval movement. J Physiol. 2011;589:3703–8.
142. Zhou Y, Cattley RT, Cario CL, Bai Q, Burton EA. Quantification of larval zebrafish motor function in multiwell plates using open-source MATLAB applications. Nat Protoc. 2014;9:1533–48.
143. Creton R. Automated analysis of behavior in zebrafish larvae. Behav Brain Res. 2009;203:127–36.
144. Pelkowski SD, Kapoor M, Richendrfer HA, Wang X, Colwill RM, Creton R. A novel high-throughput imaging system for automated analyses of avoidance behavior in zebrafish larvae. Behav Brain Res. 2011;223:135–44.
145. Rihel J, Prober DA, Arvanites A, Lam K, Zimmerman S, Jang S, Haggarty SJ, Kokel D, Rubin LL, Peterson RT, Schier AF. Zebrafish behavioral profiling links drugs to biological targets and rest/wake regulation. Science. 2010;327:348–51.
146. Mathias JR, Saxena MT, Mumm JS. Advances in zebrafish chemical screening technologies. Future Med Chem. 2012;4:1811–22.
147. Hinz FI, Aizenberg M, Tushev G, Schuman EM. Protein synthesis-dependent associative long-term memory in larval zebrafish. J Neurosci. 2013;33:15382–7.
148. Goldsmith P. Zebrafish as a pharmacological tool: the how, why and when. Curr Opin Pharmacol. 2004;4:504–12.
149. Peterson RT. Discovery of therapeutic targets by phenotype-based zebrafish screens. Drug Discov Today Technol. 2004;1:49–54.

Zebrafish Neurotoxicity Models

Julian Pittman

Abstract The field of neurotoxicology is confronted with two significant demands: the testing of an ever increasing list of chemicals, and resource limitations/ethical concerns associated with testing using traditional mammalian species. National and international government agencies have well-defined a need to reduce, refine or replace mammalian species in toxicological testing with alternative testing methods and non-mammalian models. Toxicological assays using alternative animal models may relieve some of this pressure by allowing testing of more compounds while reducing expense and using fewer mammals. Recent advances in genetic technologies and the strong conservation between human and non-mammalian genomes allow for the dissection of the molecular pathways involved in neurotoxicological responses using genetically tractable organisms such as zebrafish (*Danio rerio*). A constantly increasing database on basic developmental biology, gene transfers, and the rich foundation of molecular genetic and genomic data make zebrafish a powerful modeling system for revealing mechanisms in neurotoxicology. In contrast to the highly advanced knowledge base on molecular developmental genetics in zebrafish, the databases regarding pathologic lesions in zebrafish lag far behind the information available on most other domestic mammalian and avian species, particularly rodents. Therefore, to fully utilize the potential of zebrafish as an animal model for understanding neurotoxicological influences on human development and disease we must greatly advance our knowledge on zebrafish diseases and pathology.

Keywords Cell culture • Omics • Neuro-assays • Neurobehavioral • Toxicokinetics

1 Introduction

As the most numerous and phylogenetically diverse group of vertebrates, fish possess the power to teach us important principles about fundamental processes in vertebrate evolution, development and disease. Fish have served as useful sentinels to detect environmental hazards and as efficient, cost-effective model systems for

J. Pittman (✉)
Department of Biological and Environmental Sciences, Troy University,
210 McCall Hall, Troy, AL 36082, USA
e-mail: jtpittman@troy.edu

A.V. Kalueff (ed.), *The rights and wrongs of zebrafish: Behavioral phenotyping of zebrafish*, DOI 10.1007/978-3-319-33774-6_9

mechanistic toxicology and risk assessment for many decades [1–5]. In choosing a model system for conducting research, it is essential to appreciate that no single model is best for addressing all biomedical questions. Each model species has unique strengths and weaknesses [1]. Because of their size, zebrafish embryos and early larvae can be raised in only 100 mL of water in the wells of a 96-well plate for high-throughput whole-animal assays requiring only small amounts of compounds. Drug administration is also simple because researchers can dissolve small-molecule compounds in the water, where they diffuse into the embryos. Alternatively, researchers can microinject larger molecules, such as proteins, directly into embryos. To knock down specific genes for target validation, morpholino antisense molecules can be injected into one- or two-cell-stage embryos, resulting in uniform distribution of the oligonucleotides across the embryos in several days. Most human genes have homologues in zebrafish, and the functional domains of proteins, such as ATP-binding domains of kinases, are almost 100 % identical between homologous genes; although the similarity over the entire protein is only about 60 %. Because protein function largely resides in functional domains where drugs/toxicants often bind, the zebrafish is a highly valid model for studying gene function and drug effects in humans [6]. Indeed, many zebrafish versions of mammalian genes have been cloned and found to have similar functions, and numerous drugs tested in zebrafish have caused effects similar to those observed in humans or other mammalian models [7].

The zebrafish is the only vertebrate species for which large-scale forward genetic screens have been carried out, and many mutants obtained from these genetic screens display phenotypes that mimic human disorders, such as neurodegeneration [8, 9]. These mutants not only identify genes that may be involved in diseases but also can be used for drug/toxicant screening. Zebrafish small-molecule screening takes advantage of the small size, chemical permeability, and optical transparency of the zebrafish embryo [10]. Transgenic lines expressing fluorescent proteins in specific neuronal subpopulations have also been developed, which can facilitate screening. Cardiotoxicity is the most thoroughly tested zebrafish toxicity to date, but neurotoxicity (central nervous system) is rapidly increasing [4, 11–14].

2 Applications for Zebrafish in Neurotoxicology

Zebrafish have been shown to be amenable to high-throughput screening in applications such as small-molecule drug discovery and neurotoxicology [15]. Embryonic and larval stages of the zebrafish can be grown in 96- or 384-well assay plates, exposed to small molecules by adding the compounds to the water in the wells, and the effects can be observed in the transparent embryos using microscopy. The first small molecule screens performed employed wild-type zebrafish and visual screening to identify obvious morphological defects [16, 17]. These screens identified defects in numerous organ systems including the central nervous system. Phenotypes identified in this way were generally severe; for the CNS, phenotypes ranged from loss or expansion of brain ventricles, to truncation of the telencephalon, to severe neuronal necrosis. While these studies demonstrate the ability of this approach to identify small molecules that cause severe developmental neurotoxicities, it is doubtful that such screens could reliably identify subtle neurotoxicities that are not manifest in obvious morphological changes. More sophisticated assays will be required.

One such assay involves vital dyes, like acridine orange, which have been reported to stain apoptotic cells in zebrafish and may help detect subtle neurotoxicities [4, 18]. Transgenic lines expressing fluorescent proteins in specific neuronal subpopulations have also been developed, which could facilitate screening. Numerous functional and behavioral assays, including assays of vision, hearing, touch responsiveness, memory, anxiety, and startle habituation have been developed and could also be useful for identifying neurotoxicants that do not cause obvious developmental phenotypes [19–22]. It is possible that a panel of several high-throughput morphological and functional assays could be used to screen broadly for neurotoxicants. Nevertheless, increasing the number and sophistication of high-throughput neuronal assays for zebrafish will be of little value if zebrafish and human neurotoxicities do not correlate. Much work remains to be done to determine the extent to which zebrafish toxicities are predictive, but initial data from other organ systems are encouraging. As mentioned above, cardiotoxicity is one of the most thoroughly tested zebrafish toxicity to date. In an assay for drug-induced bradycardia, 22 of 23 compounds known to cause human QT prolongation were detected among 100 tested compounds, suggesting a high degree of correlation between zebrafish and human cardiotoxicity [23]. Similar types of studies focused on neurotoxicity would be particularly useful, but have not been reported to date. However, some individual compounds have been reported to have predictable neurotoxicities in zebrafish, including ethanol, 6-hydroxydopamine, acrylamide, MPTP, and pentylenetetrazole [4, 13, 18, 24].

Beyond the use of screening for neurotoxicants, zebrafish may have great utility with regard to high-throughput platforms for identifying neuroprotectants. Zebrafish screens have been used to discover novel compounds that suppress the effects of a genetic vascular defect. Similar screens have discovered a small molecule that suppresses the effects of a mutation that causes a cell cycle defect in zebrafish. This approach could be applied to neuroprotection by exposing thousands of zebrafish en masse to a neurotoxicant, then screening in high-throughput assays for novel small molecules that block the neurotoxic effects of the toxicant. As preliminary evidence that such an approach may be feasible, several known neuroprotectants have been shown to protect zebrafish from L-hydroxyglutaric acid neurotoxicity [4], and in a separate study, l-deprenyl and nomifensine were shown to protect zebrafish from MPTP-induced neurotoxicity [13].

With the dramatic rise in the number of potential, but poorly validated targets and preliminary hit compounds, small-animal models are increasingly important for validating targets and profiling hits. Although several model systems exist, each with its own advantages, zebrafish can bridge the gap between invertebrate and mammalian models. Wider adoption of this small-vertebrate model organism in drug discovery research, and of neurotoxicity testing in particular, could help accelerate the drug development process.

3 Distinctive Advantages of the Zebrafish Model

Zebrafish have distinct advantages as models for biomedical research, beyond the well-known lower husbandry costs as compared to mammals. Zebrafish have external fertilization and development, facilitating access for observation and manipulation of developing embryos. Zebrafish are easily housed in compact recirculating systems, breed continuously year-round, and have short generation times of approximately 3–5 months [25]. The small size of adult zebrafish allows efficient, low-cost evaluation of all major organs on a limited number of slides [26]. The small size of embryos and fry minimizes the cost and waste volume for drug/toxicant studies. Consequently, minute amounts of expensive metabolites or new targeted drugs/toxicants can be rapidly evaluated.

Among vertebrates, the zebrafish embryo has unrivaled optical clarity, allowing visual tracing of individual cells [27–29]. If inhibitors of pigment formation are included in the rearing medium, cell lineages can be traced throughout the first week of zebrafish development, and immunohistochemical or immunofluorescence studies will reveal specific cell types in whole mount preparations. Histological sections of larvae over 1 week of age are required to localize specific cell types. Alternatively, confocal microscopy can optically section these thicker larvae [30]. A wide array of histochemical markers for protein and gene expression allows identification of essentially all cell types, and often reveals functional capabilities such as synthesis of nitric oxide or specific neurotransmitters during development of the major tissues [7, 31–37].

Immunohistochemical studies in zebrafish have quantified cell proliferation and cell death in specific tissues during development [37, 38]. The optical clarity of the embryo coupled with detailed understanding of basic developmental processes and a well-established timetable for specific developmental milestones allows elegant embryonic manipulations to distinguish the relative influences of the genetic composition of a specific cell (cell autonomous effects) versus the influences of the genetic suite of its surrounding environment (non-cell autonomous effects). For instance, at a precise stage of development, specific neurons can be removed from the spinal cord using a micropipette and can be replaced by those from fish of a different genotype. Then the impact on neuronal fate and innervation of skeletal muscle can be determined [39]. Or during various time points in development, single cells or cell clusters can be removed from specific anatomic fields in the embryo and relocated to other sites to clarify the processes controlling cell fate determination, and reveal when the fate of certain cell types is irreversibly specified [40, 41]. DNA or RNA constructs can be readily microinjected into embryos at the 1-cell or 2-cell stages to study effects of transient gene expression. More uniform tissue expression can be achieved with RNA injection. With injection of RNA at the 2-cell stage, typically half of the embryo expresses the exogenous construct, with the other half acting as an internal control. Using constructs with a promoter such as that from a heat shock gene, laser probes can elicit transient expression of injected constructs in precise cell types at exact stages of development [42]. The past decade has seen intense research into molecular genetic mechanisms in cell fate determination, pattern formation, morphogenesis and functional maturation of heart, blood vessel, brain, eye, ear, nose, neural crest, muscle, cartilage, bone, skin, kidney, and gonad.

4 Nervous and Sensory Organ System Sensitivity

Zebrafish are susceptible to several mammalian neurotoxins, including dopaminergic neurotoxins, NDMA receptor antagonists, non-NMDA type glutamate receptor agonists or antagonists, and nicotinic acetylcholine receptor antagonists or acetylcholinesterase inhibitors [3]. As in mammals, neurotoxicity can be assessed at the biochemical, morphological (e.g., ototoxicity) and behavioral (e.g., locomotion) levels in zebrafish. A microplate-based ELISA assay that detects zebrafish glial fibrillary acidic protein (GFAP), a neurotoxicity marker, has been developed and was able to detect twofold increases of GFAP levels in zebrafish exposed to 2,3,7,8-tetrachlorodibenzodioxin (TCDD), an environmental neurotoxin, suggesting potential for high-throughput neurotoxicity screening [3].

Some research in the zebrafish model has focused on genetic mechanisms of cell specification and morphogenesis of the nervous system [10, 43–45]. Mutant lines with extremely specific defects in most components of the nervous system are available. Also, mutant lines with defects in neural connections are established. The ontogeny of specific behaviors is well defined [21, 46, 47], and neurologic functions such as sleep have been investigated [48]. Calcium fluxes have been visualized in individual neurons in the central nervous system of live fry during behaviors such as the escape response [49, 50]. Neuronal metabolic profiles have been well described [35]. Migration pathways of neural crest cells and sequential specification of particular cell types have been documented in zebrafish [29, 39, 51, 52]. Mutant lines with specific defects in certain aspects of neural crest development are available, with lesions including abnormal pigment patterns, abnormal jaw development, and abnormal enteric neural tissue [53, 54].

5 Neurobehavioral Assays

Behavioral assays using stereotypical patterns of locomotion in zebrafish embryos and larvae can be used as a predictive neurotoxicity endpoint. A recent study showed that exposure to chlorpyrifos, a known developmental neurotoxicant, altered the locomotor activity in zebrafish embryos and larvae. Embryos had significant increases in frequency and total duration of spontaneous tail coiling, while larvae had significant decrease in total duration of swimming activity [55]. Although only one toxicant was characterized, this study showed that these two locomotor endpoints in zebrafish can potentially be used to screen for neurotoxic agents. In another recent study, Irons et al. [56] monitored modulations in locomotor activity by acute exposure to three neuroactive compounds: ethanol, d-amphetamine sulfate, and cocaine hydrochloride. At 20 min into the exposure, locomotion was assessed for each animal for 70 min using 10 min, alternating light and dark (infrared light) periods. The findings were strikingly similar to those in mammals: low concentrations of ethanol and d-amphetamine increased locomotion activity, while higher concentrations of all three compounds decreased locomotion activity, and ethanol effects occurred predominately during the light periods, whereas the d-amphetamine and cocaine effects occurred during the dark periods [56]. Two other recent innovative studies reported an automated zebrafish screen for novel neuroactive compounds which can predict mode-of-action and neurotoxic potency in uncharacterized compounds through inferences of compounds that induce similar locomotor and behavioral activities [57, 58]. In this high-throughput screening, zebrafish embryos in microplate wells received light stimuli eliciting stereotype locomotive activity (overall amount of motion in the well, both in terms of contraction frequency and number of animals in motion) that correlates to a motion index converting into a behavioral "barcode" profile. Using this approach the authors evaluated thousands of chemical compounds and found that diverse classes of neuroactive molecules cause distinct patterns of behavioral barcode profiles that can be used to predict mode-of-action and neurotoxicity of an uncharacterized neuroactive compound. The authors also noted that although many psychotropic drugs cause reproducible behavioral phenotypes, there were drugs that failed to elicit any detectable change. However, it is not known whether these false negatives were due to inappropriate screening dose, poor absorption, or imperfect conservation of the nervous system between zebrafish and humans.

Neurobehavioral effects of a variety of drugs/toxicants have been evaluated in developing or adult zebrafish. Samson et al. [59] found that impairment of swimming and predator/prey behavior were much more sensitive indicators of toxicity in zebrafish exposed as early life stages to methylmercury than were mortality or morphologic lesions. Both early and recent studies have investigated pathologic lesions and functional impairment of developing zebrafish exposed to ethanol [60]. Gerlai et al. [61] employed zebrafish to assess genetic factors predisposing lines of fish to alcohol preference. Darland and Dowling [62] screened mutant lines of zebrafish to identify those with increased preference for exposure to cocaine and for altered responses to cocaine. Thomas [63] has studied effects of tetrahydrocannabinol (THC) from marijuana in developing zebrafish.

6 Toxicokinetics

The database on metabolism of toxicants in zebrafish is much less complete than that available for certain other highly studied fish species such as rainbow trout. More data are available regarding Phase I than Phase II metabolism in zebrafish. Several cytochrome P450 enzymes from zebrafish have been mapped, but full cDNA sequences are in the public domain only for *cyp19a*, *cyp19b*, and *cyp26*.

Cyp1a1 activity is induced in adults, as well as in early life stages of zebrafish and in liver cell cultures by 2,3,7,8-tetrachlorodibenzo-*p*-dioxin (TCDD). The aryl hydrocarbon receptor (AHR) signaling pathway has been studied extensively in zebrafish. Stage-specific expression of *ahr*, the *ahr nuclear translocator* (*arnt*), and *cyp1a1* are documented in early life stages of zebrafish. Zebrafish *cyp19b* is expressed in the brain. Keizer [64] has investigated species differences in acetylcholine esterase inhibition by diazinon in fish. They found that zebrafish are relatively resistant to diazinon compared to other fish species, because their acetylcholine esterase is relatively resistant to inhibition by this pesticide. Relatively limited data are available regarding toxicokinetics of drugs/toxicants in zebrafish. Most published zebrafish studies report exposure concentrations, but not tissue concentrations of toxicants, so quantitative comparison of sensitivity between zebrafish and other species is difficult.

7 Cell Culture and Zebrafish Resources

Cell culture methods are established to create primary and immortal cell lines from adult tissues as well as from embryos [65–67]. Significantly, explants of embryos and adult tissues, such as whole brain, can be cultured to study cell–cell interactions and metabolism [68]. The NIH has funded research in zebrafish genomics for the past decade. Among fish species, the most complete database on genomics, molecular genetics and embryology available is for the zebrafish. These data are accessible through the Zebrafish Information Network (http://zfin.org/ZFIN) coordinated in conjunction with the NIH-funded Zebrafish International Resource Center (ZIRC) at the University of Oregon.

8 Omics Application

There are vast amounts of genomic resources with regard to zebrafish, and the ability to map zebrafish genes to mammalian homologs facilitates the use of an omics approach to generate predicative and mechanistic insights into neurotoxicity. Zebrafish are amenable to various omics platforms: such as transcriptomics, proteomics, and metabolomics which generate high-content data. However, at present, most of the zebrafish omics studies have employed transcriptomics platforms, many of which are related to environmental toxicology [69]. More zebrafish omics need to be extended to drug-related studies. Phenotype-based screenings can only detect chemical-induced visible perturbations. Omics can further complement phenotype-based screenings by capturing system-wide molecular responses involving mRNA transcripts, proteins, metabolites and other biomolecules. In return, this can help to detect subtle toxic potencies, elucidate mode-of-toxicity of drugs, identify biomarkers and even predict drug toxicity that induce similar phenotypic responses [69].

Although omics can offer insightful perspective, it is more costly to implement, and interpretation of data can be more time-consuming than phenotype-based screens. Additionally, an omics approach usually requires a large amount of tissue samples that involves pooling from multiple individuals or from multiple experimental replicates in order to obtain sufficient materials for the assays. In general, depending on the platform, 30–100 mg of tissue is usually required, and contingent on the source of tissue, this amount can be obtained from pooling of >30–100 whole embryos, or pooling of specific tissue from several adult individuals. Therefore, in view of the strength and constraints of the omics approach, it should be placed strategically after phenotypic screenings to enhance the prediction of toxic potencies, and lead optimization of selected drug candidates.

9 Mechanistic Omics

Mechanistic insights into drug toxicity can be elucidated by using knowledge-based data mining and algorithms on the high-content omics data to discover perturbed molecular pathways and biological processes. Phenotypic endpoints can corroborate with omics data to characterize the mechanism of drug toxicity in a conceptual framework of cause-and-effect with verifications from known molecular interactions and phenotypic anchoring. In a study investigating genes that mediate addiction to amphetamine, the transcriptome of the adult brain mutant *no addiction* ($nad^{dne3256}$), which is unresponsive to amphetamine, were compared with wild-type siblings under normal conditions, as well as both treated with amphetamine [70]. The comparison of the transcriptome profiles identified a new network of coordinated gene regulation associated with the amphetamine-triggered phenotype known as conditioned place preference behavior. The authors found that the differentially expressed gene set was significantly enriched with transcription factor-encoding genes that are involved with vertebrate brain development. Phenotype analysis using in situ hybridization indicated that these genes are also active in adult brains. The study suggested that these genes deregulated by amphetamines were involved in neurodevelopment and also mediated behavioral addiction to amphetamines in adults. Transcriptome studies demonstrate the use of comparative transcriptome profiles coupled with chemical or genetic modifiers to generate loss or gain of function phenotype in zebrafish and elucidate valuable mechanistic insights into the toxicity of compounds.

10 Predictive Omics

Prediction modeling is used to classify drugs/toxicants into predefined/known groups using machine learning algorithms, based on the assumption that drugs with similar mode-of-toxicity have similar biological effects that can be defined by discriminatory signatures/biomarkers. A discriminatory gene set identified from omics data can be used to assay toxicity directly. For example, one study showed that the developmental expression of a subset of ten genes associated with the nervous system was perturbed in zebrafish embryos during and after exposure to sublethal concentrations of ethanol, a well-characterized developmental neurotoxicant [71]. This study also demonstrated that gene markers are more responsive and useful for predicting toxicity at sublethal concentrations before the appearance of a phenotype. Robust and sensitive predictive models can be generated using zebrafish omics to aid in predicting/identifying compounds with specific mode of toxicity.

11 Future Needs and Research Directions

The utility of early life stages of zebrafish in high-throughput screening systems for drug development is already being exploited. The small size of zebrafish embryos and fry, and their ability to be cultured during the first week of life in 96 well microtiter plates make this system ideal for drug discovery and safety testing, particularly with regard to neurotoxicity. Most hydrophilic as well as lipophilic agents are readily absorbed from the culture medium of eggs or fry, facilitating efficient testing of new agents. To advance the field of zebrafish neurotoxicologic pathology toward the state-of-the-art in mammalian neurotoxicologic pathology, much more data regarding pathologic lesions following acute, subchronic, and chronic toxicant exposure will be required. We need to develop a comprehensive database regarding spontaneous and toxicant-induced lesions in the common wild type and mutant lines of zebrafish. In addition, spontaneous aging lesions in various strains of zebrafish need to be investigated. Comprehensive data on metabolism and toxicokinetics of drugs/toxicants in various wild type and mutant lines will be essential to support sophisticated neurotoxicologic research.

12 Conclusions

In the quest for alternative models for neurotoxicity testing, zebrafish have much to offer: low cost, high-throughput, a wide range of morphological and functional assays, and a high degree of correlation with mammalian systems. The only entity missing is history. Unlike many mammalian models that have been used for decades, zebrafish do not benefit from a large reservoir of historical data establishing the system's validity and limitations. If zebrafish are to become a valuable model system, it will require a commitment to accumulating and sharing that reservoir, a process that could be accelerated by the ability to acquire data rapidly in zebrafish. Although such an effort is daunting, the competing pressures for additional testing and reduced use of mammals point toward an investment in zebrafish as a sound one.

To date, zebrafish assays for neurotoxicity have been shown to be good predictors of toxicity in humans. With recent state-of-the-art developments, zebrafish are currently the only vertebrate model system with sufficient resources amendable for medium- to high-throughput toxicity screening, while at the same time providing physiologically relevant data derived from a whole-vertebrate model. Consequently, with the increasing use of high-throughput screening and chemical libraries in the drug discovery pipeline, zebrafish are in an attractive position as first- or second-tier (after in vitro cell lines) neurotoxicity screens that could help to shift compound attrition to earlier stages of the drug development pipeline. It is expected that use of zebrafish for a variety of neurotoxicological studies will become more important and commonplace in the coming years.

References

1. Woodhead AD. Nonmammalian models in biomedical research. Boca Raton, FL: CRC; 1989.
2. Laale HW. Culture and preliminary observations of follicular isolates from adult zebra fish, *Brachydanio rerio*. Can J Zool. 1977;55:304–9.
3. McGrath P, Li CQ. Zebrafish: a predictive model for assessing drug-induced toxicity. Drug Discov Today. 2008;13:394–401.
4. Parng C, Roy NM, Ton C, Lin Y, McGrath P. Neurotoxicity assessment using zebrafish. J Pharmacol Toxicol. 2007;55:103–12.
5. Ward AC, Lieschke G. The zebrafish as a model system for human disease. Front Biosci. 2002;7:d827–33.
6. Cowley AW. The emergence of physiological genomics. J Vasc Res. 1999;36:83–90.
7. Wullimann MF, Knipp S. Proliferation pattern changes in the zebrafish brain from embryonic through early postembryonic stages. Anat Embryol. 2000;202:385–400.
8. Barut BA, Zon LI. Realizing the potential of zebrafish as a model for human disease. Physiol Genomics. 2000;2:49–51.
9. Beaman JR, Finch R, Gardner H, Hoffmann F, Rosencrance A, Zelikoff JT. Mammalian immunoassays for predicting the toxicity of malathion in a laboratory fish model. J Toxicol Environ Health. 1999;56:523–42.
10. Blader P, Strähle U. Zebrafish developmental genetics and central nervous system development. Hum Mol Genet. 2000;9:945–51.
11. Driever W, Fishman MC. The zebrafish: heritable disorders in transparent embryos. J Clin Invest. 1996;97:1788–94.
12. Lele Z, Krone PH. The zebrafish as a model system in developmental, toxicological and transgenic research. Biotechnol Adv. 1996;14:57–72.
13. McKinley ET, Baranowski TC, Blavo DO, Cato C, Doan TN, Rubinstein AL. Neuroprotection of MPTP induced toxicity in zebrafish dopaminergic neurons. Brain Res Mol Brain Res. 2005;141:128–37.
14. Rubinstein AL, McKinley E, Blavo D, Cato C. A Parkinson's disease model for drug screening. Presented at fifth international meeting on zebrafish development and genetics, Madison, WI; 2002.
15. Zon LI, Peterson RT. In vivo drug discovery in the zebrafish. Nat Rev Drug Discov. 2005;4:35–44.
16. Khersonsky SM, Jung DW, Kang TW, Walsh DP, Moon HS, Jo H, Jacobson EM, Shetty V, Neubert TA, Chang YT. Facilitated forward chemical genetics using a tagged triazine library and zebrafish embryo screening. J Am Chem Soc. 2003;125:11804–5.

17. Moon HS, Jacobson EM, Khersonsky SM, Luzung MR, Walsh DP, Xiong W, Lee JW, Parikh PB, Lam JC, Kang TW, Rosania GR, Schier AF, Chang YT. A novel microtubule destabilizing entity from orthogonal synthesis of triazine library and zebrafish embryo screening. J Am Chem Soc. 2002;124:11608–9.
18. Parng C, Ton C, Lin YX, Roy NM, McGrath P. A zebrafish assay for identifying neuroprotectants in vivo. Neurotoxicol Teratol. 2006;28:509–16.
19. Bang PI, Yelick PC, Malicki JJ, Sewell WF. High-throughput behavioral screening method for detecting auditory response defects in zebrafish. J Neurosci Methods. 2002;118:177–87.
20. Brockerhoff SE, Hurley JB, Janssen-Bienhold U, Neuhauss SC, Driever W, Dowling JE. A behavioral screen for isolating zebrafish mutants with visual system defects. Proc Natl Acad Sci. 1995;92:10545–9.
21. Fetcho JR, Liu KS. Zebrafish as a model system for studying neuronal circuits and behavior. Ann N Y Acad Sci. 1998;860:333–45.
22. Peitsaro N, Kaslin J, Anichtchik OV, Panula P. Modulation of the histaminergic system and behaviour by alpha-fluoromethylhistidine in zebrafish. J Neurochem. 2003;86:432–41.
23. Milan DJ, Peterson TA, Ruskin JN, Peterson RT, MacRae CA. Drugs that induce repolarization abnormalities cause bradycardia in zebrafish. Circulation. 2003;107:1355–8.
24. Baraban SC, Taylor MR, Castro PA, Baier H. Pentylenetetrazole induced changes in zebrafish behavior, neural activity and c-fos expression. Neuroscience. 2005;131:759–68.
25. Detrich HW, Westerfield M, Zon LI. Overview of the zebrafish system. Methods Cell Biol. 1999;59:3–10.
26. Fournie JW, Hawkins WE, Krol RM, Wolfe MJ. Preparation of whole small fish for histological evaluation. In: Ostrander GM, editor. Techniques in aquatic toxicology. Boca Raton, FL: Lewis Publishers; 1996. p. 577–88.
27. Cooper MS, D'Amico LA, Henry CA. Analyzing morphogenetic cell behaviors in vitally stained zebrafish embryos. Methods Mol Biol. 1999;122:185–204.
28. Cooper MS, D'Amico LA, Henry CA. Confocal microscopic analysis of morphogenetic movements. Methods Cell Biol. 1999;59:179–204.
29. Kimmel CB, Miller CT, Keynes RJ. Neural crest patterning and the evolution of the jaw. J Anat. 2001;199:105–20.
30. Macdonald R. Zebrafish immunohistochemistry. Methods Mol Biol. 1999;127:77–88.
31. Byrd CA, Brunjes PC. Organization of the olfactory system in the adult zebrafish: histological, immunohistochemical, and quantitative analysis. J Comp Neurol. 1995;358:247–59.
32. Caminos E, Velasco A, Jarrin M, Lillo C, Jimeno D, Aijon J, Lara J. A comparative study of protein kinase C-like immunoreactive cells in the retina. Brain Behav Evol. 2000;56:330–9.
33. Connaughton V, Behar TN, Liu W, Massey S. Immunocytochemical localization of excitatory and inhibitory neurotransmitters in the zebrafish retina. Vis Neurosci. 1999;16:483–90.
34. Kawai H, Arata N, Nakayasu H. Three-dimensional distribution of astrocytes in zebrafish spinal cord. Glia. 2001;36:406–13.
35. van Raamsdonk W, de Graaf F, van Asselt E, Diegenbach PC, Mos W, van Noorden CJ, Roberts BL, Smit-Onel MJ. Metabolic profiles of spinal motoneurons in fish as established by quantitative enzyme histochemistry. Comp Biochem Physiol Comp Physiol. 1992;102:631–6.
36. van Raamsdonk W, Tekronnie G, Pool CW, van de Laarse W. An immune histochemical and enzymic characterization of the muscle fibres in myotomal muscle of the teleost *Brachydanio rerio*. Acta Histochem. 1980;67:200–16.
37. Wullimann MF, Rink E. Detailed immunohistology of Pax6 protein and tyrosine hydroxylase in the early zebrafish brain suggests role of *Pax6* gene in development of dopaminergic diencephalic neurons. Brain Res Dev. 2001;131:173–91.
38. Korfsmeier KH. PCNA in the ovary of zebrafish (*Brachydanio rerio*, Ham.-Buch.). Acta Histochem. 2002;104:73–6.
39. Eisen JS. Motoneuronal development in the embryonic zebrafish. Dev Suppl. 1991;2:141–7.
40. Hirsinger E, Westerfield M. The progressive determination of zebrafish muscle cell lineages. Presented at fifth international meeting on zebrafish development and genetics, Madison, WI; 2002.

41. Mizuno T, Shinya M, Takeda H. Cell and tissue transplantation in zebrafish embryos. Methods Mol Biol. 1999;127:15–28.
42. Halloran MC, Sato-Maeda M, Warren JT, Su F, Lele Z, Krone PH, Kuwada JY, Shoji W. Laser-induced gene expression in specific cells of transgenic zebrafish. Development. 2000;127: 1953–60.
43. Eisen JS, Weston JA. Development of the neural crest in the zebrafish. Dev Biol. 1993;159: 50–9.
44. Moens CB, Fritz A. Techniques in neural development. Methods Cell Biol. 1999;59:253–72.
45. Moens CB, Prince VE. Constructing the hindbrain: insights from the zebrafish. Dev Dyn. 2002;224:1–17.
46. Gahtan E, Sankrithi N, Campos JB, O'Malley DM. Evidence for a widespread brain stem escape network in larval zebrafish. J Neurophysiol. 2002;87:608–14.
47. Saint-Amant L, Drapeau P. Time course of the development of motor behaviors in the zebrafish embryo. J Neurobiol. 1998;37:622–32.
48. Hendricks JC, Sehgal A, Pack AI. The need for a simple animal model to understand sleep. Prog Neurobiol. 2000;61:339–51.
49. Fetcho JR, Cox KJ, O'Malley DM. Monitoring activity in neuronal populations with single-cell resolution in a behaving vertebrate. Histochem J. 1998;30:153–67.
50. Zimprich F, Ashworth R, Bolsover S. Real-time measurements of calcium dynamics in neurons developing in situ within zebrafish embryos. Pflugers Arch. 1998;436:489–93.
51. García-Castro M, Bronner-Fraser M. Induction and differentiation of the neural crest. Curr Opin Cell Biol. 1999;11:695–8.
52. Trainor PA, Krumlauf R. Patterning the cranial neural crest: hindbrain segmentation and *hox* gene plasticity. Nat Rev Neurosci. 2000;1:116–24.
53. Dutton KA, Pauliny A, Lopes SS, Elworthy S, Carney TJ, Rauch J, Geisler R, Haffter P, Kelsh RN. Zebrafish *colourless* encodes *sox10* and specifies non-ectomesenchymal neural crest fates. Development. 2001;128:4113–25.
54. Kelsh RN, Eisen JS. The zebrafish colourless gene regulates development of non-ectomesenchymal neural crest derivatives. Development. 2000;127:515–25.
55. Selderslaghs IW, Hooyberghs J, De Coen W, Witters HE. Locomotor activity in zebrafish embryos: a new method to assess developmental neurotoxicity. Neurotoxicol Teratol. 2010;32: 460–71.
56. Irons TD, MacPhail RC, Hunter DL, Padilla S. Acute neuroactive drug exposures alter locomotor activity in larval zebrafish. Neurotoxicol Teratol. 2010;32:84–90.
57. Kokel D, Bryan J, Laggner C, White R, Cheung CY, Mateus R, Healey D, Kim S, Werdich AA, Haggarty SJ, Macrae CA, Shoichet B, Peterson RT. Rapid behavior-based identification of neuroactive small molecules in the zebrafish. Nat Chem Biol. 2010;6:231–7.
58. Rihel J, Prober DA, Arvanites A, Lam K, Zimmerman S, Jang S, Haggarty SJ, Kokel D, Rubin LL, Peterson RT, Schier AF. Zebrafish behavioral profiling links drugs to biological targets and rest/wake regulation. Science. 2010;327:348–51.
59. Samson JC, Goodridge R, Olobatuyi F, Weis JS. Delayed effects of embryonic exposure of zebrafish (Danio rerio) to methylmercury (MeHg). Aquat Toxicol. 2001;51(4):369–76.
60. Blader P, Strähle U. Ethanol impairs migration of the prechordal plate in the zebrafish embryo. Dev Biol. 1998;201:185–201.
61. Gerlai R, Lahav M, Guo S, Rosenthal A. Drinks like a fish: zebra fish (*Danio rerio*) as a behavior genetic model to study alcohol effects. Pharmacol Biochem Behav. 2000;67:773–82.
62. Darland T, Dowling JE. Behavioral screening for cocaine sensitivity in mutagenized zebrafish. Proc Natl Acad Sci. 2001;98:11691–6.
63. Thomas RJ. The toxicologic and teratologic effects of delta-9-tetrahydrocannabinol in the zebrafish embryo. Toxicol Appl Pharmacol. 1975;32(1):184–90.
64. Keizer J. Enzymological differences of AChE and diazinon hepatic metabolism: correlation of in vitro data with the selective toxicity of diazinon to fish species. Sci Total Environ. 2000;171(1–3):213–20.

65. Collodi P, Kamei Y, Ernst T, Miranda C, Buhler DR, Barnes DW. Culture of cells from zebrafish (*Brachydanio rerio*) embryo and adult tissues. Cell Biol Toxicol. 1992;8:43–61.
66. Ghosh C, Collodi P. Culture of cells from zebrafish (*Brachydanio rerio*) blastula-stage embryos. Cytotechnology. 1994;14:21–6.
67. Helmrich A, Barnes D. Zebrafish embryonal cell culture. Methods Cell Biol. 1999;59:29–37.
68. Tomizawa K, Kunieda J, Nakayasu H. Ex vivo culture of isolated zebrafish whole brain. J Neurosci Methods. 2001;107:31–8.
69. Sukardi H, Ung CY, Gong Z, Lam SH. Incorporating zebrafish omics into chemical biology and toxicology. Zebrafish. 2010;7:41–52.
70. Webb KJ, Norton WH, Trümbach D, Meijer AH, Ninkovic J, Topp S, Heck D, Marr C, Wurst W, Theis FJ, Spaink HP, Bally-Cuif L. Zebrafish reward mutants reveal novel transcripts mediating the behavioral effects of amphetamine. Genome Biol. 2009;10:81–5.
71. Fan CY, Cowden J, Simmons SO, Padilla S, Ramabhadran R. Gene expression changes in developing zebrafish as potential markers for rapid developmental neurotoxicity screening. Neurotoxicol Teratol. 2010;32:91–8.

Sleep Phenotypes in Zebrafish

David J. Echevarria and Kanza M. Khan

Abstract Sleep is a vital, highly evolutionarily conserved biological function across species. It has been extensively studied in human and rodent models, and recently, the study of sleep in zebrafish has gained momentum. Neuropeptide and neurotransmitter systems that regulate sleep and waking rhythms are fully developed in the zebrafish within 4 days post fertilization, and demonstrate genetic homology to humans and rodents. Similar to the humans and rodent models, sleep in the larval and adult zebrafish is largely consolidated to dark phases. However, these systems remain susceptible to environmental and pharmacological manipulations, with drastic changes in sleep rhythms resulting in changes in gene and protein expression. As is seen with humans and rodents, sleep deprivation in zebrafish tends to result in anxiety like responses. Further analysis of the effects of genetic and pharmacological intervention would provide a deeper understanding of this essential function, potentially paving the way for the development of pharmacological treatments for sleep related disorders.

Keywords Sleep disruption • Sleep extension • Sleep disorders • Pharmacology • Gene expression • Protein expression • Anxiety

1 The Biology of Sleep

Sleep is generally regarded as a state of reversible immobility in which consciousness and sensitivity to environmental stimuli is greatly reduced [1]. In humans, two types of sleep have been identified: rapid eye movement (REM), and non-rapid eye movement sleep (NREM), which encompasses four sub-stages. Electroencephalographic (EEG) devices may be used to discriminate various sleep states from one another, and can provide insight into sleep regulation [2, 3]. At the beginning of a sleep cycle an individual will enter stage 1 of NREM sleep which is physiologically marked by the

D.J. Echevarria (✉) • K.M. Khan
Department of Psychology, The University of Southern Mississippi,
118 College Drive, Box 5025, Hattiesburg, MS 39406, USA
e-mail: david.echevarria@usm.edu; kanza.khan@usm.edu

A.V. Kalueff (ed.), *The rights and wrongs of zebrafish: Behavioral phenotyping of zebrafish*, DOI 10.1007/978-3-319-33774-6_10

slowing of breathing and heart rate [4]. Delving deeper into sleep, the individual enters stage 2 of sleep and experiences a loss in muscle tone and a decreased sensitivity to the environment [4–6]. Using an EEG, stages 1 and 2 of sleep are distinguished from waking when theta waves (4–8 Hz) predominate over alpha (8–12.5 Hz) and beta (12.5–30 Hz) waves. Stages 3 and 4 of sleep are collectively referred to as Slow Wave Sleep (SWS) and are characterized by the onset of high-voltage, low frequency delta waves (1–8 Hz) [7]. REM is the last stage in the recursive sleep cycle, and is characterized by the rapid darting of eyes underneath the eyelids, muscle atonia, the loss of reflexes, and desynchronized brain wave activity making this stage indistinguishable from waking states on an EEG [7].

The structure of sleep, its duration, and placement within the day varies across species [1, 8]. In general, four criteria dictate this behavior: (1) the assumption of a species-specific posture, (2) behavioral quiescence, or inactivity, (3) an elevated arousal threshold, and (4) state reversibility with stimulation [1, 9]. In most vertebrate mammals (e.g., humans, rodents, cats and dogs), sleep and waking periods are propagated and maintained by two key brain regions: the suprachiasmatic nucleus (SCN) and the ventrolateral preoptic nucleus (VLPO) [10]. The SCN is seated in the lateral hypothalamus and is the organism's principal circadian 'clock'. The ability of each neuron within the SCN to function as an independent clock is key in the regulation of the transcription and translation of proteins that maintain circadian rhythms relating to vital physiological functions, namely temperature regulation and arousal [11–14]. Several proteins have been identified in the regulation of mammalian circadian rhythms: PERIOD1, PERIOD2, PERIOD3, CLOCK, BMAL1, CRYPTOCHROME1 (CRY1), CRYPTOCHROME2 (CRY2), CASEIN KINASE I-DELTA (CSNK1D) and CASEIN KINASE I-EPSILON (CSNK1E) [15–21]. Dimerization of the regulatory proteins drives the rhythmic transcription and translation of regulatory feedback loops. In one loop, the CLOCK: BMAL1 heterodimer stimulates the transcription of enhancer box (E-box) genes and cis-regulatory elements including PERIOD and CRYPTOCHROME [19]. These elements are associated with the maintenance of sleep patterns, and other physiological effects such as cell division and metabolism. The dimerization of PER:CRY drives a negative feedback loop, repressing the transcription of these regulatory elements [19].

In another regulatory loop, CLOCK:BMAL1 heterodimers drive the transcription of retinoic acid-related orphan nuclear receptors, which regulate *Bmal1* expression through positive and negative feedback. These autoregulatory loops take roughly 24 h to complete one cycle and are regulated by post-translational modification factors, *Casein kinase 1 epsilon* and *Casein kinase 1 delta*, that regulate the protein turnover rate and maintain the period of the circadian cycle (Fig. 1) [19].

The VLPO exerts an inhibitory influence on the brain regions that maintain CNS arousal [10]. Several neurotransmitter and neuropeptides play a role in the communication between brain regions to signal rest or arousal states. These include, but are not limited to orexins (also referred to as hypocretins) [22], dopamine [23], gamma aminobutyric acid (GABA) [10], galanin [24], melatonin [25], and neuropeptide Y [26] (see Table 1).

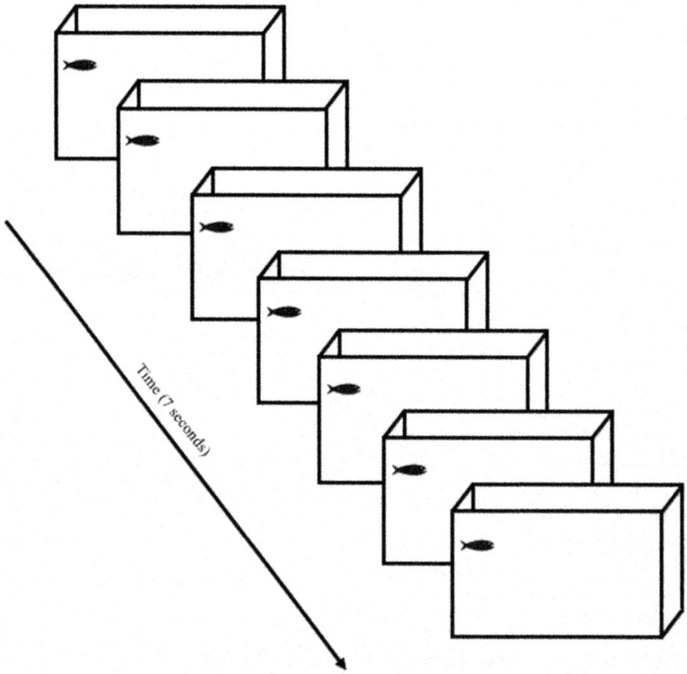

Fig. 1 Diagram of a sleeping adult zebrafish; each image depicts the passage of 1 s and the placement of the animal within the tank. During sleep, the adult zebrafish will exhibit an increased arousal threshold to aversive stimuli, and immobility for at least 6 s. Each second after the initial 6 s will be considered sleep

Table 1 Neurotransmitters and neuropeptides involved in arousal and waking

Agents	Originates in	Active during			Reference(s)
		Wake	NREM	REM	
Acetylcholine	Pedunculopontine and laterodorsal tegmental nucleus (PPT/LDT)	↑	—	↑	[118]
Noradrenaline	Locus coeruleus (LC)	↑	↓	—	[119]
Dopamine	Periaqueductal grey	↑	↓	—	[23]
Histamine	Tuberomammillary nucleus (TMN)	↑	↓	—	
Melanin concentrating hormone (MCH)	Lateral hypothalamus	?	?	↑	[117]
Hypocretin/orexin	Lateral hypothalamus	↑	?	↑	[22]
Acetylcholine	Basal forebrain	↑	↓	↑	[119], [10]
Gamma-aminobutyric acid (GABA)	Basal forebrain	↑	?	—	[64], [119]

An increase in neurotransmitter or neuropeptide release during each sleep and waking stage is signified by ↑ and a decrease is signified by ↓. If there is no change in agent release, this is marked by a dash (—), while unexplored changes are marked by a question mark (?)

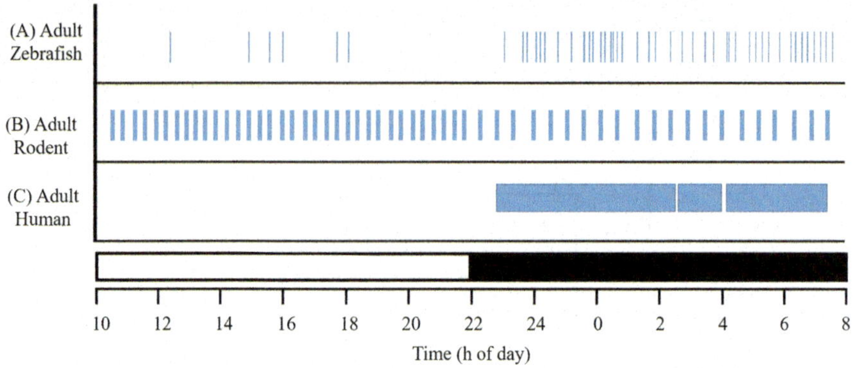

Fig. 2 Representative sleep rhythms of (**a**) adult zebrafish, (**b**) adult Long Evans strain *Rattus norvegicus*, and (**c**) a male human adult. *Blue lines/bars* depict periods of sleep. (**a**) In adult zebrafish, sleep bouts last for at least 6 s, and are largely consolidated to the night. (**b**) Adult rodents demonstrate polyphasic sleep, with fewer sleep cycles occurring during the night. (**c**) Sleep in healthy human adults is consolidated during the night hours, and presents with short intermittent wakeful periods. Adapted from [65, 115, 120]

In humans, the acute disruption of sleep rhythms (i.e., forgoing one nights rest and/or remaining awake for ~40 consecutive hours) is linked to impaired cognitive functioning and worsened physical states [27, 28]. Persistent sleep disturbances (e.g., excessive sleepiness or insomnia) are often accompanied by cognitive deficits (e.g., loss of situational awareness, slow response time, and increased compensatory effort) as well as clinical depression and anxiety [29–32].

For decades, rats and mice have been used as a valuable tool in the study of sleep and sleep-associated disorders [33–35]. Rodents demonstrate a polyphasic sleep pattern, entering several sleep bouts over the course of a 24-h day (Fig. 2). Total sleep deprivation in laboratory rats results in death after an average of 19 days [33, 36]. This is preceded by an increase in overall food intake combined with significant energy expenditure and subsequent drastic weight loss [33]. Sustained sleep deprivation also impairs immune functioning, leaving the animals susceptible to opportunistic pathogens that cause a systemic infection devoid of typical diagnostic symptoms (e.g., afebrile, absence of an inflammatory response) [36]. Along with other symptoms of total sleep deprivation, laboratory rats present with slightly elevated core body temperatures in the first half of the sleep deprivation period [37]. This is followed by a mild hypothermia (1 °C) in the second half of the deprivation period, despite an overall increase in whole body energy expenditure, suggesting impaired heat retention mechanisms [37].

Similar to humans, sleep deprivation in rodents is associated with elevated corticosterone levels, and impaired performance in cognitive tasks (e.g., t-maze) and memory deficits [38, 39]. In the short term sleep interruption (e.g., forced walk on a treadmill for 6 h [alternating 30 s periods of rest and walk]) also markedly increases the concentration of sleep promoting factors (adenosine) within the basal forebrain

[40], and disrupts hippocampal long-term potentiation [38, 41, 42] with an overall increase in hippocampal glutamate, aspartate and glycine as well as cortical glutamate, glycine and taurine levels [43]. The elevated levels of amino acids are indicative of an elevated brain metabolism, which also help to explain the increase in whole body energy expenditure [43].

Rodent models have provided invaluable insight to the study of sleep and the effects of sleep fragmentation and deprivation. Although the use of rodents as models of the human condition has been the laboratory standard for generations, there has been an increasing consideration for the use of the zebrafish (*Danio rerio*) in behavioral research. While the nervous system of the zebrafish is simpler than its rodent counterpart, zebrafish possess several of the same neuro-signaling molecules and brain regions that are conserved in rodents and humans [44]. Zebrafish produce robust phenotypes, making them a valuable model in studies of anxiety, addiction and learning [45–47]. The study of sleep in zebrafish has gained tremendous popularity in the past decade; the identification of mammalian orthologous sleep factors (e.g., orexins) and the elucidation of sleep and rest behaviors have sparked interest within the field of zebrafish research.

2 Sleep in Zebrafish

The systems modulating rest and waking in zebrafish begin to form at an early stage in larval development, and are fully functional by 4–5 days post fertilization (dpf) [48–50]. The pineal gland and retina function as the circadian clock in the zebrafish [48, 51, 52]. The pineal gland is photoreceptive and rhythmically drives the production of melatonin in the zebrafish. Peripheral and circadian clock neurons regulate rest and arousal, by making adjustments to the circadian rhythms in response to changes in environmental lighting; this process is termed entrainment [20, 53–55]. Several orthologues of mammalian clock genes have been identified in this teleost fish, including *Period* (*Per1*, *Per2*, and *Per3*), *bmal* (*zfbmal1*, *zfbmal2*) and *Cryptochrome* genes [20]. Furthermore, hormone and neurotransmitter systems involved in sleep are conserved in the zebrafish, including dopamine, GABA, serotonin, norepinephrine and orexins [56–61].

Zebrafish engage in sleep as defined by Campbell and Tobler's [1] behavioral criteria. Quiescent states are regulated in the larval and adult animal through daily fluctuations in clock genes (e.g., *per*, *cry*, *zfbmal*) and hormones (melatonin) [20, 62]. The heterodimerization of zebrafish CLOCK:BMAL1 serves as the mechanism that drives the autonomous circadian loops and transcription of per and cry genes [20]. This model functions in a similar manner to the mammalian circadian clock mechanism described above. The exact structure of sleep at each stage of zebrafish development is defined differently; once asleep, larval animals show reduced sensitivity to mechanical stimuli and changes in environmental lighting [63]. If sleep is disturbed or prevented, the animal will display sleep-rebound when introduced to a dark environment [63–65]. The relationship between amount of sleep

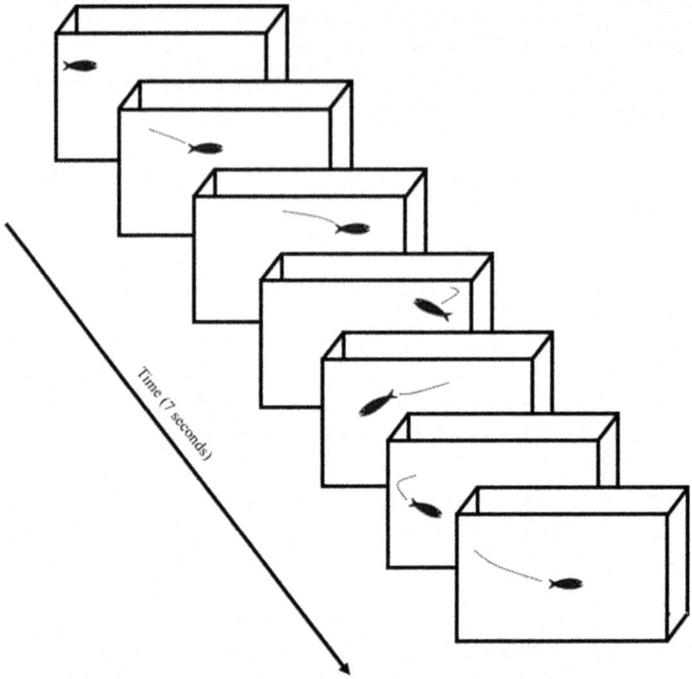

Fig. 3 Diagrams of a waking adult zebrafish. Each image depicts the passage of 1 s and the placement of the animal within the tank. During waking states, the adult zebrafish will be active and explore its environment

rebound and length of sleep disruption has not yet been tested in the larval or adult animal. Sleep in larval zebrafish is defined as a bout of immobility, lasting 2–10 min (depending on age and strain), in which the animal demonstrates a decreased sensitivity to sensory stimuli [63, 66]. This is accompanied by one of two postures: floating with the head angled downwards, or floating horizontally [62, 67].

In the adult zebrafish sleep is defined as a bout of immobility (swim speed <1 cm/s) lasting at least 6 s in which the animal has a heightened arousal threshold [65, 68, 69] (Figs. 1 and 3). Sleep is also accompanied by a reduced respiratory rate, heart rate, and mouth opening frequency [66, 67]. Sleep in both stages of development is largely consolidated to the night phase, and is associated with an increase in melatonin production (Fig. 4) [63, 65]. Zebrafish sleep may be disrupted through the presentation of environmental stimuli, such as the prolonged presentation of bright environmental lighting (e.g., >200 lx), mechanical stimulation (e.g., vibrations in the water), or a mild electric shock (e.g., 6 V/cm) [65, 68–70]. The ability of this teleost fish to enter sleep states, and the ease with which sleep may be disrupted in controlled laboratory conditions makes it an attractive model for the study of sleep-related disorders, as well as correlates of human disease that present with disrupted sleep rhythms [71].

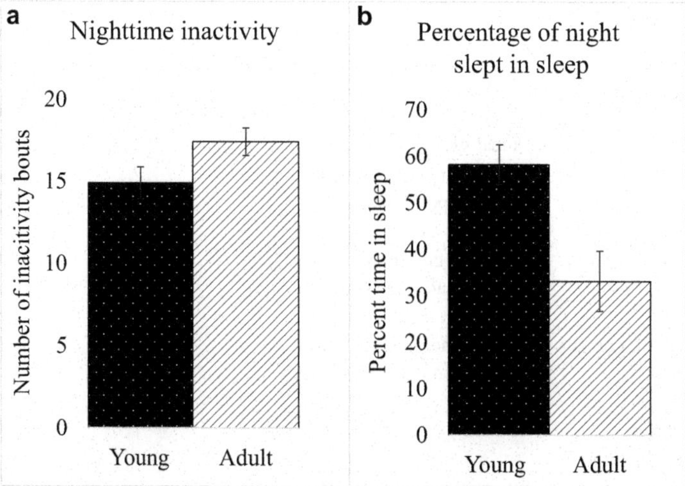

Fig. 4 A comparison of nighttime rest in young (~1 year old) and adult (~4 years old) zebrafish. (**a**) The number of inactive bouts exhibited during the dark phase (10 pm–8 am). (**b**) Total percentage of time spent asleep during the dark phase. Data adapted from [63, 116]

3 Genetic and Pharmacological Intervention on Sleep Rhythms

The neurotransmitter and neuropeptide pathways involved in the mammalian regulation of sleep are well-conserved in the zebrafish [58, 62, 72, 73]. Administration of agonists and antagonists of the sleep-regulating molecules impacts locomotor activity during the day, and alters sleep patterns throughout the night [63, 69, 70, 74] (Table 2).

3.1 Melatonin

The rate of melatonin production is regulated by the enzyme aralkylamine-N-acetyltransferase (*aanat*) [52]. The expression of the *aanat2* gene in zebrafish begins within 1dpf, and melatonin production begins within 2dpf [72]. As with humans and rodents, the presentation of ambient lights resets the circadian clock and suppresses melatonin production in the larval and adult zebrafish [51, 52]. The administration of melatonin has two main effects: it increases the rate of cell proliferation in the larval animal [63, 75]; and promotes sleep as observed through an increase in arousal threshold and a decrease in locomotion, irrespective of night or daytime [62, 63]. This effect is mediated through melatonin receptors, as the increase in rest may be blocked by pretreatment of selective melatonin receptor

Table 2 Effect of pharmacological manipulation on rest behaviors in the zebrafish

Agents	Projects to	Effect on		Reference
		Locomotion	Arousal threshold	
Melatonin	Pineal gland, retina	↓	↑	[51, 63]
Hypocretin	Telencephalon, pineal gland, hypothalamus, hindbrain	↑	?	[58, 65, 73, 76]
Gamma-aminobutyric acid (GABA) (agonist effects)	Postoptic commissure, telencephalic nucleus	↓	↑	[80, 63]
Histamine (antagonist effects)	Telencephalon, habenula, hypothalamus, optic tectum	↓	↑	[62, 88, 90]
Norepinephrine (antagonist effects)	Currently undetermined	↓	↑	[70]
Adenosine	Interrenal tissue, telencephalon, diencephalon, tegmentum, hindbrain	?	?	[98]

An increase in locomotor activity or threshold arousal (significant of rest in zf) is marked by ↑ and a decrease is marked by ↓. Unexplored changes are denoted by a question mark (?)

antagonists (e.g., luzindole), but not $GABA_A$ receptor antagonists (e.g., flumazenil) [63]. Gradual age-related declines are observed in night-time brain melatonin levels in zebrafish, and are associated with a reduction in the total sleep time during night-time [69].

3.2 *Hypocretin*

The zebrafish genome has a single orthologue for the hypocretin receptor (*hcrtr*), and is more similar to the mammalian *hcrt2* (70 % homology) than *hcrt1* (60 % homology) [73]. There are 40–45 hypocretin neurons within the zebrafish CNS located throughout the pineal gland, telencephalon, hypothalamus and hindbrain [58, 65, 73, 76, 77], that control the sleep-wake transitions in the animal. In humans, the overexpression of the neuropeptide hypocretin (HCRT) results in a narcolepsy-like state in which the affected individual is unable to consolidate sleep bouts [22]. An overexpression of HCRT in larval zebrafish greatly inhibits rest [73], while the ablation of hypocretin neurons increases the total amount of time spent asleep [78]. Additional evidence supporting the role of hypocretin neurons in the regulation of zebrafish sleep-wake transitions comes from the increase in neuroluminescence in hypocretin neurons, correspondent with an increase in spontaneous locomotion upon arousal in the morning [79].

3.3 GABA

Zebrafish have a well-developed GABA-ergic system [80]. The administration of sedative hypnotics, such as barbiturates and benzodiazepines, work in a dose-dependent manner, reducing locomotor activity and increasing arousal threshold [63]. At high doses, barbiturates (e.g., pentobarbital) and benzodiazepines (e.g., diazepam) have an anesthetic effect, resulting in a reduction of respiratory movements. A prolonged exposure to hypnotic GABA-ergic drugs may result in the death of the animal; this effect is more pronounced in the adult zebrafish [62].

3.4 Alcohol

Alcohol has a dose dependent effect in larval and adult zebrafish. Ethanol added to tank water is absorbed through the gills and affects locomotion, aggressive behavior and behavioral responses associated with anxiety [81–83]. At low doses (e.g., below 2.0 % EtOH in larval zebrafish, and 0.5 % in adult zebrafish) ethanol induces an increase in the mean swim speed and overall locomotion [82, 83]. Exposure to 1 % ethanol for roughly 20 min has anxiolytic effects in adult zebrafish [82, 84]. However, extended exposure to the same dose (e.g., 60 min) impairs locomotor activity and induces sedation by decreasing mean swim speed and hypolocomotion [82, 85–87]. A similar effect is reported in larval zebrafish following an exposure to ~4 % EtOH [83]. Recovery from the ethanol induced sedation state has not been studied, but may prove useful in the study of the effects of hypnotic properties of various drugs, including histamines and GABA receptor antagonists.

3.5 Histamine

The histamine system develops early in the zebrafish, with several receptor orthologues remaining conserved [88]. Like in rodents [89], antagonists of histamine receptors have a sedative effect in the zebrafish, and also act in a dose-dependent manner in larval zebrafish [62]. At low doses, histamine receptor antagonists (e.g., mepyramine) produce a mild sedation, while high doses act as general anesthetics [90], confirming that the anatomical similarities to human and rodent histamine receptors have a functional similarity as well.

3.6 Norepinephrine (NE)

Daily nighttime and early morning fluctuations in catecholamine levels are observed in humans [91]. Reduced plasma epinephrine and norepinephrine concentrations are detected immediately before sleep, and a corresponding spike is observed upon waking in the morning [91, 92]. Sleep deprivation in human adults increases the circulating norepinephrine levels [92], suggesting that sleep—rather than the circadian pacemaker, the SCN—regulates catecholamine levels. In humans and rodents, NE induces wakefulness and prevents NREM and REM sleep [93], while NE withdrawal or antagonists play a role in the induction of sleep states [94]. A similar effect is observed in the zebrafish; an exposure to adrenoceptor antagonists such as the alpha-1-adrenoceptor (e.g., Prazosin) modulates the effect of environmental manipulation on sleep. The presentation of ambient lights has been shown to be sufficient in preventing the onset of sleep rhythms in zebrafish [65, 70, 95]. Pretreatment of Prazosin to adult zebrafish prior to an exposure to extended light conditions increases the total amount of time spent in sleep, without altering the locomotor activity [70], suggesting a similar functionality of norepinephrine in this teleost fish.

3.7 *Adenosine*

Adenosine is the natural byproduct of metabolism of energy stores in the brain, and is produced from the metabolism of adenosine tri-phosphate (ATP) and cyclic adenosine mono-phosphate (cAMP). In humans and rodents, the accumulation of adenosine in the lateral hypothalamus has been proposed to facilitate the homeostatic drive for sleep [40, 96, 97]. Adenosine antagonists such as caffeine, work by blocking the binding of adenosine to receptors (specifically A2A), preventing the onset of the homeostatic mechanisms. Recently three distinct zebrafish genes for adenosine receptors have been identified: *adora2a.1*, *adora2a.2*, and *adora2b* [98]. Acting on the A_{2A} receptors, caffeine, a natural antagonist, is neuroprotective against 1-methyl-4-phenyl-1,2,3,6-tetrahydropyridine (MPTP) induced dopaminergic neurotoxicity in larval zebrafish [98]. Recent studies have also highlighted the anxiogenic effects of caffeine [99], and the neuroprotective benefits against scopolamine induced memory loss [100]. Presently, little is understood regarding the interaction of the adenosinergic systems with other zebrafish neurotransmitter and neuropeptide systems especially with regards to the sleep and wake systems, beckoning the need for further analysis of the behavioral effects of sleep interruption on sleep-modulating pathways.

4 Sleep Disruption

Similar to mammals, the deprivation of sleep causes differential gene expression, protein expression and altered behaviors (e.g., anxiety) in zebrafish. For example, keeping the animals under extended light conditions (>14 h) reduces the total amount of time animals spent in rest [65, 68, 70]. An extended exposure to bright environmental conditions results in the up-regulation of roughly 279 gene transcripts (after controlling for ambient light effects), while keeping the animals under extended dark conditions results in the up-regulation of just one gene transcript

[68]. A differential protein expression is also found following extended light and extended dark conditions [95]. Manipulation of environmental conditions cause a 1.5-fold increase in the expression of protein transcripts that maintain the circadian rhythms, including increased transcription of the *clock*, *cry*, and *per* genes [95], similar to the effect seen in humans and rodents following sleep-cycle disruption. Administration of low voltage electric shocks (6 V/cm) is effective in disrupting sleep rhythms in adult animals; however, its impact on gene and protein transcription has not yet been evaluated.

The impact of sleep loss on behavioral measures of anxiety in adult zebrafish has produced mixed results. Measures of anxiety typically focus on one of three tests: exposure to a novel environment (e.g., novel tank test, the open field task) and/or light-dark preference (e.g., the light dark box) are highly similar to rodent anxiety novelty-based tests [47, 101]. In each of these tasks, the position of the animal within the tank is measured along with several other behavioral parameters (e.g., total distance traveled, average swim speed, latency to enter different environments, freezing behavior, erratic swimming serve as indicators of an anxiety-like response [67, 101]). Following exposure to a stressful stimulus, zebrafish tend to remain in the lower regions of their environment in the novel tank test [47]. In the light dark preference task, the animal will spend a greater proportion of their time in the darker region, demonstrating "scototaxic" behavior [47, 101]. The zebrafish open field task works in the same way as the rodent task [102]. The tank is typically devoid of environmental stimuli; stressed animals will circle the periphery of the tank, a behavior dubbed "thigmotaxis", in search of an opportunity for escape or shelter [47, 101]. Physiological markers of a stress response are evaluated through the collection and quantification of cortisol levels, since higher anxiety-like behavior positively correlates with whole body cortisol concentrations [103].

Sleep disturbances (i.e., disruption or prevention) in humans and rodents are associated with the presentation of anxiety like symptoms [104–106]. In humans, chronic partial sleep deprivation (<4 h/night) increases blood pressure and decreases the parasympathetic tone [107], accompanied by an increase in evening blood cortisol levels and pro-inflammatory cytokines [108, 109]. Anxiety disorders are often comorbid with a disruption in the individual's sleep patterns (i.e., a reduction or extension of total sleep time) [29]. Although there is a clear relationship between the disruption of sleep rhythms and anxiety, the cause and effect relationship between the two is not well understood. In such cases, the study of animal models is particularly advantageous as it allows for greater experimental control. As has been discussed, the disruption of sleep in rodents results in impaired cognitive functioning and elevated stress hormone levels [39, 110]. Extended (e.g., 96 h) paradoxical sleep deprivation (e.g., via forced swim, or forced exercise) disrupts the acquisition, consolidation and retrieval of an avoidance learning task [111]. The use of animals provides insight into the relationship between sleep and anxiety related symptoms at a deeper level than is possible with human subjects. As is discussed below, sleep disruption in zebrafish has produced mixed physiological and behavioral results.

Following sleep deprivation (e.g., via the presentation of bright ambient lighting for at least 6 h), adult zebrafish present interesting behaviors in the anxiety tests. Animals will prefer to spend more of their time in the darker regions of a light/dark tank [70]—the behavior typically suggesting an anxiety-like response (albeit it is unknown if the animals remained in this region of the tank so as to rest). Previous research has demonstrated sleep rebound in larval and adult zebrafish, although this effect was less pronounced in the adult animal [63].

In a different test of anxiety, animals were deprived of sleep by the presentation of bright lights or low-voltage electric shocks, resulting in reduced place preference (top versus bottom) in the novel tank test, and unaltered body cortisol [68]. It has previously been documented that cortisol release follows daily circadian fluctuations in several teleost species [112]. Thus, zebrafish deprived of sleep would likely not have gone through the typical hormone and steroid fluctuations throughout the day. However, this speculation necessitates additional analysis of typical behavior to enhance our understanding of the interaction between sleep and anxiety.

5 Discussion and Conclusion

The study of sleep in humans, rodents and teleost fishes has propelled our understanding of this vital and necessary behavior. Sleep deprivation can result in adverse reactions in humans and rodents. These reactions can manifest as cognitive deficits, immune dysfunction, and mood dysregulation [27, 28, 32, 33, 36, 37]. Within the realm of sleep research, zebrafish hold a tremendous potential as a translational model, and can serve to complement the existing understanding of sleep behavior as well as sleep disorders [44, 113, 114].

As a diurnal animal, zebrafish sleep behavior is thought to be driven by a circadian rhythm. Adult zebrafish sleep bouts are consolidated during the nighttime, with few sleep bouts occurring during the day [68]. The circadian rhythms guiding sleep and arousal are maintained by circadian and peripheral oscillators [20, 48, 53]. Daily fluctuations in signaling molecules [20], and hormones [63] control the transition between, and subsequently maintain rest and wakeful states. As is the case with humans and rodents, an exposure to oscillating light and dark phases in the first few days of life is essential for the entrainment of circadian rhythms. Though the zebrafish circadian clock is well equipped to maintain rhythms in cell cultures ex vivo, the systems remain susceptible to genetic, pharmacological and environmental manipulation [63, 65, 68, 70].

Our understanding of zebrafish sleep structure and function is still in its infancy, but this branch of research is very promising. Studies to date have largely examined the effects of various pharmacological manipulations (e.g. orexin, melatonin, GABA) and environmental conditions (e.g., prolonged darkness, brightness and electrical stimulation) on swim activity in adult zebrafish throughout the night [62, 63, 68, 70]. Genetic models elucidate the importance of neuropeptide agents (e.g., hypocretin) on the transition between rest and waking states in zebrafish [65, 73, 76,

78]. However, few transgenic strains have been studied with respect to sleep regulating in the zebrafish, and the effect of genetic ablation or overexpression across systems (i.e., exploring the interaction between sleep-arousal NP or NT regulating systems).

Keeping zebrafish under extended light or dark conditions effectively disrupts typical sleep rhythms and produces a sleep rebound, signifying the presence of homeostatic drives for sleep and rest in the animal [63–65]. Disrupted sleep rhythms are associated with differential gene and protein transcription of the regulatory molecules that control synapse plasticity, energy balance and circadian rhythms [68, 95]. However, these effects are specific to sleep rhythm disruption or extension by way of extending light or dark phases. The impact of sleep disruption via electric shock, mechanical stimulation (i.e., vibrations within the water), or pharmacological manipulation on gene and protein transcription remains to be explored.

The zebrafish model presents a unique combination of neural simplicity and behavioral complexity that allows for the translation of behaviors and results to rodents and humans. We know that sleep deprivation is correlated with the onset of anxiety related symptoms in humans and the propagation of anxiety related symptoms, and faulty memory in rodents [106–108, 110, 111]. The effects of sleep deprivation, and extended sleep on anxiety related behaviors in zebrafish has been recently explored and the results seem comparable to both rodent and human data [68, 70]. Along with this, evaluating the effects of environmental and pharmacological manipulations during light and dark cycles would be crucial in establishing the zebrafish sleep model. Establishing this model would then unlock the potential of pharmacological screens for the treatments of sleep disorders, due to the ease of behavioral screening, genetic manipulation and drug exposure.

References

1. Campbell SS, Tobler I. Animal sleep: a review of sleep duration across phylogeny. Neurosci Biobehav Rev. 1984;8(3):269–300.
2. Kohman MH, Carney PR. Sleep-related disorders in neurologic disease during childhood. Pediatr Neurol. 2000;23(2):107–13.
3. Lauer CJ, Riemann D, Wiegand M, Berger M. From early to late adulthood changes in EEG sleep of depressed patients and healthy volunteers. Biol Psychiatry. 1991;29(10):979–93.
4. Andrillon T, Nir Y, Staba RJ, Ferrarelli F, Cirelli C, Tononi G, Fried I. Sleep spindles in humans: insights from intracranial EEG and unit recordings. J Neurosci. 2011;31(49):17821–34.
5. Gais S, Mölle M, Helms K, Born J. Learning-dependent increases in sleep spindle density. J Neurosci. 2002;22(15):6830–4.
6. Tamminen J, Payne JD, Stickgold R, Wamsley EJ, Gaskell G. Sleep spindle activity is associated with the integration of new memories and existing knowledge. J Neurosci. 2010;30(43):14356–60.
7. Carskadon MA, Dement WC. Monitoring and staging human sleep. In: Kryger MH, Roth T, Dement WC, editors. Principles and practice of sleep medicine. 5th ed. St. Louis: Elsevier Saunders; 2011. p. 16–26.
8. Siegel JM. Do all animals sleep? Trends Neurosci. 2008;31(4):208–13.

9. Tobler I. Is sleep fundamentally different between mammalian species? Behav Brain Res. 1995;69(1):35–41.
10. Saper, C. B., Scammel, T. E., & Lu, J. (2005). Hypothalamic regulation of sleep and circadian Nature, 437(7063), 1257–1263.
11. Jin X, Shearman LP, Weaver DR, Zylka MJ, De Vries GJ, Reppert SM. A molecular mechanism regulating rhythmic output from the suprachiasmatic circadian clock. Cell. 1999;96:57–68.
12. Liu C, Weaver DR, Strogatz SH, Reppert SM. Cellular construction of a circadian clock: period determination in the suprachiasmatic nuclei. Cell. 1997;91:855–60.
13. Reppert SM. A Clockwork Explosion! Neuron. 1998;21(1):1–4.
14. Zietzer JM, Dijk DJ, Kronauer RE, Brown EN, Czeisler CA. Sensitivity of the human circadian pacemaker to nocturnal light: melatonin phase resetting and suppression. J Physiol. 2000;526(3):695–702.
15. Albrecht U, Sun ZS, Eichele G, Lee CC. A differential response of two putative mammalian circadian regulators, mper1 and mper2, to light. Cell. 1997;91:1055–64.
16. Griffin EA, Staknis D, Weitz CJ. Light-independent role of CRY1 and CRY2 in the mammalian circadian clock. Science. 1999;286(5440):768–71.
17. Honma S, Ikeda M, Abe H, Tanahashi Y, Namihira M, Honma K, Nomura M. Circadian oscillation of BMAL1, a partner of a mammalian clock gene Clock, in rat suprachiasmatic nucleus. Biochem Biophys Res Commun. 1998;250(1):83–7.
18. King DP, Zhao Y, Sangoram AM, Wilsbacher LD, Tanaka M, Antoch MP, Steeves TD, Vitaterna MH, Kornhauser JM, Lowrey PL, Turek FW, Takahashi JS. Positional cloning of the mouse circadian clock gene. Cell. 1997;89:641–53.
19. Ko CH, Takahashi JS. Molecular components of the mammalian circadian clock. Hum Mol Genet. 2006;15 Suppl 2:R271–7.
20. Pando M, Sassone-Corsi P. Unraveling the mechanisms of the vertebrate circadian clock: zebrafish may light the way. BioEssays. 2002;24:419–26.
21. Zylka MJ, Shearman LP, Weaver DR, Reppert SM. Three period homologs in mammals: differential light responses in the suprachiasmatic circadian clock and oscillating transcripts outside of brain. Neuron. 1998;20:1103–10.
22. Nishino S, Ripley B, Overeem S, Lammers GJ, Mignot E. Hypocretin (orexin) deficiency in human narcolepsy. Lancet. 2000;335(9197):39–40.
23. Corsini GU, Del Zompo M, Manconi S, Piccardi MP, Onali PL, Mangoni A, Gessa GL. Evidence for dopamine receptors in the human brain mediation sedation and sleep. Life Sci. 1977;20(9):1613–8.
24. Gaus SE, Strecker RE, Tate BA, Parker RA, Saper CB. Ventrolateral preoptic nucleus contains sleep-active, galaninergic neurons in multiple mammalian species. Neuroscience. 2002;115(1):285–94.
25. Cajochen C, Kräuchi K, Wirz-Justice A. Role of melatonin in the regulation of human circadian rhythms and sleep. J Neuroendocrinol. 2003;15(4):432–7.
26. Antonijevic IA, Murck H, Bohlhalter S, Frieboes RM, Holsboer F, Stieger A. Neuropeptide Y promotes sleep and inhibits ACTH and cortisol release in young men. Neuropharmacology. 2000;39(8):1474–81.
27. Drummond SP, Brown GG. The effects of total sleep deprivation on cerebral responses to cognitive performance. Neuropsychopharmacology. 2001;25:S68–73.
28. Jung CM, Melanson EL, Frydendall EJ, Perreault L, Eckel RH, Wright KP. Energy expenditure during sleep, sleep deprivation and sleep following sleep deprivation in adult humans. J Physiol. 2011;589(1):235–44.
29. American Psychiatric Association. Diagnostic and statistical manual of mental disorders: DSM-5. Washington, DC: American Psychiatric Association; 2013.
30. Durmer JS, Dinges DF. Neurocognitive consequences of sleep deprivation. Semin Neurol. 2005;25(1):117–29.

31. Harrison Y, Horne JA. The impact of sleep deprivation on decision making: a review. J Exp Psychol Appl. 2000;6(3):236.
32. Riemann D. Insomnia and comorbid psychiatric disorders. Sleep Med. 2007;8:S15–20.
33. Everson CA, Bergmann BM, Rechtschaffen A. Sleep deprivation in the rat: III. Total sleep deprivation. Sleep. 1989;12(1):13–21.
34. Pires GN, Tufik S, Andersen ML. Sleep deprivation and anxiety in humans and rodents—translational considerations and hypotheses. Behav Neurosci. 2015;129(5):621–33.
35. Rechtschaffen A, Bergmann BM, Everson CA. Sleep deprivation in the rat: X. Integration and discussion of the findings. Sleep. 2002;25(1):68–87.
36. Everson CA. Sustained sleep deprivation impairs host defense. Am J Phys Regul Integr Comp Phys. 1993;265(5):R1148–54.
37. Everson CA. Functional consequences of sustained sleep deprivation in the rat. Behav Brain Res. 1995;69(1):43–54.
38. Campbell IG, Guinan MJ, Horowitz JM. Sleep deprivation impairs long-term potentiation in rat hippocampal slices. J Neurophysiol. 2002;88(2):1073–6.
39. Silva RH, Abilio VC, Takatsu AL, Kameda SR, Grassl C, Chehin AB, et al. Role of hippocampal oxidative stress in memory deficits induced by sleep deprivation in mice. Neuropharmacology. 2004;46(6):895–903.
40. McKenna JT, Tartar JL, Ward CP, Thakkar MM, Cordeira JW, McCarley RW, Strecker RE. Sleep fragmentation elevates behavioral, electrographic and neurochemical measures of sleepiness. Neuroscience. 2007;146:1462–73.
41. Smith CT, Conway JM, Rose GM. Brief paradoxical sleep deprivation impairs reference, but not working, memory in the radial arm maze task. Neurobiol Learn Mem. 1998;69(2):264–78.
42. Tartar JL, McKenna JT, Ward CP, McCarley RW, Strecker RE, Brown RE. Sleep fragmentation reduces hippocampal CA1 pyramidal cell excitability and response to adenosine. Neurosci Lett. 2010;469:1–5.
43. Mohammed HS, Ezz HS, Khadrawy YA, Noor NA. Neurochemical and electrophysiological changes induced by paradoxical sleep deprivation in rats. Behav Brain Res. 2011;225(1):39–46.
44. Howe K, Clark MD, Torroja CF, Torrance J, Berthelot C, Muffato M, et al. The zebrafish reference genome sequence and its relation to the human genome. Nature. 2013;496(7446):498–503.
45. Best JD, Berghmans S, Hunt JJ, Clarke SC, Fleming A, Goldsmith P, Roach AG. Non-associative learning in larval zebrafish. Neuropsychopharmacology. 2008;33(5):1206–15.
46. Collier AD, Echevarria DJ. The utility of the zebrafish model in conditioned place preference to assess the rewarding effects of drugs. Behav Pharmacol. 2013;24(5–6):375–83.
47. Egan RJ, Bergner CL, Hart PC, Cachat JM, Canavello PR, Elegante MF, Elkhayat SI, Bartels BK, Tien AK, Tien DH, Mohnot S, Beeson E, Glasgow E, Amri H, Zukowska Z, Kalueff AV. Understanding behavioral and physiological phenotypes of stress and anxiety in zebrafish. Behav Brain Res. 2009;205(1):38–44.
48. Ben-Moshe Z, Foulkes N, Gothilf Y. Functional development of the circadian clock in the zebrafish pineal gland. Biomed Res Int. 2014;2014:235781.
49. Hirayama J, Kaneko M, Cardone L, Cahill GM, Sassone-Corsi P. Analysis of circadian rhythms in zebrafish. Methods Enzymol. 2005;393:186–204.
50. Hurd MW, Cahill GM. Entraining signals initiate behavioral circadian rhythmicity. J Biol Rhythm. 2002;17(4):307–14.
51. Cahill GM. Circadian regulation of melatonin production in cultured zebrafish pineal and retina. Brain Res. 1996;708(1):177–81.
52. Cahill GM. Clock mechanisms in zebrafish. Cell Tissue Res. 2002;309(1):27–34.
53. Vatine G, Vallone D, Gothilf Y, Foulkes N. It's time to swim! Zebrafish and the circadian clock. FEBS Lett. 2011;4(1):19.
54. Whitmore D, Foulkes N, Sassone-Corsi P. Light acts directly on organs and cells in culture to set the vertebrate circadian clock. Nature. 2000;404:87–91.

55. Whitmore D, Foulkes N, Strähle U, Sassone-Corsi P. Zebrafish clock rhythmic expression reveals independent peripheral circadian oscillators. Nat Neurosci. 1998;1(8):701–7.
56. Alsop D, Vijayan MM. Development of the corticosteroid stress axis and receptor expression in zebrafish. Am J Phys Regul Integr Comp Phys. 2008;294(3):R711–9.
57. Falcon J, Miguad H, Munoz-Cueto J, Carrillo M. Current knowledge on the melatonin system in teleost fish. Gen Comp Endocrinol. 2010;165:469–82.
58. Faraco JH, Appelbaum L, Marin W, Gaus SE, Mourrain P, Mignot E. Regulation of hypocretin (orexin) expression in embryonic zebrafish. J Biol Chem. 2006;281(40):29753–61.
59. Kim YJ, Nam RH, Yoo YM, Lee CJ. Identification and functional evidence of GABAergic neurons in parts of the brain of adult zebrafish (*Danio rerio*). Neurosci Lett. 2004;335(1):29–32.
60. Rinkwitz S, Mourrain P, Becker T. Zebrafish; an integrative system for neurogenomics and neurosciences. Prog Neurobiol. 2011;93(2):231–43.
61. Schweitzer J, Driever W. Development of the dopamine systems in zebrafish. Adv Exp Med Biol. 2009;651:1–14.
62. Zhdanova I. Sleep and its regulation in zebrafish. Rev Neurosci. 2011;22(1):27–36.
63. Zhdanova I, Wang SY, Leclair OU, Danilova NP. Melatonin promotes sleep-like state in zebrafish. Brain Res. 2001;903(1):263–8.
64. Chiu C, Prober D. Regulation of zebrafish sleep and arousal states: current and prospective approaches. Front Neural Circuits. 2013;7, 58.
65. Yokogawa T, Marin W, Faraco JH, Pézeron G, Appelbaum L, Zhang J, Rosa F, Mourrain P, Mignot E. Characterization of sleep in zebrafish and insomnia in hypocretin receptor mutants. PLoS Biol. 2007;5(10), e277.
66. Zhdanova I. Sleep in zebrafish. Zebrafish. 2006;3(2):215–26.
67. Kalueff AV, Gebhardt M, Stewart AM, Cachat JM, Brimmer M, Chawla JS, Craddock C, Kyzar EJ, Roth A, Landsman S, Gaikwad S, Robinson K, Baatrup E, Tierney K, Shamchuk A, Norton W, Miller N, Nicolson T, Braubach O, Gilman CP, Pittman J, Rosemberg DB, Gerlai R, Echevarria D, Lamb E, Neuhauss SC, Weng W, Bally-Cuif L, Schneider H, Zebrafish Neuroscience Research Consortium. Towards a comprehensive catalog of zebrafish behavior 1.0 and beyond. Zebrafish. 2013;10(1):70–86.
68. Sigurgeirsson B, Þorsteinsson H, Sigmundsdóttir S, Lieder R, Sveinsdóttir H, Sigurjónsson Ó. Sleep-wake dynamics under extended light and extended dark conditions in adult zebrafish. Behav Brain Res. 2013;256:377–90.
69. Zhdanova I, Yu L, Lopez-Patino M, Shang E, Kishi S, Guelin E. Aging of the circadian system in zebrafish and the effects of melatonin on sleep and cognitive performance. Brain Res Bull. 2008;75(2–4):433–41.
70. Singh A, Subhashini N, Sharma S, Mallick BN. Involvement of the α1-adrenoceptor in sleep-waking and sleep loss-induced anxiety behavior in zebrafish. Neuroscience. 2013;245:136–47.
71. Nishimura Y, Okabe S, Sasagawa S, Murakami S, Ashikawa Y, Yuge M, Kawaguchi K, Kawase R, Tanaka T. Pharmacological profiling of zebrafish behavior using chemical and genetic classification of sleep-wake modifiers. Front Pharmacol. 2015;6, 257.
72. Elbaz I, Foulkes N, Gothilf Y, Appelbaum L. Circadian clocks, rhythmic synaptic plasticity and the sleep-wake cycle in zebrafish. Front Neural Circuits. 2013;7, 9.
73. Prober D, Rihel J, Onah AA, Sung RJ, Schier AF. Hypocretin/orexin overexpression induces an insomnia-like phenotype in zebrafish. J Neurosci. 2006;26(51):13400–10.
74. Rihel J, Prober D, Schier AF. Monitoring sleep and arousal in zebrafish. Methods Cell Biol. 2010;100:281–94.
75. Danilova NP, Krupnik VE, Sugden D, Zhdanova I. Melatonin stimulates cell proliferation in zebrafish embryo and accelerates its development. FASEB J. 2004;18(6):751–3.
76. Appelbaum L, Wang GX, Maro GS, Mori R, Tovin A, Marin W, et al. Sleep-wake regulation and hypocretin-melatonin interaction in zebrafish. Proc Natl Acad Sci. 2009;106(51):21942–7.

77. Sundvik M, Panula P. Interactions of the orexin/hypocretin neurones and the histaminergic system. Acta Physiol. 2015;213:321–33.
78. Elbaz I, Yelin-Bekerman L, Ncenboim J, Vatine G, Appelbaum L. Genetic ablation of hypocretin neurons alters behavioral state transitions in zebrafish. J Neurosci. 2012;32(37):12961–72.
79. Naumann EA, Kampff AR, Prober D, Schier AF, Engert F. Monitoring neural activity with bioluminescence during natural behavior. Nat Neurosci. 2010;13(4):513–20.
80. Doldan MJ, Prego B, Holmquist BI, de Miguel E. Distribution of GABA-immunolabeling in the early zebrafish (*Danio rerio*) brain. Eur J Morphol. 1999;37(2-3):126–9.
81. Echevarria DJ, Hammack CM, Jouandot DJ, Toms CN. Does acute alcohol exposure modulate aggressive behaviors in the zebrafish (*Danio rerio*), or is the bark worse than the bite? Int J Comp Psychol. 2010;23:62–9.
82. Gerlai R, Lahav M, Guo S, Rosenthal A. Drinks like a fish: zebra fish (*Danio rerio*) as a behavior genetic model to study alcohol effects. Pharmacol Biochem Behav. 2000;67:773–82.
83. Lockwood B, Bjerke S, Kobayashi K, Guo S. Acute effects of alcohol on larval zebrafish: a genetic system for large-scale screening. Pharmacol Biochem Behav. 2004;77:647–54.
84. Mathur P, Guo S. Differences of acute versus chronic ethanol exposure on anxiety-like behavioral responses in zebrafish. Behav Brain Res. 2011;219(2):234–9.
85. Pannia E, Tran S, Rampersad M, Gerlai R. Acute ethanol exposure induces behavioural differences in two zebrafish (*Danio rerio*) strains: a time course analysis. Behav Brain Res. 2014;259:174–85.
86. Rosemberg DB, Braga MM, Rico EP, Loss CM, Córdova SD, Mussulini BH, Blaser RE, Leite CE, Campos MM, Dias RD, Calcagnotto ME, de Oliveira DL, Souza DO. Behavioral effects of taurine pretreatment in zebrafish acutely exposed to ethanol. Neuropharmacology. 2012;63(4):613–23.
87. Tran S, Gerlai R. Time-course of behavioural changes induced by ethanol in zebrafish (*Danio rerio*). Behav Brain Res. 2013;252:204–13.
88. Panula P, Sundvik M, Karlstedt K. Developmental roles of brain histamine. Trends Neurosci. 2014;37(3):159–68.
89. Passani MB, Lin J, Hancock A, Crochet S, Blandina P. The histamine H3 receptor as a novel therapeutic target for cognitive and sleep disorders. Trends Pharmacol Sci. 2004;25(12):618–25.
90. Renier C, Faraco JH, Bourgin P, Motley T, Bonaventure P, Rosa F, Mignot E. Genomic and functional conservation of sedative-hypnotic targets in the zebrafish. Pharmacogenet Genomics. 2007;17(4):237–53.
91. Dodt C, Breckling U, Fehm HL, Born J. Plasma epinephrine and norepinephrine concentrations of healthy humans associated with nighttime sleep and morning arousal. Hypertension. 1997;30(1):71–6.
92. Irwin M, Thompson J, Miller C, Gillin JC, Ziegler M. Effects of sleep and sleep deprivation on catecholamine and interleukin-2 levels in humans: clinical implications. J Clin Endocrinol Metab. 1999;84(6):1979–85.
93. Mitchell HA, Weinshenker D. Good night and good luck: norepinephrine in sleep pharmacology. Biochem Pharmacol. 2010;79(6):801–9.
94. Mallick BN, Majumdar S, Faisal M, Yadav V, Madan V, Pal D. Role of norepinephrine in the regulation of rapid eye movement sleep. J Biosci. 2002;28(5):539–51.
95. Purushothaman S, Saxena S, Meghah V, Meena Lakshmi MG, Singh SK, Brahmendra Swamy CV, Idris MM. Proteomic and gene expression analysis of zebrafish brain undergoing continuous light/dark stress. J Sleep Res. 2015;24(4):458–65.
96. Basheer R, Strecker RE, Thakkar MM, McCarley RW. Adenosine and sleep-wake regulation. Prog Neurobiol. 2004;73(6):379–96.
97. Porkka-Heiskanen T, Alanko L, Kalinchuk A, Stenberg D. Adenosine and sleep. Sleep Med Rev. 2002;64(4):321–32.

98. Boehmler W, Petko J, Woll M, Frey C, Thisse B, Canfield V, Levenson R. Identification of zebrafish A2 adenosine receptors and expression in developing embryos. Gene Expr Patterns. 2009;9(3):144–51.
99. Maximino C, Lima MG, Olivera KR, Picanco-Diniz DL, Herculano AM. Adenosine A1, but not A2, receptor blockade increases anxiety and arousal in zebrafish. Basic Clin Pharmacol Toxicol. 2011;109(3):203–7.
100. Bortolotto JW, de Melo GM, Cognato G, Vianna MRM, Bonan CD. Modulation of adenosine signaling prevents scopolamine-induced cognitive impairment in zebrafish. Neurobiol Learn Mem. 2015;118:113–9.
101. Stewart A, Gaikwad S, Kyzar E, Green J, Roth A, Kalueff AV. Modeling anxiety using adult zebrafish: a conceptual review. Neuropharmacology. 2012;62(1):135–43.
102. Champagne DL, Hoefnagels CC, de Kloet RE, Richardson MK. Translating rodent behavioral repertoire to zebrafish (*Danio rerio*): relevance for stress research. Behav Brain Res. 2010;214(2):332–42.
103. Canavello PR, Cachat JM, Beeson EC, Laffoon AL, Grimes C, Haymore WA, Elegante M, Bartels B, Hart P, Elkhayat S, Tien D, Mohnot S, Amri H, Kalueff AV. Measuring endocrine (cortisol) responses of zebrafish to stress. Totowa, NJ: Humana Press; 2011.
104. Kjelsberg FN, Ruud EA, Stavem K. Predictors of symptoms of anxiety and depression in obstructive sleep apnea. Sleep Med. 2005;6:341–6.
105. Neckelmann D, Mykletun A, Dahl AA. Chronic insomnia as a risk factor for developing anxiety and depression. Sleep. 2007;30(7):873–80.
106. Silva RH, Kameda SR, Carvalho RC, Takatsu-Coleman AL, Niigaki ST, Abílio VC, Tufik S, Frussa-Filho R. Anxiogenic effect of sleep deprivation in the elevated plus-maze test in mice. Psychopharmacology. 2004;176:115–22.
107. McEwen BS. Sleep deprivation as a neurobiologic and physiologic stressor: allostasis and allostatic load. Metab Clin Exp. 2006;55 Suppl 2:S20–3.
108. Leproult R, Copinschi G, Buxton O, Van Cauter E. Sleep loss results in an elevation of cortisol levels the next evening. Sleep. 1997;20:865–70.
109. Vgontzas AN, Zoumakis E, Bixler EO, Lin HM, Follett H, Kales A, Chrousos GP. Adverse effects of modest sleep restriction on sleepiness, performance, and inflammatory cytokines. J Clin Endocrinol Metab. 2004;89(5):2119–26.
110. Tartar JL, Ward CP, Cordeira JW, Legare SL, Blanchette AJ, McCarley RW, Strecker RE. Experimental sleep fragmentation and sleep deprivation in rats increases exploration in an open field test of anxiety while increasing plasma corticosterone levels. Behav Brain Res. 2009;197(2):450–3.
111. Alvarenga TA, Patti CL, Andersen ML, Silva RH, Calzavara MB, Lopez GB, et al. Paradoxical sleep deprivation impairs acquisition, consolidation, and retrieval of a discriminative avoidance task in rats. Neurobiol Learn Mem. 2008;90:624–32.
112. Peter RE, Hontela A, Cook AF, Paulencu CR. Daily cycles in serum cortisol levels in the goldfish: effects of photoperiod, temperature, and sexual condition. Can J Zool. 1978;56(11):321–32.
113. Dodd A, Curtis PM, Williams LC, Love DR. Zebrafish: bridging the gap between development and disease. Hum Mol Genet. 2000;9(16):2443–9.
114. Mathur P, Guo S. Use of zebrafish as a model to understand mechanisms of addiction and complex neurobehavioral phenotypes. Neurobiol Dis. 2010;40(1):66–72.
115. van Twyer H. Sleep patterns of five rodent species. Physiol Behav. 1969;4:901–5.
116. Sorribes A, Porsteinsson H, Arnardottir H, Johannesdottir I, Sigurgeirsson B, Polavieja G, Karlsson K. The ontogeny of sleep-wake cycles in zebrafish: a comparison to humans. Front Neural Circuits. 2013;7, 178.
117. Verret, L., Goutagny, R., Fort, P., Cagnon, L., Salvert, D., Leger, L., Boissard, R., Salin, P., Peyron, C., & Luppi, P. H. (2003). A role of melanin-concentrating hormone producing neurons in the central regulation of paradoxical sleep. BMC Neuroscience, 4(19).

118. Hallanger, A. E., Levey, A. I., Lee, H. J., Rye, D. B., & Wainer, B. H. (1987). The origins of cholinergic and other subcortical afferents to the thalamus in the rat. Journal of Comparative Neurology, 262(1), 105–124.
119. Gallopin, T., Fort, P., Eggermann, E., Cauli, B., Luppi, P. H., Rossier, J.,Audinat, E. Mühlethaler, M., & Serafin, M. (2000). Identificationof sleep-promoting neurons in vitro. Nature, 404(6781), 992–995.
120. Carskadon, M. A., & Dement, W. C. (2005). Normal human sleep: An overview. In M. H. Kryger, T. Roth, & W. C. Dement (Eds.), Principles and Practice of Sleep Medicine. 4th ed. (pp. 13-23). Philadelphia: Elsevier Saunders.

Zebrafish Behavioral Models of Ageing

Alistair J. Brock, Ari Sudwarts, Matthew O. Parker, and Caroline H. Brennan

Abstract With lifespans rapidly increasing worldwide there has been a marked increase in age-related diseases—particularly those affecting cognition—that place a major socioeconomic burden on society. Despite this, much of what occurs during the aging process at a molecular level is poorly understood, facilitating the need for a greater understanding of the processes involved. In recent years, zebrafish have proved a useful model for the identification of genetic and cellular mechanisms affecting a variety of disease processes. Here we review the potential of zebrafish as a model for the study of cognitive ageing.

Keywords Zebrafish • Cognition • Ageing • Memory • Degeneration

1 Introduction

Over the past 150 years, average lifespans have doubled in much of the world [1]. With this increase in population age, we have seen an increase in age-related diseases such as atherosclerosis, cataracts and Alzheimer's disease. Thus, whilst many continue with good health well into their 80s and 90s, others suffer increasingly debilitating cognitive decline and disease. Although one's life experiences and environment are considered to play the major role in determining healthy ageing, as much as 25 % of the variation in cognitive stability (the maintenance of mental ability into old age) within the population can be explained by genetic variation [2].

With advances in sequencing power, human genome wide association studies (GWAS) have been able to identify genes affecting numerous disease processes [3, 4]. Such genetic studies have made some progress with regard to alleles linked to cognitive disease phenotypes (e.g., presenilin, Amyloid Beta (A4) Precursor Protein

A.J. Brock • A. Sudwarts • C.H. Brennan (✉)
Queen Mary University of London, London, UK
e-mail: C.H.Brennan@qmul.ac.uk; CHBrennan@me.com

M.O. Parker
School of Health Sciences and Social Work, University of Portsmouth, Portsmouth, UK

A.V. Kalueff (ed.), *The rights and wrongs of zebrafish: Behavioral phenotyping of zebrafish*, DOI 10.1007/978-3-319-33774-6_11

(APP) and their link with dementia) but it has proved difficult to identify alleles linked to more normal age-related cognitive decline. This is in part due to the difficulty in conducting longitudinal studies that span the 70 or so years from childhood to old age as is required [2]. In addition, clinical studies must deal with a wide range of environmental and social factors which cannot be controlled and make identification of genetic associations difficult. Mutational analysis in model species with shorter lifespans and where environmental factors can be controlled, can avoid these issues faced by human studies. In comparison to naturally occurring variations, mutagenesis approaches often generate very strong phenotypes making identification of effects possible from a relatively small number of individuals. Although the very severe phenotypes may not exist within the normal human population (having been selected against through evolution), there may exist variants within the same gene that influence gene function to a lesser degree. Thus, in addition to identifying cellular processes affected, mutagenesis studies can direct human research by telling researchers where to look.

A wide range of species from worms and flies to mice and non-human primates are currently used for biomedical research into the mechanisms of ageing. However, there is increasing interest in the use of zebrafish, an established model system widely used for developmental genetic screening, as a promising model for ageing research [5–7].

Here we review behavioural assays used to assess zebrafish cognitive performance and their application to studies of age-related cognitive decline.

2 Zebrafish Ageing

During the normal aging process, animals experience age-related deterioration in cognitive ability [8]. Historically, it was thought that massive cell loss and deterioration of dendritic branching was the primary cause. However, it is now clear that cellular changes that occur are more subtle, affecting dendritic morphology, cellular connectivity, Ca2+ regulation as well as gene expression [8]. Changes in these and other cellular processes affect neuronal plasticity to ultimately affect the network dynamics of neural assemblies that support cognition [8].

Recently, zebrafish have emerged as a useful model for the cellular mechanisms of age-related phenotypes [7, 9–12]. They display gradual senescence similar to humans over a much shorter lifespan of approximately 3 years [6] with documented changes in senescence, circadian rhythmicity and onset of melatonin deficiency occurring at around 2 years of age [12, 13]. Zebrafish also display a number of other age-related phenotypes, including the aggregation of lipofusin in the liver and in the retinal pigment epithelium [5] and oxidized protein accumulation in the muscle [5, 11, 13]. Fish show a decline in the capacity for epithelial regeneration, with increases in wound lesions and impaired fin regeneration displayed in older fish [13]. Zebrafish show similar progressive declines in response (up regulation Hsp90alpha, Hsp90beta, and heat shock factor 1a) to heat shock as seen in humans [14, 15], as well as displaying comparable oxidative stress responses [16], collectively underlying the usefulness of zebrafish for studies on aging.

Fig. 1 Zebrafish with age. Images are taken at 6-months (**a**), 1-year (**b**) and 2.5 years (**c**)

While recent medical advances, better disease management and improved diets have contributed to greater life-expectancy, it has also shone the spotlight on age-related decreases in cognition [17]. The brain naturally undergoes gradual structural and functional changes, even in the absence of neurodegenerative disease, and age-related cognitive decline has become more prevalent with an ageing population [18]. Cognitive decline is one of the most common reasons for institutional care and among the most feared aspects of growing old. It is also the most costly, in terms of the financial, personal and societal burdens. It is important not only, because it often precedes dementia, illness and death, but also because of the psychological and financial burden placed on the ageing individuals and their families. It is widely agreed that more research is needed to understand the mechanisms of cognitive ageing and the factors that contribute to individual differences in rates of decline.

Age-related cognitive decline has been observed in zebrafish, with significant decreases in zebrafish learning and memory observed with age over a 4-year lifespan [7]. Comparing these behaviours in wild-type fish and mutants with altered acetylcholinesterase activity (*acheb55/+*) showed cholinergic signalling to be important to cognitive performance in ageing zebrafish [7, 19] as in humans. In humans, transcriptional profiling of ageing brains has shown a number of neurotransmitter pathways to be down-regulated [20]. Identifying which molecules and pathways are responsible for the acquisition and retention of memory will help in the understanding of how these faculties decrease with age (Fig. 1).

3 Behavioural Assays Used to Assess Cognitive Decline and Their Zebrafish Counterparts

In humans, the extent to which 'normal' ageing affects the brain is not consistent across neuroanatomical regions. For example, the hippocampi and prefrontal cortices are particularly vulnerable, with behaviours that require processing in these regions most severely affected [8, 21]. Examples of these behaviours include spatial, episodic and working memory as well as aspects of executive function.

The CANTAB Alzheimer's battery is a commonly used group of five tests designed to assess mild cognitive decline in humans and includes measures of memory and learning (spatial working memory, delayed matching to sample (DMTS) or non-matching to sample (DNMS), pair associates learning), sustained attention and reaction time. Rodent and zebrafish versions of tests designed to measure these features have been developed [22–24].

4 Spatial Memory

In humans, a 2-choice forced discrimination paradigm is commonly used as a test of visual spatial recognition memory [25]. In rodents, spatial associative learning is typically assayed using a radial arm maze [26] or the Morris water navigation task [27] which measure the subjects' ability to use, learn and memorise spatial queues in the apparatus to locate and retrieve hidden food stores. Spatial learning has been attributed through lesion studies to the function of the mammalian hippocampus [28]. Despite the typical hippocampal anatomy being missing from the teleost brain [29] spatial learning has been demonstrated in zebrafish [30] where fish were trained to find a food reward in differing arms of a four-arm radial maze using fixed environmental queues.

Another example of a behavioural assay of spatial memory in adult zebrafish has been demonstrated [7] in a model where food was administered to one side of a tank (right or left) and associated with a red-coloured wall. After 6 days of conditioning, the food anticipatory reaction (measured as the percentage of time spent in the food administered side 30-min prior to food administration) was significantly higher than baseline [7] (Fig. 2).

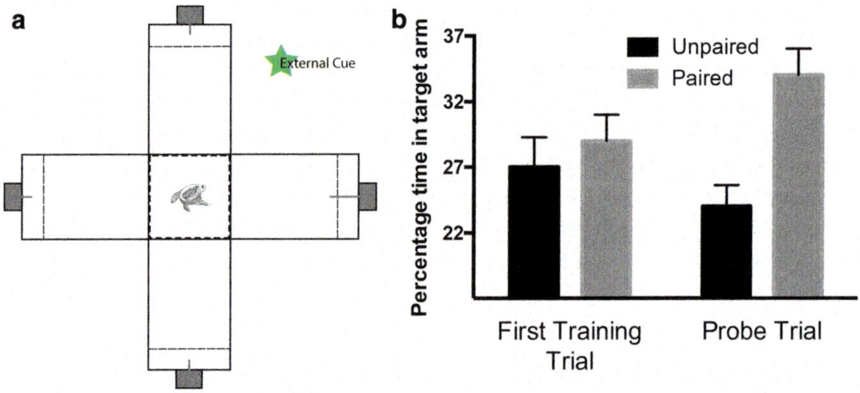

Fig. 2 The four-arm radial maze. Panel (**a**) displays a schematic of the apparatus. Fish start in the center partition where the barrier is raised and the fish must navigate to the correct arm based on environmental cues (external to the tank). Each arm is identical aside from Teflon feeding tubing (*grey lines*) being placed just in-front or just behind perforated sheets so fish only have access to food if the correct arm is chosen. The entire apparatus can be rotated between trials to give access to different arms so fish cannot rely on queues within the apparatus [30]. Panel (**b**) provides performance data of subjects at both the first training trial and at the probe trial (after 20 training trials). Zebrafish show a significant increase in time spent travelling down the arm with which they were trained to receive food, using only external cues as a means to locate the correct arm. Note: graphs reproduced from Sison and Gerlai [30]

5 Sustained Attention and Reaction Time

Sustained attention and reaction time can be assayed in humans using a continuous performance task (CPT; [31]). In animals, a variety of tests have been used (e.g., stop-signal task [32]; go no-go task [33]), but arguably the most useful has been the 5-choice serial reaction time task (5-CSRTT; [34, 35]), owing to the rich variety of parameters measurable in this assay. The task itself is designed to measure sustained attention and response inhibition by requiring the animal to detect the presence of a brief stimulus light in one of five apertures, and 'nose-poke' into that aperture to signal recognition. Responses in the correct location during a limited time following the stimulus presentation (limited hold; LH) are conditionally reinforced with illumination of a food signal light, and subsequently food reinforcement, at the opposite end of the apparatus. There is a pause (called an inter-trial interval or ITI) prior to the onset of the next stimulus light, during which a 'premature' response can be interpreted as 'impulsivity'. Once the animal has learned the task (demonstrated as >70% correct responses per session), and is performing reliably, it is possible to introduce long ITI sessions in order to probe motor impulsivity or to reduce the length of presentation of the light stimulus, or add distractors, to probe attention. Sustained attention has been successfully demonstrated in zebrafish using 3 [36, 37] and 5 [38] choice systems. It has been shown that zebrafish are able to learn to perform the 5 choice serial reaction time task reaching levels of correct performance comparable with rodents [39]. As in rodents increasing the ITI increases premature responses and, critical for genetic screening, we saw significant variation in individual responses in terms of both numbers correct and premature [38] (Figs. 3 and 4).

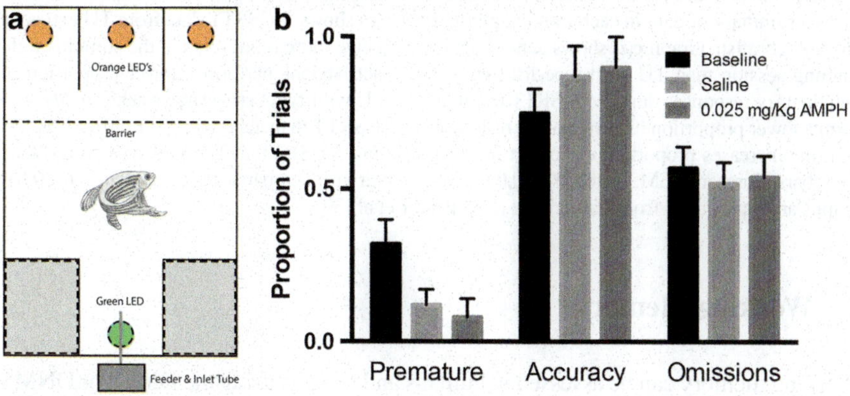

Fig. 3 3-Choice discrimination assay. Panel (**a**) shows a diagram of the testing tank with lights comprising 1× *green* (feeding) and 3× *yellow* (choice) LED's. Fish are exposed to three spatially distinct, but perceptually identical, stimuli, presented in a random order after a fixed-time inter-trial interval (ITI). Entries to the correct response are reinforced with illumination of the magazine light and delivery of a small food reward. Panel (**b**) shows performance data from the long ITI probe trials during baseline, saline and 0.025 mg/kg AMPH. There appeared to be little difference between the three conditions (i.e., baseline, saline, 0.025 mg/kg AMPH) for omissions, but a slight increase in accuracy rates. There did, however, appear to be an effect for premature responding, with the highest level of premature responding in the baseline probe trial and lowest during the AMPH treatment. Note: data is reproduced from Parker et al. [40]

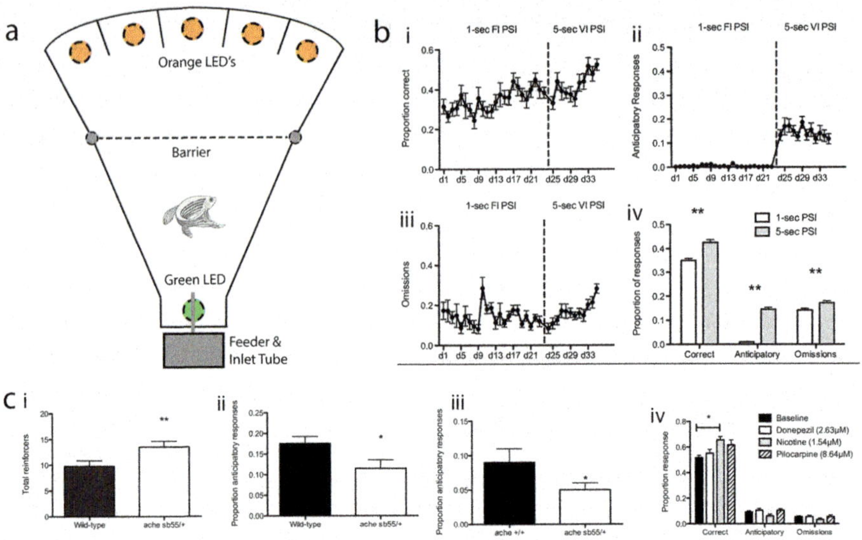

Fig. 4 5-Choice serial reaction time task (5-CSRTT). Panel (**a**) depicts the testing tank. Zebrafish are trained to swim into the lit aperture when the barrier opens at which point the green feeder light is lit and food provided. By varying the pre-stimulus interval (PSI) as well as the duration, it is possible to probe differing aspects of executive function such as attention and impulsivity [38, 39]. Panel (**b**) shows training data from 1-s fixed interval PSI and 5-s variable interval PSI of 5-CSRTT. Criterion for moving from phase 1 to 2 was ≥20 trials per session for three consecutive sessions. In graph **b**(*i*) correct responses increase steadily throughout training and significantly increased between phases 1 and 2; **b**(*ii*) shows anticipatory responses increased on initiation of the 5-s variable interval PSI; **b**(*iii*) shows that omission errors increased significantly in phase 2; **b**(*iv*) gives a summary of data in each training phase. Panel (**c**) shows 5-CSRTT data from TU and ache sb55/+ zebrafish. Graph **c**(*i*) shows ache sb55/+ fish receive more reinforcers in the stimulus-light training session than TU wild-type fish; **c**(*ii*) shows achesb55/+ perform a lower proportion of anticipatory responses during 5-CSRTT training than TU wild-type; **c**(*iii*) shows ache sb55/+ perform a lower proportion of anticipatory responses in 5-CSRTT than ache +/+; c(iv) shows 1.54 uM nicotine increases proportion of correct responses during 5-CSRTT in TU wild-type fish. (Note: error bars represent SEM. Note: $^{**}p<0.01$, post-hoc pairwise comparisons. $^{*}P<0.05$; $^{**}P<0.01$; graphs are reproduced from Parker et al [38], Parker et al [39])

6 Working Memory

Working memory can be assessed in humans and rodents using the DMTS or DNMS task where a sample stimulus is presented to the subject and after a delay, the sample is presented again together with a new stimulus. The subject is rewarded for choosing the matching (DMTS) or new (DNMTS) stimulus. Aged rats and NHP show time dependent deficits on either of these tests, with the magnitude of the deficit increasing as the delay increases [41–47]. This test can also be used to assess reaction times-latency to make a choice. Although DMTS has not been demonstrated in zebrafish, matching to sample has been documented in other fish species including goldfish [48, 49].

Novel object recognition is another established measure of short-term memory in mammalian models. In rodents, the novel object recognition test is a one-trial memory task. In this test, memory of a familiar object is manifested as preferential exploration of novel objects when given a choice. This task quantifies a naturalistic rodent behaviour in a non-stressful environment without primary reinforcing stimuli and is similar to visual recognition tests used in non-human primates. Measures of novel object recognition in zebrafish larvae have been established [50, 51], and as with all vertebrates that show lateralisation, zebrafish demonstrate differences in the use of the left vs. the right eye systems when viewing novel environments or reflections for the first time [52]. Briefly, the left eye system is used to assess novelty. When viewing a reflection or novel object for the first time, 8-day zebrafish larvae preferentially use the left eye for the initial viewing period before switching to the right eye and back. On subsequent exposure, the left eye preference is reduced presumably as the larvae can establish more readily if the object/image is novel. Although the left vs. right eye preference is useful in larval novel object recognition assays, and Miklosi [53] demonstrated a preference for left eye use when adults view conspecifics, eye preference has not yet been used as a measure of novel object recognition in adults. The novel object recognition tests used in adult zebrafish (see Fig. 5) assesses latency to approach, and time spent in the vicinity of, a previously seen 'novel' object placed in a familiar environment (e.g. home tank), or behavior on re-exposure to a 'novel' environment [54, 55]. These behavioral expressions, including thigmotaxis (preferring the edges of a test chamber), freezing, hyperactivity, erratic movement, and accelerated movements, are all characteristics used when describing individual stress coping [54, 56–62]. It could therefore be difficult to assign these behaviors to altered memory deficits as opposed to stress, boldness and anxiety when using the traditional NOR in adult zebrafish.

7 Behavioural Flexibility

People routinely encounter situations requiring them to deal with unexpected changes in their environments. The ability to adapt behaviour to environmental challenges is termed behavioural flexibility. Human and non-human animals near the end of their lifespans show impaired behavioural flexibility as measured by tests of reversal learning [63, 64]. The typical procedure for examining reversal learning involves, first, an animal being trained on a simple discrimination between two stimuli, $A+$ $B-$. When the animal reaches a set criterion (e.g., 80 % correct in a given number of choices, or *n* consecutive correct choices) the reinforced alternative is switched with the previously unreinforced alternative (i.e., $A-$ $B+$). In many species (e.g., rats, pigeons, monkeys and humans), despite a high number of errors typically during the initial reversal, the number of errors decreases as a function of subsequent reversals [65–67] (Fig. 6).

Assays of reversal learning have been performed in zebrafish [40, 68]. One study trained fish on a 2-choice colour discrimination task, before switching to the previously unreinforced alternative. Consequently, upon reaching criterion, the cues

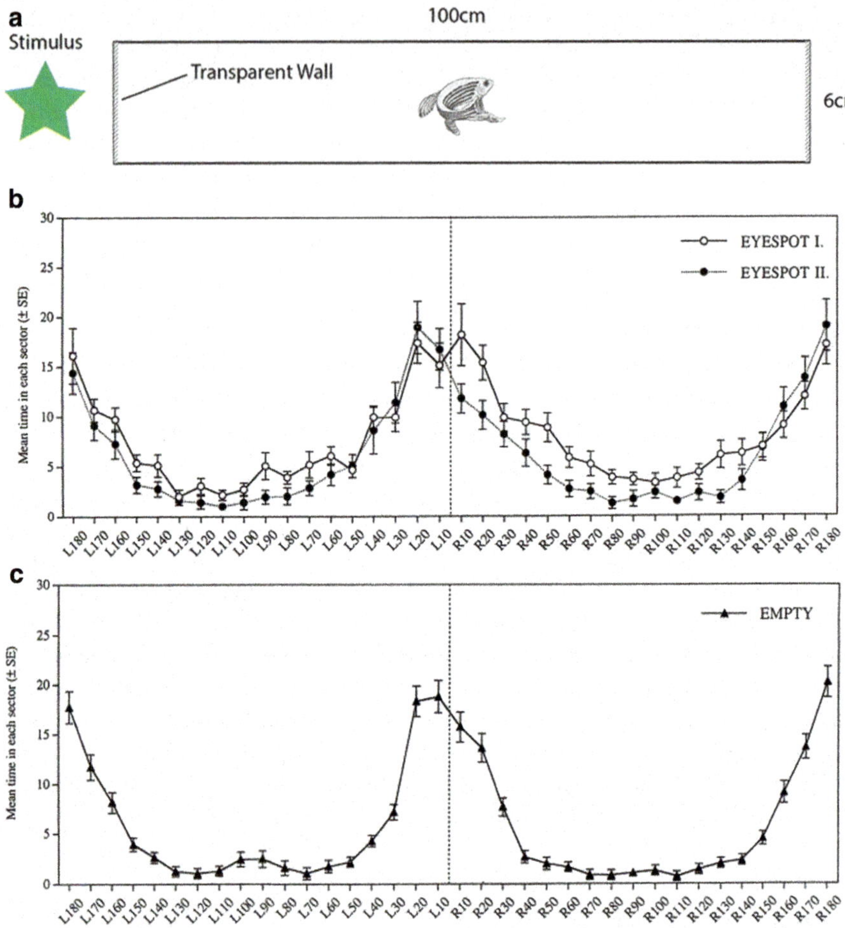

Fig. 5 Novel object recognition (NOR) assay. Panel (**a**) depicts a simple tank used in a type of NOR assay in adult zebrafish. Fish are placed in the 100 × 6 cm tank with removeable mirrors at each end. Fish are allowed to get used to swimming up and down the tank. Then the mirror is removed at one end and a novel object placed outside so it can be seen through the transparent wall at the end of the tank. By filming and assessing frames of the video it is possible to assay the angle of the fish in regard to the novel stimulus and determine which eye is being used to assess the novel object at any given point as the fish approaches the end of the tank. Panel (**b**) depicts the mean time (in seconds) spent in each 10° sector is shown for the first presentation of the viewing area, but without any objects (empty scene). In panel (**c**) the graph shows the mean time spent in each 10° sector for the first and second trials with a novel stimulus (a white ball with two dots to simulate eyes). In the first trial there is a slight tendency to use R10 (right 10°) over both L10 (left 10°) and L20 whereas in the second trial they display preferential use of L10 and L20 showing the later in this instance the left eye system was used to view familiar conspecifics and the right eye is used when it is necessary to inhibit premature response. Note: graphs are reproduced from Miklosi et al. [53]

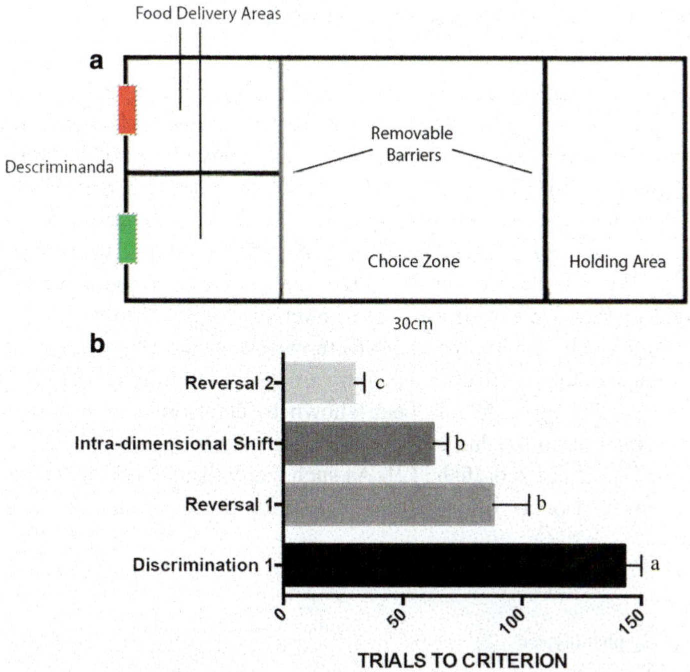

Fig. 6 Reversal learning tank. Panel (**a**) shows a schematic of a typical tank used in this type of assay. The subject is first restricted to the holding area, and at the start of each trial allowed access to the rest of the tank. A choice is made when the fish swam into either of the compartments and approached the stimulus therein. The graph in panel (**b**) denotes the trials to criterion during each of the four phases of the experiment different letters signify significant difference ($P<0.05$) [40]

were replaced with a novel pair of colours (intra-dimensional shift) and reversed again on reaching criteria [40]. Using this assay zebrafish were found to show progressive improvement in response, with the number of trials to criterion reducing as they progressed through the stages of the experiment. A different, arguably more ethological approach [7] used measures of entrainment to a temporal cue to assess behavioural flexibility in adult zebrafish. Physiological functions depends on both the intrinsic biological clock and the entrainment to environmental cues [69, 70]. To determine zebrafish acquisition of food entrainment, at the start of the adaptation period their feeding schedule was changed to a new time and restricted to 1 h/day. Following 7 days of new feeding, locomotor activity 30 min prior to food administration, was compared to the preceding 90-min period to determine acquisition of food entrainment [7]. Fish that showed increased locomotion in the 30 min prior to feeding were considered to show entrainment [7].

With the establishment of these assays of cognition, memory and learning researchers are now poised to exploit the genetic tractability of zebrafish to gain understanding of the cell biology underlying individual differences in rates of cognitive decline.

8 Other Assays of Ageing in Zebafish

In addition to the above measures of cognitive performance, zebrafish assays of physical ability (Table 1) have been developed, paralleling the fact that humans and other mammals exhibit a decline in whole-organism physical performance with age. To model this in zebrafish, sprint and endurance swimming abilities have been observed as well as the associated kinematics [77]. Zebrafish turning frequency has also been observed, as this can be an indicator of the ability to perform routine behaviours [78]. Swimming ability can be measured in a swim tunnel with an adjustable flow that forces fish to swim to maintain their position. The test section is large enough to allow zebrafish to perform various swimming gaits, which can be recorded using behavioural tracking software [77]. Locomotor activity has also been recorded with age and has been shown to deteriorate in zebrafish, with a significant reduction in the daily amplitude of activity in 3-year old and 4-year-old, compared to 1-year old zebrafish [12]. As such, zebrafish provide a useful model to investigate other aspects of age related deterioration, not just those involving cognition.

Table 1 Ageing phenotypes

Human aging phenotypes	Mouse aging phenotypes	Zebrafish aging phenotypes
Skin becomes less elastic, the oil glands gradually produce less oil [71]	Chronological skin aging [72]	Impaired wound healing and fin regeneration [13]
Hair loss/thinning	Mouse models for human hair loss disorders [73, 74]	n/a
Bone compression, joint compression, muscle weakening	Muscle aging [75] and age-related bone loss [76]	Curved spine [11], reduced swim performance [77, 78] and muscle degeneration [9]
Hearing loss	Age-related hearing loss is a observed [79, 80]	Not tested
Vision loss, cataracts	Reduced visual function [81, 82], cataracts [83]	Increased cataracts [84]
Sleep pattern/circadien rhythm	Decline in circadian output [85, 86]	Altered circadian rhythmicity [12] and onset of melatonin deficiency [13]
Anxiety-related symptoms	Changes in anxiety [87–89]	Models of anxiety exist [54], though have not been examined with regards to age
Cognitive decline	Gradual senescence [90, 91]	Age-related cognitive decline observed [7]
Memory loss, rates of learning	Decreased memory [92, 93]	Reduced rate of learning [7]

Typical aging phenotypes observed in aging humans according to the World Health Organisation (http://www.who.int/mediacentre/factsheets/fs404/en/) with and their mouse and zebrafish counterparts

8.1 Ageing Phenotypes Observed?

The majority of behavioural assays described above have yet to be fully adapted to zebrafish ageing research. The exceptions to this are studies which looked at the effect of aging (1, 2 and 3 years) as well as cholinergic signalling on spatial learning and behavioural flexibility [7]. Yu et al [7] found that in the case of entrainment to a temporal food cue, young age correlated with the ability of wild type zebrafish to anticipate a meal. Several middle-aged (2 year) and old fish (3 year) failed to significantly entrain to the new time of feeding [7]. In the same assay, the effect of increased cholinergic signalling on food entrainment in young and ageing zebrafish was assessed using fish heterozygous for an acetylcholinesterase loss of function mutation (*acheb55/+)* [7]. In these experiments middle-aged *acheb55/+* showed no difference in food anticipation to young *acheb55/+* fish, with increased cholinergic signalling attenuating the effect seen in wild-type. Similarly, in the spatial entrainment task, only young wild-type fish demonstrated both anticipation and preference of the food-associated side. Both 1- and 2-year-old fish demonstrated spatial entrainment in the *acheb55/+* when compared with baseline indicating reduced decline in this mutant line. Studies looking at the same acheb55/+ mutants in the 5-choice serial reaction time task showed them to show far fewer anticipatory responses when compared with wild-type as well as showing increased learning rates during training [19]. These studies are consistent with findings in humans that anticholinesterase drugs improve cognitive performance in both young and ageing individuals.

Finally, using zebrafish assays of physical ability, Gilbert et al. [77] observed an age associated decline in endurance and sprint performance. Further to this, they found swimming performance to be improved with exercise training in young and middle-aged zebrafish, but not in old zebrafish (Fig. 7).

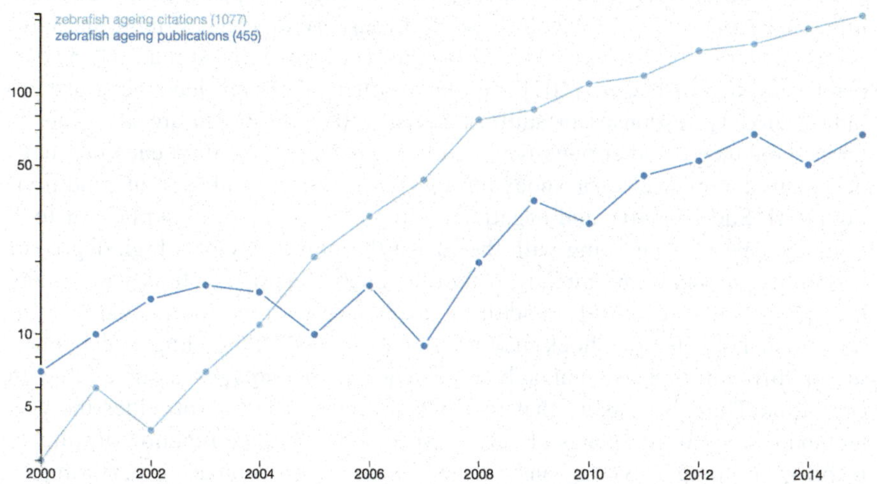

Fig. 7 Graph of number of publications (*dark blue*) and citations (*light blue*) pertaining to the zebrafish ageing research by year starting in 2000 until present. The number of publications has largely grown year on year

9 Limitations and Future Directions

The utility of zebrafish as a tool for researching aging and age-related diseases has substantially increased in recent years. As zebrafish are vertebrate species that share many cellular and neural processes with humans and other mammals, any mechanisms affecting age-related cognition will be of significant translational relevance. With an estimated 70 % of translated human genes—and around 85 % of human disease-associated genes—found in zebrafish, this organisms provides a highly valuable in vivo method of genetic and pharmacological screening [94]. For example, forward genetic screens that assess the relative rate of decline of performance of wildtype and mutant lines in cognitive tasks can be used to inform human studies. Similarly, the ability to generate loss of function lines using CRISPR or TALEN gene editing allows for the effect of candidate genes on task performance and decline to be assessed [95–98]. In addition, numerous mutant resources are available [99, 100]. Thus, zebrafish have great potential to contribute to understanding of the ageing brain. However, currently one limitation to screening approaches using the assays described above is the time and resources required to screen the numbers necessary. The possible answer lies in development of automated behavioural assays for specific aspects of cognition where large numbers of fish can be screened with limited researcher interaction and handling, minimising variability. A number of automated systems have been described [19, 38, 101] however, these have yet to be applied to the problem of cognitive ageing.

The use of food as the reward in these learning tasks could also be an issue. As is seen in other species, it may be possible that zebrafish appetite decreases with age [7], which may make comparisons in learning between young adults and older subjects problematic. Indeed, as described above, it has been shown that older wild-type zebrafish are slower in establishing an anticipatory locomotor response to regular food administration when compared with their juvenile counterparts [7]. Zebrafish respond favourably to social stimuli and have been shown to preferentially shoal with conspecific artificial stimuli when compared with heterospecific [102]. Similarly, artificial stimuli have been used to induce anxiety responses in adult zebrafish [103]. Zebrafish are also able to learn avoidance of other negative stimuli, for example avoiding cues that have been paired with an electric shock through both classical and operant conditioning [104]. Such positive and negative reinforcers may be used in place of food to assay aspects of learning with the caveat that there may be a high degree of variability in social and anxiety responses due to factors such as within-tank hierarchal structures [105]. In addition, avoidance learning from negative reinforcers such as electric shock may provide problems with ceiling effects making it difficult to tease out subtle behavioural phenotypes associated with cognition. Thus, it is likely that food-reward based learning provides the best method of identifying genes affecting aspects of executive function despite an inability to compare young and old fish directly; by comparing performance relative to age matched young and old wild-type populations.

Despite the importance of healthy ageing to society, the genes involved in cognitive ageing remain largely unknown [2, 106]. *In zebrafish, age-related changes in cognition manifest as early as 2 years of age and then further deteriorate* [7, 12]. *Use of mutant and environmentally-affected zebrafish showed potential advantages of studying the genetic and environmental factors that modulate the rate of ageing in this animal model. Their relatively long life-spans allow for the use of forward genetic screens to identify the effect of genes on senescence over an entire 3-year period—which is not possible in many vertebrate models—and may allow for the discovery of novel candidate genes for study in mammals, including humans.* Identifying these genes and the cellular processes they affect will not only help us understand how environmental factors may interact with genetics to regulate cognitive stability, but by doing so will also aid the development of strategies to minimize the impact of ageing on society.

References

1. Mathers CD, Stevens GA, Boerma T, White RA, Tobias MI. Causes of international increases in older age life expectancy. Lancet. 2015;385(9967):540–8. doi:10.1016/S0140-6736(14)60569-9.
2. Deary IJ, Yang J, Davies G, Harris SE, Tenesa A, Liewald D, et al. Genetic contributions to stability and change in intelligence from childhood to old age. Nature. 2012;482(7384):212–5. doi:10.1038/nature10781.
3. McCarthy MI, Abecasis GR, Cardon LR, Goldstein DB, Little J, Ioannidis JP, Hirschhorn JN. Genome-wide association studies for complex traits: consensus, uncertainty and challenges. Nat Rev Genet. 2008;9(5):356–69. doi:10.1038/nrg2344.
4. Zeggini E, Scott LJ, Saxena R, Voight BF, Marchini JL, Hu T, et al. Meta-analysis of genome-wide association data and large-scale replication identifies additional susceptibility loci for type 2 diabetes. Nat Genet. 2008;40(5):638–45. doi:10.1038/ng.120.
5. Kishi S, Bayliss PE, Uchiyama J, Koshimizu E, Qi J, Nanjappa P, et al. The identification of zebrafish mutants showing alterations in senescence-associated biomarkers. PLoS Genet. 2008;4(8), e1000152. doi:10.1371/journal.pgen.1000152.
6. Kishi S, Slack BE, Uchiyama J, Zhdanova IV. Zebrafish as a genetic model in biological and behavioral gerontology: where development meets aging in vertebrates—a mini-review. Gerontology. 2009;55(4):430–41. doi:10.1159/000228892.
7. Yu L, Tucci V, Kishi S, Zhdanova IV. Cognitive aging in zebrafish. PLoS One. 2006;1, e14. doi:10.1371/journal.pone.0000014.
8. Burke SN, Barnes CA. Neural plasticity in the ageing brain. Nat Rev Neurosci. 2006;7(1):30–40. doi:10.1038/nrn1809.
9. Gerhard GS, Kauffman EJ, Wang X, Stewart R, Moore JL, Kasales CJ, et al. Life spans and senescent phenotypes in two strains of Zebrafish (*Danio rerio*). Exp Gerontol. 2002;37(8–9):1055–68.
10. Keller ET, Murtha JM. The use of mature zebrafish (*Danio rerio*) as a model for human aging and disease. Comp Biochem Physiol C Toxicol Pharmacol. 2004;138(3):335–41. doi:10.1016/j.cca.2004.04.001.
11. Kishi S, Uchiyama J, Baughman AM, Goto T, Lin MC, Tsai SB. The zebrafish as a vertebrate model of functional aging and very gradual senescence. Exp Gerontol. 2003;38(7):777–86.
12. Zhdanova IV, Yu L, Lopez-Patino M, Shang E, Kishi S, Guelin E. Aging of the circadian system in zebrafish and the effects of melatonin on sleep and cognitive performance. Brain Res Bull. 2008;75(2–4):433–41. doi:10.1016/j.brainresbull.2007.10.053.

13. Tsai SB, Tucci V, Uchiyama J, Fabian NJ, Lin MC, Bayliss PE, et al. Differential effects of genotoxic stress on both concurrent body growth and gradual senescence in the adult zebrafish. Aging Cell. 2007;6(2):209–24. doi:10.1111/j.1474-9726.2007.00278.x.
14. Keller JM, Escara-Wilke JF, Keller ET. Heat stress-induced heat shock protein 70 expression is dependent on ERK activation in zebrafish (*Danio rerio*) cells. Comp Biochem Physiol A Mol Integr Physiol. 2008;150(3):307–14. doi:10.1016/j.cbpa.2008.03.021.
15. Murtha JM, Keller ET. Characterization of the heat shock response in mature zebrafish (*Danio rerio*). Exp Gerontol. 2003;38(6):683–91.
16. Malek RL, Sajadi H, Abraham J, Grundy MA, Gerhard GS. The effects of temperature reduction on gene expression and oxidative stress in skeletal muscle from adult zebrafish. Comp Biochem Physiol C Toxicol Pharmacol. 2004;138(3):363–73. doi:10.1016/j.cca.2004.08.014.
17. Salthouse TA. When does age-related cognitive decline begin? Neurobiol Aging. 2009;30(4):507–14. doi:10.1016/j.neurobiolaging.2008.09.023.
18. Driscoll I, Davatzikos C, An Y, Wu X, Shen D, Kraut M, Resnick SM. Longitudinal pattern of regional brain volume change differentiates normal aging from MCI. Neurology. 2009;72(22):1906–13. doi:10.1212/WNL.0b013e3181a82634.
19. Parker MO, Brock AJ, Sudwarts A, Teh M-T, Combe F, Brennan CH. Developmental role of acetylcholinesterase in impulse control in zebrafish. Front Behav Neurosci. 2015;9. doi:10.3389/fnbeh.2015.00271.
20. Lu T, Pan Y, Kao SY, Li C, Kohane I, Chan J, Yankner BA. Gene regulation and DNA damage in the ageing human brain. Nature. 2004;429(6994):883–91. doi:10.1038/nature02661.
21. Maillet D, Rajah MN. Association between prefrontal activity and volume change in prefrontal and medial temporal lobes in aging and dementia: a review. Ageing Res Rev. 2013;12(2):479–89. doi:10.1016/j.arr.2012.11.001.
22. Gerlai R. Phenomics: fiction or the future? Trends Neurosci. 2002;25(10):506–9.
23. Keeler JF, Robbins TW. Translating cognition from animals to humans. Biochem Pharmacol. 2011;81(12):1356–66. doi:10.1016/j.bcp.2010.12.028.
24. Weed MR, Taffe MA, Polis I, Roberts AC, Robbins TW, Koob GF, et al. Performance norms for a rhesus monkey neuropsychological testing battery: acquisition and long-term performance. Brain Res Cogn Brain Res. 1999;8(3):185–201.
25. Kuzmickiene J, Kaubrys G. Selective ability of some CANTAB battery test measures to detect cognitive response to a single dose of donepezil in Alzheimer disease. Med Sci Monit. 2015;21:2572–82. doi:10.12659/msm.895381.
26. Crusio WE, Schwegler H, Lipp HP. Radial-maze performance and structural variation of the hippocampus in mice: a correlation with mossy fibre distribution. Brain Res. 1987;425(1): 182–5.
27. Morris RGM. Spatial localization does not require the presence of local cues. Learn Motiv. 1981;12(2):239–60. doi:10.1016/0023-9690(81)90020-5.
28. Dusek JA, Eichenbaum H. The hippocampus and memory for orderly stimulus relations. Proc Natl Acad Sci U S A. 1997;94(13):7109–14.
29. Salas C, Broglio C, Duran E, Gomez A, Ocana FM, Jimenez-Moya F, Rodriguez F. Neuropsychology of learning and memory in teleost fish. Zebrafish. 2008;3(2):157–71. doi:10.1089/zeb.2006.3.157.
30. Sison M, Gerlai R. Associative learning in zebrafish (*Danio rerio*) in the plus maze. Behav Brain Res. 2010;207(1):99–104. doi:10.1016/j.bbr.2009.09.043.
31. Beck LH, Bransome Jr ED, Mirsky AF, Rosvold HE, Sarason I. A continuous performance test of brain damage. J Consult Psychol. 1956;20(5):343–50.
32. Logan GD, Schachar RJ, Tannock R. Impulsivity and inhibitory control. Psychol Sci. 1997;8(1):60–4. doi:10.1111/j.1467-9280.1997.tb00545.x.
33. Finn PR, Justus A, Mazas C, Steinmetz JE. Working memory, executive processes and the effects of alcohol on Go/No-Go learning: testing a model of behavioral regulation and impulsivity. Psychopharmacology (Berl). 1999;146(4):465–72.

34. Carli M, Robbins TW, Evenden JL, Everitt BJ. Effects of lesions to ascending noradrenergic neurones on performance of a 5-choice serial reaction task in rats; implications for theories of dorsal noradrenergic bundle function based on selective attention and arousal. Behav Brain Res. 1983;9(3):361–80.
35. Robbins TW. The 5-choice serial reaction time task: behavioural pharmacology and functional neurochemistry. Psychopharmacology (Berl). 2002;163(3–4):362–80. doi:10.1007/s00213-002-1154-7.
36. Bilotta J, Risner ML, Davis EC, Haggbloom SJ. Assessing appetitive choice discrimination learning in zebrafish. Zebrafish. 2005;2(4):259–68. doi:10.1089/zeb.2005.2.259.
37. Parker MO, Millington ME, Combe FJ, Brennan CH. Development and implementation of a three-choice serial reaction time task for zebrafish (*Danio rerio*). Behav Brain Res. 2012;227(1):73–80. doi:10.1016/j.bbr.2011.10.037.
38. Parker MO, Ife D, Ma J, Pancholi M, Smeraldi F, Straw C, Brennan CH. Development and automation of a test of impulse control in zebrafish. Front Syst Neurosci. 2013;7:65. doi:10.3389/fnsys.2013.00065.
39. Parker MO, Brock AJ, Sudwarts A, Brennan CH. Atomoxetine reduces anticipatory responding in a 5-choice serial reaction time task for adult zebrafish. Psychopharmacology (Berl). 2014;231(13):2671–9. doi:10.1007/s00213-014-3439-z.
40. Parker MO, Gaviria J, Haigh A, Millington ME, Brown VJ, Combe FJ, Brennan CH. Discrimination reversal and attentional sets in zebrafish (*Danio rerio*). Behav Brain Res. 2012;232(1):264–8. doi:10.1016/j.bbr.2012.04.035.
41. Andrews JS, Jansen JH, Linders S, Princen A. Effects of disrupting the cholinergic system on short-term spatial memory in rats. Psychopharmacology (Berl). 1994;115(4):485–94.
42. Buccafusco JJ. Estimation of working memory in macaques for studying drugs for the treatment of cognitive disorders. J Alzheimers Dis. 2008;15(4):709–20.
43. Bushnell PJ, Levin ED, Overstreet DH. Spatial working and reference memory in rats bred for autonomic sensitivity to cholinergic stimulation: acquisition, accuracy, speed, and effects of cholinergic drugs. Neurobiol Learn Mem. 1995;63(2):116–32. doi:10.1006/nlme.1995.1012.
44. Deacon RMJ. Pharmacological studies of a rat spatial delayed nonmatch-to-sample task as an animal-model of dementia. Drug Dev Res. 1991;24(1):67–79. doi:10.1002/ddr.430240106.
45. Rodriguez JS, Zurcher NR, Bartlett TQ, Nathanielsz PW, Nijland MJ. CANTAB delayed matching to sample task performance in juvenile baboons. J Neurosci Methods. 2011;196(2):258–63. doi:10.1016/j.jneumeth.2011.01.012.
46. Ruotsalainen S, MacDonald E, Miettinen R, Puumala T, Riekkinen Sr P, Sirvio J. Additive deficits in the choice accuracy of rats in the delayed non-matching to position task after cholinolytics and serotonergic lesions are non-mnemonic in nature. Psychopharmacology (Berl). 1997;130(4):303–12.
47. Zola-Morgan S, Squire LR. The neuropsychology of memory. Parallel findings in humans and nonhuman primates. Ann N Y Acad Sci. 1990;608:434–50. discussion 450–6.
48. Goldman M, Shapiro S. Matching-to-sample and oddity-from-sample in goldfish. J Exp Anal Behav. 1979;31(2):259–66. doi:10.1901/jeab.1979.31-259.
49. Zerbolio DJ, Royalty J. Matching and oddity conditional discrimination in the goldfish as avoidance responses: evidence for conceptual avoidance learning. Anim Learn Behav. 1983;11(3):341–8. doi:10.3758/BF03199786.
50. Andersson MA, Ek F, Olsson R. Using visual lateralization to model learning and memory in zebrafish larvae. Sci Rep. 2015;5:8667. doi:10.1038/srep08667.
51. Miklosi A, Andrew RJ. The zebrafish as a model for behavioral studies. Zebrafish. 2006;3(2):227–34. doi:10.1089/zeb.2006.3.227.
52. Sovrano VA, Andrew RJ. Eye use during viewing a reflection: behavioural lateralisation in zebrafish larvae. Behav Brain Res. 2006;167(2):226–31. doi:10.1016/j.bbr.2005.09.021.
53. Miklosi A, Andrew RJ, Savage H. Behavioural lateralisation of the tetrapod type in the zebrafish (*Brachydanio rerio*). Physiol Behav. 1997;63(1):127–35.

54. Maximino C, de Brito TM, Colmanetti R, Pontes AA, de Castro HM, de Lacerda RI, et al. Parametric analyses of anxiety in zebrafish scototaxis. Behav Brain Res. 2010;210(1):1–7. doi:10.1016/j.bbr.2010.01.031.
55. Wright D, Nakamichi R, Krause J, Butlin RK. QTL analysis of behavioral and morphological differentiation between wild and laboratory zebrafish (*Danio rerio*). Behav Genet. 2006;36(2):271–84. doi:10.1007/s10519-005-9029-4.
56. Blaser R, Gerlai R. Behavioral phenotyping in zebrafish: comparison of three behavioral quantification methods. Behav Res Methods. 2006;38(3):456–69.
57. Cachat J, Canavello P, Elegante M, Bartels B, Hart P, Bergner C, et al. Modeling withdrawal syndrome in zebrafish. Behav Brain Res. 2010;208(2):371–6. doi:10.1016/j.bbr.2009.12.004.
58. Egan RJ, Bergner CL, Hart PC, Cachat JM, Canavello PR, Elegante MF, et al. Understanding behavioral and physiological phenotypes of stress and anxiety in zebrafish. Behav Brain Res. 2009;205(1):38–44. doi:10.1016/j.bbr.2009.06.022.
59. Gerlai R, Fernandes Y, Pereira T. Zebrafish (*Danio rerio*) responds to the animated image of a predator: towards the development of an automated aversive task. Behav Brain Res. 2009;201(2):318–24. doi:10.1016/j.bbr.2009.03.003.
60. Lopez-Patino MA, Yu L, Cabral H, Zhdanova IV. Anxiogenic effects of cocaine withdrawal in zebrafish. Physiol Behav. 2008;93(1-2):160–71. doi:10.1016/j.physbeh.2007.08.013.
61. Lopez Patino MA, Yu L, Yamamoto BK, Zhdanova IV. Gender differences in zebrafish responses to cocaine withdrawal. Physiol Behav. 2008;95(1–2):36–47. doi:10.1016/j.physbeh.2008.03.021.
62. Wong K, Elegante M, Bartels B, Elkhayat S, Tien D, Roy S, et al. Analyzing habituation responses to novelty in zebrafish (*Danio rerio*). Behav Brain Res. 2010;208(2):450–7. doi:10.1016/j.bbr.2009.12.023.
63. Barense MD, Fox MT, Baxter MG. Aged rats are impaired on an attentional set-shifting task sensitive to medial frontal cortex damage in young rats. Learn Mem. 2002;9(4):191–201. doi:10.1101/lm.48602.
64. Gallagher M, Stocker AM, Koh MT. Mindspan: lessons from rat models of neurocognitive aging. ILAR J. 2011;52(1):32–40.
65. Crutch SJ, Warrington EK. Contrasting patterns of comprehension for superordinate, basic-level, and subordinate names in semantic dementia and aphasic stroke patients. Cogn Neuropsychol. 2008;25(4):582–600. doi:10.1080/02643290701862290.
66. Gonzalez RC, Behrend ER, Bitterman ME. Reversal learning and forgetting in bird and fish. Science. 1967;158(3800):519–21.
67. Rajalakshmi R, Jeeves MA. The relative difficulty of reversal learning (reversal index) as a basis of behavioural comparisons. Anim Behav. 1965;13(2):203–11.
68. Colwill RM, Raymond MP, Ferreira L, Escudero H. Visual discrimination learning in zebrafish (*Danio rerio*). Behav Processes. 2005;70(1):19–31. doi:10.1016/j.beproc.2005.03.001.
69. Bolles RC, Stokes LW. Rat's anticipation of diurnal and a-diurnal feeding. J Comp Physiol Psychol. 1965;60(2):290–4.
70. Schibler U, Ripperger J, Brown SA. Peripheral circadian oscillators in mammals: time and food. J Biol Rhythms. 2003;18(3):250–60.
71. Gilchrest BA. Age-associated changes in the skin. J Am Geriatr Soc. 1982;30(2):139–43.
72. Hwang KA, Yi BR, Choi KC. Molecular mechanisms and in vivo mouse models of skin aging associated with dermal matrix alterations. Lab Anim Res. 2011;27(1):1–8. doi:10.5625/lar.2011.27.1.1.
73. Porter R, Jahoda C, Lunny D, Henderson G, Ross J, McLean W, et al. 26 Mouse models for human hair loss disorders. J Anat. 2002;201(5):424.
74. Porter RM. Mouse models for human hair loss disorders. J Anat. 2003;202(1):125–31.
75. Musaro A, Rosenthal N. Transgenic mouse models of muscle aging. Exp Gerontol. 1999;34(2):147–56.
76. Jilka RL. The relevance of mouse models for investigating age-related bone loss in humans. J Gerontol A Biol Sci Med Sci. 2013;68(10):1209–17. doi:10.1093/gerona/glt046.

77. Gilbert MJ, Zerulla TC, Tierney KB. Zebrafish (*Danio rerio*) as a model for the study of aging and exercise: physical ability and trainability decrease with age. Exp Gerontol. 2014;50:106–13. doi:10.1016/j.exger.2013.11.013.
78. Plaut I. Critical swimming speed: its ecological relevance. Comp Biochem Physiol A Mol Integr Physiol. 2001;131(1):41–50.
79. Kujawa SG, Liberman MC. Acceleration of age-related hearing loss by early noise exposure: evidence of a misspent youth. J Neurosci. 2006;26(7):2115–23. doi:10.1523/Jneurosci.4985-05.2006.
80. Someya S, Xu JZ, Kondo K, Ding DL, Salvi RJ, Yamasoba T, et al. Age-related hearing loss in C57BL/6 J mice is mediated by Bak-dependent mitochondrial apoptosis. Proc Natl Acad Sci U S A. 2009;106(46):19432–7. doi:10.1073/pnas.0908786106.
81. Kolesnikov AV, Fan J, Crouch RK, Kefalov VJ. Age-related deterioration of rod vision in mice. J Neurosci. 2010;30(33):11222–31. doi:10.1523/Jneurosci.4239-09.2010.
82. Lehmann K, Schmidt KF, Lowel S. Vision and visual plasticity in ageing mice. Restor Neurol Neurosci. 2012;30(2):161–78. doi:10.3233/Rnn-2012-110192.
83. Wolf N, Penn P, Pendergrass W, Van Remmen H, Bartke A, Rabinovitch P, Martin GM. Age-related cataract progression in five mouse models for anti-oxidant protection or hormonal influence. Exp Eye Res. 2005;81(3):276–85. doi:10.1016/j.exer.2005.01.024.
84. Greiling TM, Houck SA, Clark JI. The zebrafish lens proteome during development and aging. Mol Vis. 2009;15:2313–25.
85. Nakamura TJ, Nakamura W, Yamazaki S, Kudo T, Cutler T, Colwell CS, Block GD. Age-related decline in circadian output. J Neurosci. 2011;31(28):10201–5. doi:10.1523/JNEUROSCI.0451-11.2011.
86. Valentinuzzi VS, Scarbrough K, Takahashi JS, Turek FW. Effects of aging on the circadian rhythm of wheel-running activity in C57BL/6 mice. Am J Physiol. 1997;273(6 Pt 2):R1957–64.
87. Chen GH, Wang C, Yangcheng HY, Liu RY, Zhou JN. Age-related changes in anxiety are task-specific in the senescence-accelerated prone mouse 8. Physiol Behav. 2007;91(5):644–51. doi:10.1016/j.physbeh.2007.03.023.
88. Miyamoto M. Characteristics of age-related behavioral changes in senescence-accelerated mouse SAMP8 and SAMP10. Exp Gerontol. 1997;32(1–2):139–48.
89. Miyamoto M, Kiyota Y, Nishiyama M, Nagaoka A. Senescence-accelerated mouse (SAM): age-related reduced anxiety-like behavior in the SAM-P/8 strain. Physiol Behav. 1992;51(5):979–85.
90. Gower AJ, Lamberty Y. The aged mouse as a model of cognitive decline with special emphasis on studies in NMRI mice. Behav Brain Res. 1993;57(2):163–73. doi:10.1016/0166-4328(93)90132-A.
91. Murphy GG, Rahnama NP, Silva AJ. Investigation of age-related cognitive decline using mice as a model system: behavioral correlates. Am J Geriatr Psychiatr. 2006;14(12):1004–11. doi:10.1097/01.JGP.0000209405.27548.7b.
92. Marighetto A, Brayda-Bruno L, Etchamendy N. Studying the impact of aging on memory systems: contribution of two behavioral models in the mouse. Curr Top Behav Neurosci. 2012;10:67–89. doi:10.1007/7854_2011_151.
93. Miyamoto M, Kiyota Y, Yamazaki N, Nagaoka A, Matsuo T, Nagawa Y, Takeda T. Age-related changes in learning and memory in the senescence-accelerated mouse (SAM). Physiol Behav. 1986;38(3):399–406.
94. Langheinrich U. Zebrafish: a new model on the pharmaceutical catwalk. Bioessays. 2003;25(9):904–12. doi:10.1002/bies.10326.
95. Hwang WY, Fu YF, Reyon D, Maeder ML, Tsai SQ, Sander JD, et al. Efficient genome editing in zebrafish using a CRISPR-Cas system. Nat Biotechnol. 2013;31(3):227–9. doi:10.1038/nbt.2501.
96. Ma AC, Lee HB, Clark KJ, Ekker SC. High efficiency in vivo genome engineering with a simplified 15-RVD GoldyTALEN design. PLos One. 2013;8(5), e65259. doi:10.1371/journal.pone.0065259.

97. Shah AN, Davey CF, Whitebirch AC, Miller AC, Moens CB. Rapid reverse genetic screening using CRISPR in zebrafish. Nat Methods. 2015;12(6):535–40. doi:10.1038/nmeth.3360.
98. Zu Y, Tong XJ, Wang ZX, Liu D, Pan RC, Li Z, et al. TALEN-mediated precise genome modification by homologous recombination in zebrafish. Nat Methods. 2013;10(4):329–31. doi:10.1038/Nmeth.2374.
99. Clark KJ, Argue DP, Petzold AM, Ekker SC. zfishbook: connecting you to a world of zebrafish revertible mutants. Nucleic Acids Res. 2012;40(D1):D907–11. doi:10.1093/nar/gkr957.
100. Kettleborough RNW, Busch-Nentwich EM, Harvey SA, Dooley CM, de Bruijn E, van Eeden F, et al. A systematic genome-wide analysis of zebrafish protein-coding gene function. Nature. 2013;496(7446):494–7. doi:10.1038/nature11992.
101. Mueller KP, Neuhauss SC. Automated visual choice discrimination learning in zebrafish (*Danio rerio*). J Integr Neurosci. 2012;11(1):73–85. doi:10.1142/S0219635212500057.
102. Saverino C, Gerlai R. The social zebrafish: behavioral responses to conspecific, heterospecific, and computer animated fish. Behav Brain Res. 2008;191(1):77–87. doi:10.1016/j.bbr.2008.03.013.
103. Luca RM, Gerlai R. In search of optimal fear inducing stimuli: differential behavioral responses to computer animated images in zebrafish. Behav Brain Res. 2012;226(1):66–76. doi:10.1016/j.bbr.2011.09.001.
104. Valente A, Huang KH, Portugues R, Engert F. Ontogeny of classical and operant learning behaviors in zebrafish. Learn Mem. 2012;19(4):170–7. doi:10.1101/lm.025668.112.
105. Dahlbom SJ, Lagman D, Lundstedt-Enkel K, Sundstrom LF, Winberg S. Boldness predicts social status in zebrafish (*Danio rerio*). PLoS One. 2011;6(8), e23565. doi:10.1371/journal.pone.0023565.
106. Harris, Sarah E., and Ian J. Deary. "The genetics of cognitive ability and cognitive ageing in healthy older people." *Trends in cognitive sciences* 15.9 (2011):388–394

Integrating Morphological and Behavioral Phenotypes in Developing Zebrafish

Guozhu Zhang, Lisa Truong, Robert L. Tanguay, and David M. Reif

Abstract The zebrafish, with its prolific reproduction, rapid development, and genetic homology to humans, provides an ideal model to efficiently characterize behavioral, developmental, and morphological phenotypes in biomedical research. By designing experiments to take advantage of these properties, behavioral phenotypes can be interpreted within the context of relevant neuromorphological phenotypes in order to present a truly integrative analysis of the rich biological data. These experiments may take the form of targeted studies or scaled up to a discovery mode that can keep pace with in vitro high-throughput screening (HTS) experimental systems. Achieving the goal of data integration will require appropriate application and adaptation of traditional statistical approaches, as well as the development of novel methods implemented in concert with new experimental approaches. In this chapter, we survey experimental designs and statistical methods for behavioral studies in developing zebrafish, then highlight experimental factors and analysis strategies that facilitate integration of morphological and behavioral phenotypes. We conclude that appropriate design and analysis of integrated morphological and behavioral studies using zebrafish can elucidate new chemical bioactivity pathways, identify compounds eliciting a broad range of effects, and paint a more comprehensive picture of development than either class of phenotype alone.

Keywords Data integration • High-throughput • Zebrafish • Aquatic models • Integrated testing • Development

---

G. Zhang • D.M. Reif (✉)
Department of Biological Sciences, Bioinformatics Research Center, North Carolina State University, Raleigh, NC, USA
e-mail: gzhang6@ncsu.edu; dmreif@ncsu.edu

L. Truong • R.L. Tanguay
Department of Environmental and Molecular Toxicology, The Sinnhuber Aquatic Research Laboratory and the Environmental Health Sciences Center at Oregon State University, Corvallis, OR, USA
e-mail: lisa.truong@oregonstate.edu; robert.tanguay@oregonstate.edu

A.V. Kalueff (ed.), *The rights and wrongs of zebrafish: Behavioral phenotyping of zebrafish*, DOI 10.1007/978-3-319-33774-6_12

1 Introduction

There is a growing number of chemicals in the environment, with new entities added commercially at an ever-increasing pace [1], however, the baseline toxicity and environmental health hazard information is still limited to a small number of these chemicals. Thus, characterizing the potential adverse impacts of these chemicals poses a challenge. Traditional animal testing using mammalian models is very expensive and time consuming, thus limiting its usefulness for screening applications. Applications of HTS in vitro chemical screening approaches are widely used to speed the pace of identifying chemical bioactivity using a suite of targeted individual assays [2, 3].

These target-specific in vitro assays fail to take into account the complexity of the vertebrate nervous system; thus, these assays may not identify neurotoxic chemicals. Phenotype-based assays could improve screening coverage to better identify neuroactive small molecules with no prior understanding of individual chemical molecular targets or their underling mechanism of action [4, 5]. HTS in vivo assay using zebrafish have proven useful for predicting different modes of action, including cellular differentiation, proliferation, migration, synapse formation, and apoptosis related to human diseases [6–10].

Zebrafish have emerged as a premier alternative model organism for chemical testing because of their small size, rapid development, low cost, and overall genomic similarity with human genes [11, 12]. Additionally, the central nervous system (CNS) of zebrafish is homologous to those of humans in many key features [13]. Behavioral assessment in zebrafish can be captured through spontaneous tail contractions as early as 19–29 hpf [14] or assessed at later developmental stages, where higher-level behaviors, such as aggression or response to novel situations, can be assessed [15].

In this chapter, we survey experimental designs for behavioral profiling; discuss experimental factors that facilitate integrated, high-throughput testing; evaluate statistical methods and common pitfalls for behavioral data analysis; discuss examples of data integration; and discuss translational applications within an integrated testing framework. An overview of experimental designs and associated statistical analysis methods is provided in Fig. 1.

2 Experimental Methods

2.1 *Survey of Behavioral Tests*

There are hundreds of behaviors that have been measured using the zebrafish model, covering aspects of behavior such as response to external stimuli, locomotion, changing light conditions, learning ability, and aggression [15]. Here, we introduce broad classes of tests selected as readily amenable to scale-up for integrated testing in larval (<6 dpf) zebrafish, though the candidate test list is evolving as new

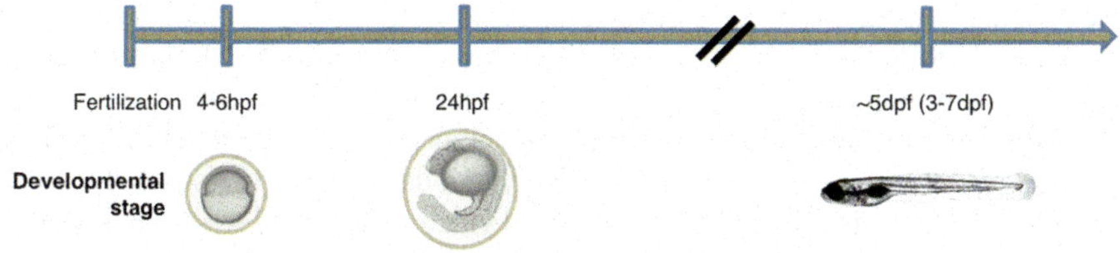

	24hpf	~5dpf (3-7dpf)
Morphological phenotypes	• Mortality • Developmental delay • Notochord • Spontaneous movement	• Mortality • Edemas (Yolk sac, Pericardial) • Malformations (Axis, Brain, Fin(s), Eye, Heart, Jaw, Notochord, Otic, Snout, Somite, Trunk) • Pigmentation • Circulation • Swim bladder
Behavioral tests	• Photomotor response (PMR)	• Light/Dark transition • Locomotor
Behavioral phenotypes (output)	• Motion (spontaneous contraction) index	• Distance moved • Speed • Time spent at different locations
Statistical methods	• Behavioral barcodes (pseudo z-scores) • Kruskal-Wallis test* • Kolmogorov-Smirnov plus relative change	• ANOVA models (+/- repeated measures) • Student's t-test • Wilcoxon test* • Kruskal-Wallis test* • MANOVA/Hotelling's T^2 test

Fig. 1 Summary of morphological and behavioral data collected at larval zebrafish developmental stages highlighted in chapter text. The 4–6 hpf timepoint is noted as the dechorionation decision-point, although the images depict intact chorions. Non-parametric alternative tests are denoted by an *asterisk* (*)

technologies and experiments are invented. Choosing the appropriate test depends on developmental stage, feasibility, treatment conditions (e.g. chemical stressors or biomolecular intervention), and study goals.

At only 24 hpf, the photomotor response (PMR) assay is one of the earliest measures of reproducible behavioral phenomena. In the PMR assay, the embryos experience two strong light stimuli (~1 s each), the first causing excitation and the second causing no discernable change in activity. The experimental period of less than 1 min can be divided into multiple phases, with the main phases defined according to light stimuli as background (prior to first light stimulus), latency (immediately following first light stimulus), excitatory (post-latency, following first light stimulus), and refractory (following second light stimulus) [16]. The response of the embryos is recorded using custom-built recording devices, such as the Photomotor Response Analysis Tool (PRAT) [17]. A movement index of the embryos is then estimated by analyzing video and compared to expected responses in each experimental phase. While the potential biological mechanisms underlying this test are not yet fully understood, the sensitivity to changing light conditions is thought to stem from photoreceptors in the developing hindbrain [18]. Due to the sensitivity of the normal PMR to chemical perturbation, this system has already been used to classify small molecules and diverse sets of compounds [16, 19, 20].

The locomotor test measures swimming activity in larval zebrafish [21]. This response can be measured in embryos as early as 4 days post fertilization (dpf). The swimming activity is tracked by an automated video system, and several parameters are chosen for analysis (e.g., total distance covered, swimming speed, location, and related measures). This test can be used in the analysis of compounds for potential effects on developmental neurotoxicity and neuromotor activity [22].

The light/dark transition test is used to analyze free swimming activity of zebrafish larvae in response to alternating light and dark periods or a repeated light and dark transition cycle [23]. This behavioral test is also usually performed in zebrafish larvae that are older than 4 dpf. The expectation for normally-developing (i.e. unexposed) zebrafish larvae is to show little or no movement when exposed to light, with increased swimming activity when transitioned to the dark phase. An automated video system is used to assess locomotion over the light cycle(s).

2.2 *Experimental Design for Integrated Testing*

The zebrafish model is amendable to rapid throughput testing, but to fully take advantage of the model, protocols need to be streamlined and standardized. There are a number of points to consider, ranging from sample preparation to exposure protocols to data recording and management.

The first consideration when conducting HTS chemical screening studies is controlling the exposure. First, to remove a potential exposure barrier, the chorions should be removed [24]. Once embryos are dechorionated (from 5 hpf), they are delicate and extremely sensitive to physical damage when handling. Although the transfer of dechorionated embryos into individual wells of multiwell plates can be done manually, rapid and careful robotic handling will support HTS. For example, the automated embryonic placement systems (AEPS) ensures reproducibility and minimizes damage [24]. Most importantly, AEPS allows for the ability to scale up without sacrificing quality for speed. The vulnerability of 5–6 hpf dechorionated embryos and the lack of widespread availability of robotics has led to many groups initiating exposures at later developmental stages, where embryos are markedly less sensitive to manual manipulations. However, the potential downside for starting exposures later in development is that critical windows may be missed due to the speed of zebrafish embryonic development. For example, if exposures are delayed for just 24 h, primary organogenesis will be nearly complete, thus precluding identification of chemicals targeting important gene products that were exclusively expressed in the first 24 h of embryonic development [25]. This scenario could increase the number of false negatives in screening.

To reduce confounding factors, embryos can be placed one-per-well, allowing individual assessments of both endpoint tests and chemical (or other stressor) exposure. Once embryos are placed into individual wells, the next critical step is controlling chemical delivery and exposure. The solvent of choice most commonly used for toxicity studies with the embryonic zebrafish is dimethyl sulfoxide (DMSO), but embryos can develop until 5 dpf normally in reverse osmosis water [26]. There are also opportunities to optimize chemical delivery and exposures. Most HTS approaches use liquid handling systems to dispense chemicals. These systems may use stainless steel or plastic pipettes, but there is a chance of cross-chemical contamination and adsorption. The newest technology in chemical delivery to solve these problems are digital dispensing devices such as the HP D300e (http://www8.hp.com). By using this system, it is possible to precisely dispense directly from stock solutions, eliminating costly dilution steps which often increase the possibility of chemical loss or handling errors. After dispensing chemicals, there is an intense mixing to ensure availability to the embryos. Exposed plates with embryos are sealed tightly and placed in the dark to minimize evaporation and photodegradation.

To fully maximize throughput of zebrafish behavioral testing, handling of experimental animals should be minimized and standardized. For this, it is preferred that chemical exposures be static over the course of the study, and not exchanged. The renewal of chemical solutions could protect against losses of bioactivity due to chemical instability or metabolism, but rarely are these chemical properties known in priori. In conditions where the chemical is highly stable and widely bioavailable, reintroduction of fresh exposure solutions into the system could lead to an exceedingly high chemical body burdens as the chemical would accumulate into the zebrafish because of the additive chemical mass. These high body burden doses may increase the false positive rate. In the absence of knowledge regarding individual chemical uptake kinetics, metabolism and elimination (which are not routinely measured) it may be advisable to utilize multiple exposure regiments and to experimentally determine if the concentration relationship is impacted by the different exposure protocols.

Data recording and management is another major consideration. All the data collected from HTS assays must be stored in a database to allow for ease of data retrieval and analysis. To accomplish this, a laboratory information management system (LIMS) should be used to house the chemical information, zebrafish strain used, exposure protocol, and methods/equipment that were used to acquire behavioral and morphological data [10].

3 Analysis Methods

3.1 Traditional Analysis Methods

The traditional methods that have been used in analyzing behavioral data can be broadly partitioned into parametric and non-parametric methods [27, 28]. In parametric analysis, the Student's t-test or linear models (both fixed and mixed effects)

[29–31] have been used to determine whether the exposure group(s) differ significantly from the control group. The linear models include one-way or two-way Analysis of Variance (ANOVA) or repeated-measures ANOVA, which should be followed by appropriate post hoc tests, such as the Bonferroni correction, Tukey's procedure, Dunnett's correction, or Fisher's least significant difference (LSD) [32].

However, a recent review of zebrafish behavioral assays [28] found that post hoc tests (adjustments) were conducted in less than half of the total reviewed studies. This presents a potential issue related to the general problem of multiple testing, which refers to simultaneous empirical tests of more than one hypothesis [33]. In zebrafish studies, multiple testing may arise in experiments comparing several doses of chemical, treatment groups, or molecular interventions. First, it should be decided whether to adjust for multiple tests, which may not be warranted if (1) the study is limited to a small number of pre-planned comparisons; (2) the study is exploratory in nature; or (3) avoidance of Type II (false negative) errors is the primary concern [34, 35]. Next, the choice of adjustment should consider both the statistical analysis framework (e.g. Dunnett's test for planned comparisons of multiple treatment groups versus a reference control group) and the goals of the particular study. Adjustment methods range from strict control of the family-wise error rate (FWER) via the Bonferroni method through less-conservative, false discovery rate (FDR) methods [36]. A strict Bonferroni correction should be considered when avoidance of Type I (false positive) errors are a primary concern in a study without preplanned hypotheses, but its incorrect use (e.g. correcting for irrelevant null hypotheses) may result in decreased statistical power [37, 38]. For large-scale experiments where concerns must be balanced between Type I and Type II errors, controlling the FDR may be more appropriate [39].

Particular applications will differ with respect to the consequences of failing to address multiple testing issues and violating assumptions of a particular statistical procedure. However, failing to properly check assumptions may give incorrect or misleading results. Potential violations of assumptions in a general ANOVA framework may relate to: (1) lack of independence within or between samples; (2) non-normality of the entire sample (or influential outliers); (3) heterogeneity of variances; (4) small or unbalanced samples; (5) inappropriate or absent post hoc correction. There are many options for dealing with violations within an ANOVA framework [40, 41]. For dealing with non-normality of the data, one can (1) remove or replace outliers with imputed values (e.g. sample mean); (2) transform data using variance-stabilizing algorithms; and/or (3) choose non-parametric test alternatives. For ANOVA alternatives, the Kruskal–Wallis test does not require the assumption of normality and can be applied if the assumption of homogeneity of variances has been violated [41]. For dealing with small or unbalanced samples, bootstrapping (i.e. resampling with replacement from within the data) may be appropriate [42]. For standard models, lack of independence presents a serious problem, but the addition of blocking factors (i.e. stratification) may provide a solution [32]. Non-parametric t-tests alternatives include the Wilcoxon rank sum test (non-paired) or Wilcoxon signed-rank test (paired).

3.2 Advanced Analysis Methods

The complexity of time-series data generated by modern, high-throughput experiments presents statistical challenges arising from shared correlation, unusual distributions, and sheer data volume. Moreover, when it comes to developmental neurotoxicological studies, individual zebrafish can behave very differently, resulting in unusual behavioral patterns that do not follow the most common statistical distributions. Thus, new analysis methods are needed to deal with high-volume data, determine outliers, and model behavioral data for different study goals.

The Multivariate Analysis of Variance (MANOVA) and Hotelling's T-squared test are examples of advanced methods to look at multiple factors that affect the zebrafish larval activity [43]. These factors include biological variations (zebrafish strains and developmental stages), treatment effects (repeated measurements and different light/dark conditions), and experimental variations (physical location and independent replication), as well as interactions between these factors. Hotelling's T-squared test can be applied to perform pairwise comparisons of any individual model parameter to identify the key factor affecting response. MANOVA can be applied when one or more independent variables have an effect on a group of dependent response variables. The assumptions underlying both methods include multivariate normality of response variables and homogeneity of covariance matrices. Compared to a set of univariate tests, formulating models within the frameworks of Hotelling's T-square test or MANOVA can reduce the Type I error rate [43].

Advanced analysis methods may also recombine elements of novel and traditional approaches, such as the behavioral barcoding method presented in [16] to identify compound-similarity clusters from a high-throughput PMR assay using pooled-embryo wells. Prior to statistical analysis, the behavioral profiles were normalized to each other to adjust for potential artifacts. The data were then log transformed and normalized after the transformation by using a global profile, which was the average of all the plate profiles. The total experimental period (~40 s) was divided into seven phases according to light stimuli. For each phase, in each well, the first and third quartile motion index values were computed. A pseudo z-score was calculated by subtracting the selected quartile motion index value from the median motion index value of the control group and then dividing by the control median absolute deviation (MAD). These pseudo z-scores were characterized as behavioral barcodes. The reproducibility of the barcodes was confirmed via permutation and by looking at the similarity of independent wells that were treated with the same compound.

In order to identify compounds eliciting aberrant behavioral effects in concentration-response PMR data from embryos in individual wells, [20] developed a robust approach that used global estimates to inform local statistical tests of differential activity. First, a detailed quality control pipeline was implemented that took advantage of the magnitude of data (tested using a diverse set of 1060 compounds in five-point, concentration-response mode) to check for batch effects, identify outliers, and quantify expected (normal) responses. For analysis, the ends of the experimental period were truncated, then divided into three features surrounding the two PMR light pulses: background, excitatory, and refractory. For each experimental interval, the behavioral pattern of chemical-treated groups was compared to its corresponding negative (vehicle) control. The statistical significance was defined using a combination of change-in-activity threshold (either hyper- or hypo-active compared to control) and a non-parametric Kolmogorov-Smirnov test. Since the significant call of the movement is a ratio-based comparison to local controls, this method can be applied to smaller scale screening of varying experimental designs. The reproducibility of the data treatment and statistical methods were validated using sets of separately-sourced, blinded chemical replicates.

The advantages of these advanced statistical frameworks over more traditional alternatives include the ability to model multiple factors and robustness in the face of changing, nonstandard distributions [16, 20]. While methods such as MANOVA are also attractive for their relative familiarity to those used to an ANOVA framework, these parametric methods come with the additional assumptions of multivariate normality of response and homogeneity of covariance matrices. For a spontaneous response phenotype such as the PMR, the zero-heavy distributions may confound these assumptions, so it is vital that underlying response patterns be verified prior to modeling [20]. Additionally, experimental goals should determine the balance between fine-scale modeling of paired, second-to-second responses on one extreme, versus statistically robust modeling of responses analyzed across a longer experimental intervals on the other extreme.

3.3 Integration of Morphological and Behavioral Data

Application of appropriate experimental design and analysis methods affords the opportunity to explore integrated testing strategies by linking various neurological and behavioral phenotypes. Such data integration can be achieved via meta-analysis across external datasets or across endpoints within an experimental system.

Meta-analysis of zebrafish developmental toxicity studies reveals how chemicals can affect a range of developmental outcomes and can be used to estimate rankings of chemical teratogenicity [44]. As information on basic genetics continues to grow, meta-analysis across species offers another avenue for integration. For example, *Drosophila melanogaster* has been used for primary screens and secondary validation for drug discovery for a range of human diseases [45]. With appropriate bioinformatical mapping of related genetic elements across species, results from similar behavioral tests that measure locomotion, aggression, and learning in *Drosophila* can be used to inform zebrafish studies and help understand genetic underpinnings related to behavior.

Bioinformatically mapping to emerging HTS in vitro systems may provide additional evidence for biological targets associated with chemical-elicited behavioral changes in zebrafish. For example, a recent study [20] found that chemicals eliciting neurobehavioral phenotypes in zebrafish were significantly enriched for relevant in vitro biochemical assay activity from data reported previously [46]. The in vitro assays that were activated by the same chemicals associated with hypoactivity in zebrafish included neurologically-relevant assays probing aminergic G-protein coupled receptor binding, ion channel binding, and cholinesterase enzyme inhibition. Conversely, chemicals that were negative in the PMR assay had "protective" odds ratios (odds ratios <1), meaning that the compounds inactive in this behavioral assay were less likely to "hit" the given targeted assays. Thus, chemicals that were negative in this 24 hpf behavioral assay displayed an inverse relationship with in vitro assays targeting plausible neuroactivity pathways.

Integration across endpoints within an experimental system can directly probe linkages between morphological and behavioral endpoints [47]. Given appropriate design, these morphology-behavior linkages can even be explored across developmental stages. For example, Reif et al. [20] found that chemicals eliciting an early, 24 hpf, hypoactive responses in a PMR assay were predictive of distinct developmental abnormalities (notochord defects, edema, and structural defects) and mortality measured at 120 hpf. Moreover, by measuring all endpoints in the same five-point, chemicals concentration range (from 6.4 nM to 64 μM), the authors observed that critical concentrations (lowest effect level, LEL) in the 24 hpf behavioral assay could be more potent than those for gross morphological effects, as with all three neurotoxicant avermectins [48] tested: abamectin, milbemectin, and emamectin benzoate. Beyond canonical neurotoxic modes of action, the PMR assay also identified chemicals disrupting gross structural development, such as tributyltin chloride, tributyltin methacrylate, and triphenyltin hydroxide, at concentrations lower than those at which morphological abnormalities were observed [20]. These findings illustrate the vital importance of a concentration-response design when integrating behavioral and morphological data, where observed behavior at a given concentration may be hyperactive, hypoactive, or neither. In the extreme case, a single concentration may happen to coincide with the point at which a behavioral response is transitioning between hyper- and hypoactivity (i.e. neither or "no change"), providing no information with respect to relative potency of observed morphological versus behavioral consequences. Therefore, single concentration data on behavior is of limited utility for inferring mechanism, as it provides no information on as to whether observed behavioral differences arise from altered neurological signaling or simply the result of impaired structural developmental.

4 Interpretation and Translation

Zebrafish have shown to be a powerful model for studying brain disorders, with assayable behaviors that increase in complexity with life stage [49, 50]. Behavioral profiling in zebrafish has also been used to both identify biological targets and identify small molecules as prospective neurological drugs [16, 30, 51]. Emerging data also indicate that the zebrafish is a powerful model for studying mechanisms by which environmental compounds may perturb neurodevelopment or contribute to Parkinson-like symptoms [10, 50].

In toxicology testing applications, because of the unknown space of possible targets that must be probed, new biological and computational strategies are needed to adequately evaluate toxicity potential and identify biological pathways of interest. Building adverse outcome pathways (AOPs) that include behavioral endpoints could advance development of integrated testing strategies [52]. Depending on the toxicological outcome of interest, in vitro screening results can be used to construct putative AOPs [53]; however, for outcomes that depend on coordinated function of an entire organism, such as behavior, purely in vitro assays will be insufficient. Therefore, behavioral profiling using zebrafish, especially if implemented on an HTS scale, provides a useful, complementary strategy for developing AOPs related to behavioral outcomes.

Behavioral data from zebrafish offer significant, potential value when integrated with morphological data in a testing context employing a comprehensive suite of endpoints. This is especially true in early development, where a delicate cascade of events leading to normal development may be interrupted by environmental stressors yet not manifest in an observable endpoint of a particular type (or measured at a particular time in a given experiment). For example, in screening a diverse set of 1060 compounds using a PMR assay, [20] found 102 chemicals eliciting behavioral changes at 24 hpf yet no statistically significant morphological alterations when examined at 5 dpf. This situation may be due to a number of factors, from experimental artifacts to random chance, but the possibility that early behavioral alterations may be sensitive to otherwise undetectable chemical bioactivity should be considered when designing integrated testing strategies.

5 Conclusions

Because in vivo zebrafish assays can provide rapid, yet comprehensive, measures of normal development and behavior, they offer many advantages when integrated with data from morphological screens, targeted in vitro assays, or phenotypic measurements in mammalian models. Integrated analysis that links behavioral and morphological phenotypes may avoid false-negatives that would arise when evaluating only a single type of data. For example, behavioral assessments may identify chemicals associated with unmeasured morphological phenotypes, or observed neuromorphological aberrations may help establish empirical thresholds for behavioral changes. Furthermore, non-destructive behavioral assays may be used to assess multiple stages of development in the same individuals that can later be assessed for apical, morphological phenotypes, shedding light on developmental bases of adult disease. The observation that chemicals show significant behavioral responses in the absence of known modes of action may suggest modifications to existing AOPs or demand development of entirely new pathways. Functionally, realizing the promise of data integration will necessitate investment in methods development and associated software infrastructure, as new methods must also have the flexibility to accommodate new data as experimental technology continues to progress [54]. In conclusion, appropriate design and analysis of integrated morphological and behavioral studies using zebrafish can paint a more comprehensive picture of development than either class of phenotype alone.

References

1. Wambaugh JF, Setzer RW, Reif DM, Gangwal S, Mitchell-Blackwood J, Arnot JA, Joliet O, Frame A, Rabinowitz J, Knudsen TB, Judson RS, Egeghy P, Vallero D, Cohen Hubal EA. High-throughput models for exposure-based chemical prioritization in the ExpoCast project. Environ Sci Technol. 2013;47(15):8479–88.
2. Collins FS, Gray GM, Bucher JR. Toxicology: transforming environmental health protection. Science. 2008;319(5865):906–7.
3. Judson RS, Houck KA, Kavlock RJ, Knudsen TB, Martin MT, Mortensen HM, Reif DM, Rotroff DM, Shah I, Richard AM, Dix DJ. In vitro screening of environmental chemicals for targeted testing prioritization: the ToxCast project. Environ Health Perspect. 2010;118(4):485–92.
4. Burns CJ, McIntosh LJ, Mink PJ, Jurek AM, Li AA. Pesticide exposure and neurodevelopmental outcomes: review of the epidemiologic and animal studies. J Toxicol Environ Health B Crit Rev. 2013;16(3–4):127–283.
5. Selderslaghs IW, Hooyberghs J, Blust R, Witters HE. Assessment of the developmental neurotoxicity of compounds by measuring locomotor activity in zebrafish embryos and larvae. Neurotoxicol Teratol. 2013;37:44–56.
6. Cao Y, Semanchik N, Lee SH, Somlo S, Barbano PE, Coifman R, Sun Z. Chemical modifier screen identifies HDAC inhibitors as suppressors of PKD models. Proc Natl Acad Sci U S A. 2009;106(51):21819–24.
7. Makris N, Biederman J, Monuteaux MC, Seidman LJ. Towards conceptualizing a neural systems-based anatomy of attention-deficit/hyperactivity disorder. Dev Neurosci. 2009;31(1–2):36–49.
8. Padilla S, Corum D, Padnos B, Hunter DL, Beam A, Houck KA, Sipes N, Kleinstreuer N, Knudsen T, Dix DJ, Reif DM. Zebrafish developmental screening of the ToxCast Phase I chemical library. Reprod Toxicol. 2012;33(2):174–87.
9. Sanes DH, Reh TA, Harris WA, editors. Development of the nervous system. New York: Elsevier Science; 2005.
10. Truong L, Reif DM, St Mary L, Geier MC, Truong HD, Tanguay RL. Multidimensional in vivo hazard assessment using zebrafish. Toxicol Sci. 2014;137(1):212–33.
11. Barbazuk WB, Korf I, Kadavi C, Heyen J, Tate S, Wun E, Bedell JA, McPherson JD, Johnson SL. The syntenic relationship of the zebrafish and human genomes. Genome Res. 2000;10(9):1351–8.
12. Howe K, Clark MD, Torroja CF, Torrance J, Berthelot C, Muffato M, Collins JE, Humphray S, McLaren K, Matthews L, McLaren S, Sealy I, Caccamo M, Churcher C, Scott C, Barrett JC, Koch R, Rauch GJ, White S, Chow W, Kilian B, Quintais LT, Guerra-Assuncao JA, Zhou Y, Gu Y, Yen J, Vogel JH, Eyre T, Redmond S, Banerjee R, Chi J, Fu B, Langley E, Maguire SF, Laird GK, Lloyd D, Kenyon E, Donaldson S, Sehra H, Almeida-King J, Loveland J, Trevanion S, Jones M, Quail M, Willey D, Hunt A, Burton J, Sims S, McLay K, Plumb B, Davis J, Clee C, Oliver K, Clark R, Riddle C, Eliott D, Threadgold G, Harden G, Ware D, Mortimer B, Kerry G, Heath P, Phillimore B, Tracey A, Corby N, Dunn M, Johnson C, Wood J, Clark S, Pelan S, Griffiths G, Smith M, Glithero R, Howden P, Barker N, Stevens C, Harley J, Holt K, Panagiotidis G, Lovell J, Beasley H, Henderson C, Gordon D, Auger K, Wright D, Collins J, Raisen C, Dyer L, Leung K, Robertson L, Ambridge K, Leongamornlert D, McGuire S, Gilderthorp R, Griffiths C, Manthravadi D, Nichol S, Barker G, Whitehead S, Kay M, Brown

J, Murnane C, Gray E, Humphries M, Sycamore N, Barker D, Saunders D, Wallis J, Babbage A, Hammond S, Mashreghi-Mohammadi M, Barr L, Martin S, Wray P, Ellington A, Matthews N, Ellwood M, Woodmansey R, Clark G, Cooper J, Tromans A, Grafham D, Skuce C, Pandian R, Andrews R, Harrison E, Kimberley A, Garnett J, Fosker N, Hall R, Garner P, Kelly D, Bird C, Palmer S, Gehring I, Berger A, Dooley CM, Ersan-Urun Z, Eser C, Geiger H, Geisler M, Karotki L, Kirn A, Konantz J, Konantz M, Oberlander M, Rudolph-Geiger S, Teucke M, Osoegawa K, Zhu B, Rapp A, Widaa S, Langford C, Yang F, Carter NP, Harrow J, Ning Z, Herrero J, Searle SM, Enright A, Geisler R, Plasterk RH, Lee C, Westerfield M, de Jong PJ, Zon LI, Postlethwait JH, Nusslein-Volhard C, Hubbard TJ, Roest Crollius H, Rogers J, Stemple DL. The zebrafish reference genome sequence and its relationship to the human genome. Nature. 2013;496(7446):498–503.

13. Tropepe V, Sive HL. Can zebrafish be used as a model to study the neurodevelopmental causes of autism? Genes Brain Behav. 2003;2(5):268–81.
14. Kimmel CB, Ballard WW, Kimmel SR, Ullmann B, Schilling TF. Stages of embryonic development of the zebrafish. Dev Dyn. 1995;203(3):253–310.
15. Kalueff AV, Gebhardt M, Stewart AM, Cachat JM, Brimmer M, Chawla JS, Craddock C, Kyzar EJ, Roth A, Landsman S, Gaikwad S, Robinson K, Baatrup E, Tierney K, Shamchuk A, Norton W, Miller N, Nicolson T, Braubach O, Gilman CP, Pittman J, Rosemberg DB, Gerlai R, Echevarria D, Lamb E, Neuhauss SC, Weng W, Bally-Cuif L, Schneider H, C. Zebrafish Neuroscience Research . Towards a comprehensive catalog of zebrafish behavior 1.0 and beyond. Zebrafish. 2013;10(1):70–86.
16. Kokel D, Bryan J, Laggner C, White R, Cheung CY, Mateus R, Healey D, Kim S, Werdich AA, Haggarty SJ, Macrae CA, Shoichet B, Peterson RT. Rapid behavior-based identification of neuroactive small molecules in the zebrafish. Nat Chem Biol. 2010;6(3):231–7.
17. Noyes PD, Haggard DE, Gonnerman GD, Tanguay RL. Advanced morphological—behavioral test platform reveals neurodevelopmental defects in embryonic zebrafish exposed to comprehensive suite of halogenated and organophosphate flame retardants. Toxicol Sci. 2015;145(1):177–95.
18. Kokel D, Dunn TW, Ahrens MB, Alshut R, Cheung CY, Saint-Amant L, Bruni G, Mateus R, van Ham TJ, Shiraki T, Fukada Y, Kojima D, Yeh JR, Mikut R, von Lintig J, Engert F, Peterson RT. Identification of nonvisual photomotor response cells in the vertebrate hindbrain. J Neurosci. 2013;33(9):3834–43.
19. Raftery TD, Isales GM, Yozzo KL, Volz DC. High-content screening assay for identification of chemicals impacting spontaneous activity in zebrafish embryos. Environ Sci Technol. 2014;48(1):804–10.
20. Reif DM, Truong L, Mandrell D, Marvel S, Zhang G, Tanguay RL. High-throughput characterization of chemical-associated embryonic behavioral changes predicts teratogenic outcomes. Arch Toxicol. 2016;90(6):1459–70.
21. Lockwood B, Bjerke S, Kobayashi K, Guo S. Acute effects of alcohol on larval zebrafish: a genetic system for large-scale screening. Pharmacol Biochem Behav. 2004;77(3):647–54.
22. Baraban SC, Taylor MR, Castro PA, Baier H. Pentylenetetrazole induced changes in zebrafish behavior, neural activity and c-fos expression. Neuroscience. 2005;131(3):759–68.
23. MacPhail RC, Brooks J, Hunter DL, Padnos B, Irons TD, Padilla S. Locomotion in larval zebrafish: influence of time of day, lighting and ethanol. Neurotoxicology. 2009;30(1):52–8.
24. Mandrell D, Truong L, Jephson C, Sarker MR, Moore A, Lang C, Simonich MT, Tanguay RL. Automated zebrafish chorion removal and single embryo placement: optimizing throughput of zebrafish developmental toxicity screens. J Lab Autom. 2012;17(1):66–74.
25. Westerfield M. The zebrafish book: a guide for the laboratory use of zebrafish (*Danio rerio*). Eugene: University of Oregon Press; 2007.
26. Truong L, Zaikova T, Richman EK, Hutchison JE, Tanguay RL. Media ionic strength impacts embryonic responses to engineered nanoparticle exposure. Nanotoxicology. 2012;6(7):691–9.
27. Jarema KA, Hunter DL, Shaffer RM, Behl M, Padilla S. Acute and developmental behavioral effects of flame retardants and related chemicals in zebrafish. Neurotoxicol Teratol. 2015;52(Pt B):194–209.

28. Legradi J, El Abdellaoui N, van Pomeren M, Legler J. Comparability of behavioural assays using zebrafish larvae to assess neurotoxicity. Environ Sci Pollut Res Int. 2015;22(21):16277–89.
29. Parker MO, Ife D, Ma J, Pancholi M, Smeraldi F, Straw C, Brennan CH. Development and automation of a test of impulse control in zebrafish. Front Syst Neurosci. 2013;7:65.
30. Rihel J, Prober DA, Arvanites A, Lam K, Zimmerman S, Jang S, Haggarty SJ, Kokel D, Rubin LL, Peterson RT, Schier AF. Zebrafish behavioral profiling links drugs to biological targets and rest/wake regulation. Science. 2010;327(5963):348–51.
31. Selderslaghs IW, Hooyberghs J, De Coen W, Witters HE. Locomotor activity in zebrafish embryos: a new method to assess developmental neurotoxicity. Neurotoxicol Teratol. 2010;32(4):460–71.
32. Box GEP, Hunter JS, Hunter WG, editors. Statistics for experimenters: design, innovation, and discovery. Hoboken, NJ: Wiley-Interscience; 2005.
33. Noble WS. How does multiple testing correction work? Nat Biotechnol. 2009;27(12):1135–7.
34. Armstrong RA. When to use the Bonferroni correction. Ophthalmic Physiol Opt. 2014;34(5):502–8.
35. Schulz KF, Grimes DA. Multiplicity in randomised trials I: endpoints and treatments. Lancet. 2005;365(9470):1591–5.
36. Benjamini Y, Hochberg Y. Controlling the false discovery rate: a practical and powerful approach to multiple testing. J R Stat Soc Ser B Methodol. 1995;57(1):289–300.
37. Perneger TV. What's wrong with Bonferroni adjustments. BMJ. 1998;316(7139):1236–8.
38. Rothman KJ. No adjustments are needed for multiple comparisons. Epidemiology. 1990;1(1):43–6.
39. Nakagawa S. A farewell to Bonferroni: the problems of low statistical power and publication bias. Behav Ecol. 2004;15(6):1044–5.
40. Fox J, editor. Applied regression analysis and generalized linear models. London: Sage; 2008.
41. Higgins JJ, editor. Introduction to modern nonparametric statistics. Belmont, CA: Duxbury Press; 2003.
42. Krishnamoorthy K, Lu F, Mathew T. A parametric bootstrap approach for ANOVA with unequal variances: fixed and random models. Comput Stat Data Anal. 2007;51(12):5731–42.
43. Liu Y, Carmer R, Zhang G, Venkatraman P, Brown SA, Pang CP, Zhang M, Ma P, Leung YF. Statistical analysis of zebrafish locomotor response. PLoS One. 2015;10(10), e0139521.
44. Ducharme NA, Peterson LE, Benfenati E, Reif D, McCollum CW, Gustafsson JA, Bondesson M. Meta-analysis of toxicity and teratogenicity of 133 chemicals from zebrafish developmental toxicity studies. Reprod Toxicol. 2013;41:98–108.
45. Pandey UB, Nichols CD. Human disease models in Drosophila melanogaster and the role of the fly in therapeutic drug discovery. Pharmacol Rev. 2011;63(2):411–36.
46. Sipes NS, Martin MT, Kothiya P, Reif DM, Judson RS, Richard AM, Houck KA, Dix DJ, Kavlock RJ, Knudsen TB. Profiling 976 ToxCast chemicals across 331 enzymatic and receptor signaling assays. Chem Res Toxicol. 2013;26(6):878–95.
47. Beker van Woudenberg A, Snel C, Rijkmans E, de Groot D, Bouma M, Hermsen S, Piersma A, Menke A, Wolterbeek A. Zebrafish embryotoxicity test for developmental (neuro)toxicity: demo case of an integrated screening approach system using anti-epileptic drugs. Reprod Toxicol. 2014;49:101–16.
48. Lumaret JP, Errouissi F, Floate K, Rombke J, Wardhaugh K. A review on the toxicity and non-target effects of macrocyclic lactones in terrestrial and aquatic environments. Curr Pharm Biotechnol. 2012;13(6):1004–60.
49. Gerlai R, Lahav M, Guo S, Rosenthal A. Drinks like a fish: zebra fish (*Danio rerio*) as a behavior genetic model to study alcohol effects. Pharmacol Biochem Behav. 2000;67(4):773–82.
50. Kalueff AV, Stewart AM, Gerlai R. Zebrafish as an emerging model for studying complex brain disorders. Trends Pharmacol Sci. 2014;35(2):63–75.
51. MacRae C, Peterson R. Zebrafish-based small molecule discovery. Chem Biol. 2003;10:901–8.

52. Vinken M. The adverse outcome pathway concept: a pragmatic tool in toxicology. Toxicology. 2013;312:158–65.
53. Kleinstreuer NC, Yang J, Berg EL, Knudsen TB, Richard AM, Martin MT, Reif DM, Judson RS, Polokoff M, Dix DJ, Kavlock RJ, Houck KA. Phenotypic screening of the ToxCast chemical library to classify toxic and therapeutic mechanisms. Nat Biotechnol. 2014;32(6):583–91.
54. Meisner M, Reif DM. Computational methods used in systems biology. In: Fry RC, editor. Systems biology in toxicology and environmental health. 1st ed. Amsterdam: Academic; 2015. p. 85–115.

Neuroimaging Phenotypes in Zebrafish

Jeremy F.P. Ullmann and Andrew L. Janke

Abstract The zebrafish has become an established model in neuroscience due to the ease with which gene discovery, chemical screening, behaviour, and disease modelling can be performed. More recently, neuroimaging, a crucial pre-clinical technique for probing tissue structure, examining volumetric changes, and studying in vivo brain activity has also been applied to zebrafish. The zebrafish brain is particularly attractive for neuroimaging due to its small size, numerous translucent strains, and distinct forebrain organization. In this chapter we discuss the range of imaging techniques which have been utilized to examine the zebrafish brain. While many of these methods have only begun to be utilized in zebrafish, correlating neuroimaging phenotypes with behaviour in zebrafish has a bright future.

Keywords Zebrafish • Brain • Magnetic resonance imaging • Calcium imaging • Optical projection tomography • Transmission electron microscopy

1 Introduction

Zebrafish are a premier organism for neuroscience research. While initially developed as a genetic model for developmental biology, a large behavioural repertoire [1], a similar brain archetype [2], and a wide range of powerful genetic techniques has led to the explosion of zebrafish as a model in neuroscience [3]. Many features of the adult and larval zebrafish brain have led to it being extensively imaged, including a small brain size (the larval brain is 0.5 mm thick and 1.5 mm long, with the total number of neurons on the order of 10^5 [4] and the adult brain is 2 mm thick

J.F.P. Ullmann (✉)
Centre for Advanced Imaging, The University of Queensland, Brisbane, QLD, Australia

Epilpesy Genetics Program, Department of Neurology, Boston Children's Hospital and Harvard Medical School, Boston, MA, USA
e-mail: jeremy.ullmann@childrens.harvard.edu

A.L. Janke
Centre for Advanced Imaging, The University of Queensland, Brisbane, QLD, Australia

A.V. Kalueff (ed.), *The rights and wrongs of zebrafish: Behavioral phenotyping of zebrafish*, DOI 10.1007/978-3-319-33774-6_13

and 4.5 mm long, with about 10^7 neurons [5]); transparency of larvae and adults; inversion of the forebrain during development resulting in the dorsal location of subcortical structures such as the hippocampus (Fig. 1); and, finally, cutaneous breathing at early stages of development facilitates embedding in agarose and subsequent imaging without surgical procedures to expose the brain. In this chapter we will discuss the range of approaches to image both the larval and adult zebrafish brain, and their relevance to zebrafish neurophenotyping (Table 1).

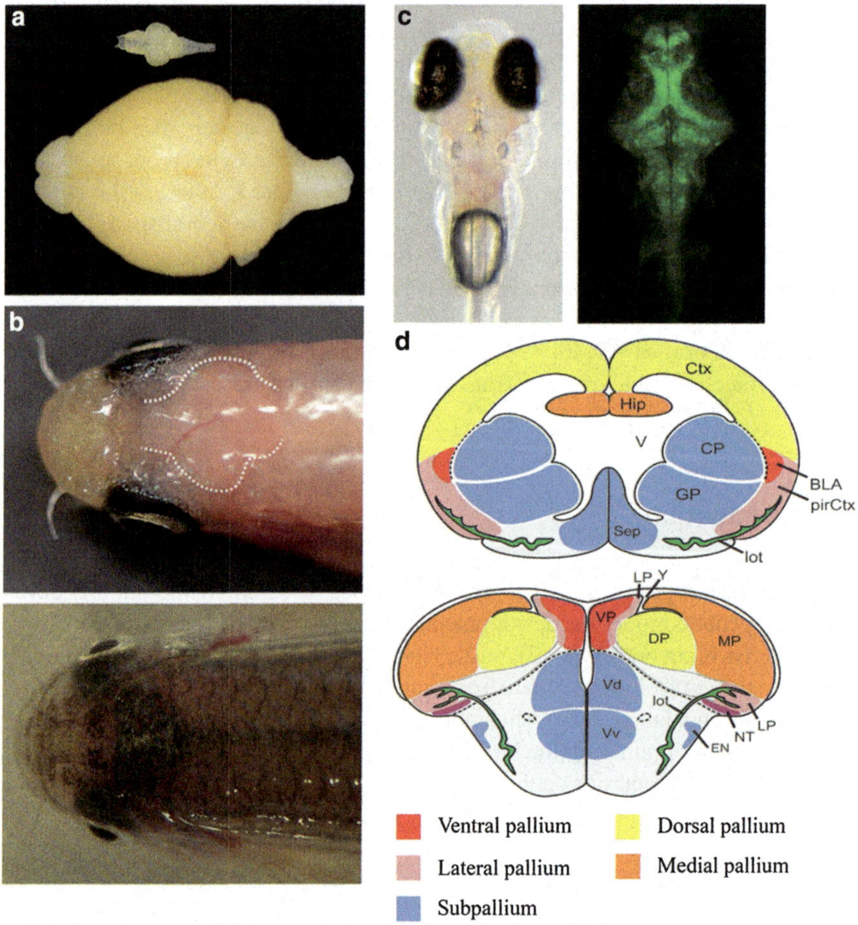

Fig. 1 Advantages of the zebrafish brain for imaging. (**a**) Size comparison between the adult mouse and adult zebrafish brains. (**b**) Ability to visualize the brain in a non-pigmented (*absolute*) adult zebrafish and a wild-type zebrafish. (**c**) Transparency of non-pigmented larval zebrafish and expression of pan neuronal GFP (*elavl3:H2B-GCamp6S*). (**d**) Schematic drawing illustrating the topological differences between zebrafish and rodents from [6]

Table 1 Comparison of imaging modalities for imaging the zebrafish brain

	Age	In/ex vivo	Field of view	Resolution	Analysis outcomes	References
Serial electron microscopy	All ages	Ex vivo	>3 nm	3 nm	Cellular microstructure	Yet to be done for zebrafish
					Circuit analysis	
Calcium imaging	5–7 dpf although older ages are possible. See [25]	In vivo	Objective dependent	Single cell ~4 μm	Structural analysis of cleared brains or functional analyses of behaviour	[26–30]
			For zebrafish usually 800 × 600 μm			
OPT/OCT/optical acoustic imaging	Adult	In vivo	5 mm	Single cell ~4 μm	3D view of micro-structure and morphometry as seen in traditional microscopy	[21, 22, 24, 31]
Magnetic resonance imaging	Adult	Ex vivo	Coil dependent. >2 mm	7 μm	Meso-structural volume and morphometry	[11, 12]
Diffusion weighted imaging	Adult	Ex vivo	Col dependent >2 mm	48 μM	Meso-structural morphometry	[32]

2 Magnetic Resonance Imaging

Magnetic resonance imaging (MRI) is a widely used non-invasive pre-clinical and clinical imaging modality. MRI visualizes the anatomy of the brain by exploiting differences in the relaxation values of various microstructures. By altering repetition times and echo times contrast can be optimized and different neuroanatomical structures visualized. MR imaging was devised by adding the ability to determine spatial location to NMR spectroscopic imaging that is typically used in chemical analysis of structures. MR imaging was initially developed for human brain but subsequent improvements in coil design and increases in magnetic field strength have enabled a range of species including fish to be imaged. MRI permits the acquisition of in-vivo three-dimensional volumes of the whole brain eliminating the need for tedious sectioning. By imaging the whole brain many histological artifacts such as shrinkage, tearing, and variations in labeling are minimized and instead 're-slicing' of the data in any arbitrary orientation is possible. Different contrasts are also achievable by using a variety of contrast agents or MRI pulse sequences (see [7] for review).

In zebrafish, MRI was first used in to visualize the entire anatomy of the adult zebrafish [8]. Ex vivo MRI performed on a 9.4 T magnet with an anisotropic field of view of 20×35 mm resulted in a spatial resolution of 137 μm, and visualization of all of the major organs of the zebrafish including the brain. In the same study in vivo experiments using a flow-through chamber produced images with an in-plane resolution of 78 μm. While studies in other teleost species have been able to obtain high resolution and good contrast to noise ratios in an in situ preparation [9, 10] more recent zebrafish MRI studies have dissected the brain out of the skull to minimize the field of view and maximize resolution. The concurrent development of zebrafish-specific fixation and incubation protocols with gadolinium-based contrast agents led to the acquisition of 10 μm^3 images [11] and the creation of a single-brain high-resolution atlas [12]. Although this single-brain atlas describes many brain regions in the adult zebrafish brain a probabilistic atlas that minimizes individual differences and instead is based upon a large population is still needed. The resultant data set would only exhibit structures present throughout the population and generate mean morphometric measures that represent the population [7, 13] (Fig. 2). MRI has also been utilized for examining zebrafish models of neurological disease. MRI of OCRL1 mutant zebrafish demonstrated white matter irregularities in the periventricular region that were later identified as areas with gliosis [14].

3 Diffusion Weighted Imaging

Diffusion weighed imaging (DWI) is the analysis of water diffusion using MRI. By using specialized gradients and sequences, the diffusivity of water can be quantified by a loss of signal related to movement of water in a particular direction and the orientation of internal structures inferred [15–18]. Specifically, neuroanatomical

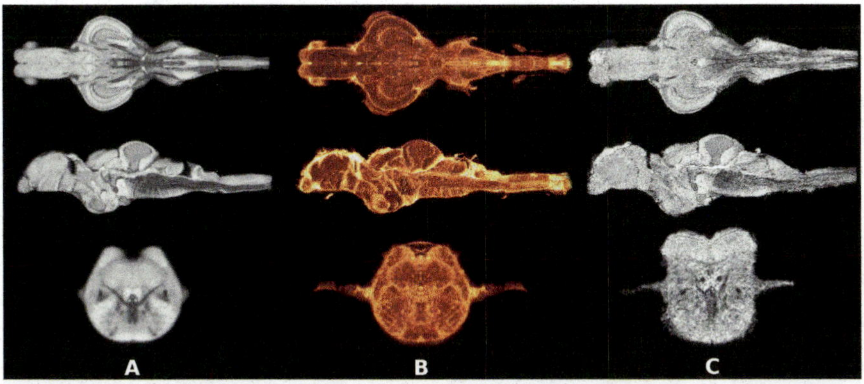

Fig. 2 Comparison of resolution and contrast achieved between a minimum deformation model (**a**) and a single brain data set (**c**). The minimum deformation model minimizes individual differences and only exhibits structures present throughout the population. (**b**) Standard deviation map with areas in *yellow* highly variable between individual brains and areas in *red* very consistent

regions that do not preferentially restrict the movement of water in any direction, such as grey matter, have relatively uniform or isotropic levels of diffusivity, while regions that are highly structured and restrict movement in particular directions, such as white matter, have anisotropic levels of diffusivity. This is most readily manifest in myelinated axonal bundles as in these areas water movement is restricted by both the longitudinal axonal bodies and the surrounding myelin sheaths. The differences in diffusivity between white and grey matter can be visualized in parameter maps (e.g. diffusion weighted imaged, apparent diffusion coefficient map, or fractional anisotropy map) to assist in the delineation of neuronal structures.

Despite its significant utility in other preclinical models diffusion weighted imaging has only been minimally performed on zebrafish. Freidlin et al. [19] used in vitro diffusion tensor microscopy visualize the spinal cord and muscle fibre groups in the adult zebrafish. Ex vivo DWI has also used to examine the adult zebrafish brain. Using a 16.4 tesla vertical magnet and a custom-made 5 mm solenoid coil a three-dimensional data set with an isotropic resolution of 48 μm was acquired [20]. Despite the limited resolution, numerous commissures, fiber tracts, and lamination previously described with MRI could be identified (Fig. 3).

4 OPT

Optical projection tomography (OPT) is a non-invasive in-vivo imaging technique that allows for longitudinal studies of live models. It has been most used in zebrafish as the size of a fish is not prohibitively large and thus light dispersion is within reasonable limits. The technique uses the same technique as traditional CT imaging. Multiple images are taken at varying angles around the sample by moving either the object

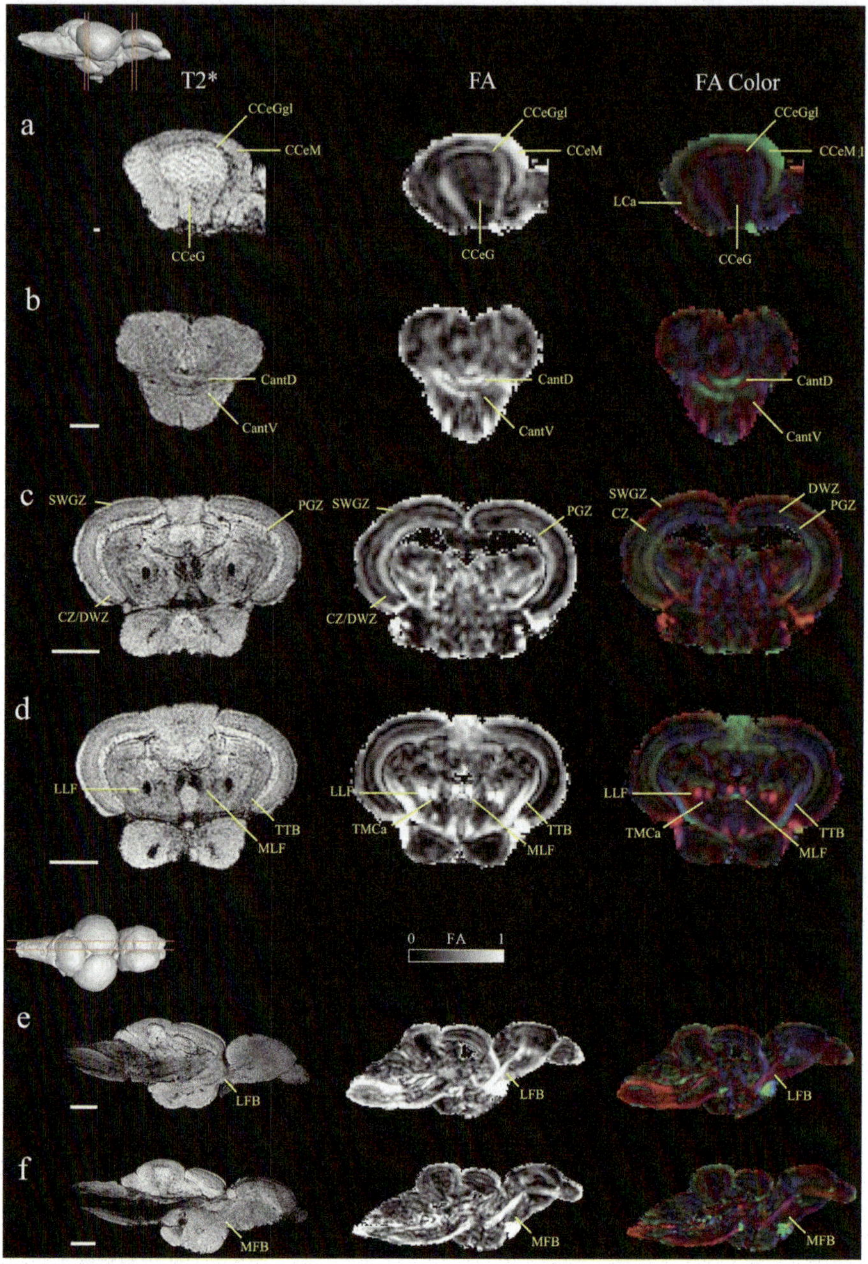

Fig. 3 Comparison of contrast obtained in T_2* 10 µm resolution images from, *left hand column*, FA *middle column* and FA color images (24 µm resolution), *right hand column*. The location of the planes of sections (**a–d**) is demonstrated as orange lines across the lateral brain, while the planes of sections (**e**, **f**) are in the dorsal brain. *Red*, *green*, *blue* in FA color maps indicate rostral-caudal, medial-lateral and dorsal-ventral structural orientation, respectively. *CantD* dorsal anterior

being imaged or the light source and detector. The 3D image is subsequently reconstructed from the series of 2D images via filtered back projection (FBP) with a Radon transform. As with all FBP techniques the larger the number of projections the better the resolution of the final image. The number of projections is typically on the order of several hundred as this provides a good balance between imaging time (several minutes) and resolution. The resulting 3D image is typically resolvable to the order of micron [21]. Imaging in the spectral domain via OPT has also been achieved [22]

The projections are typically acquired as transmission images, using a wide-field imaging system and a light source aligned to the optical axis of the lens. If a fluorescent marker has been used the direction of the light source is far less critical. An OPT setup requires a depth of field that covers the entire imaging field such that a 3D volume can be reconstructed from the 2D projections. As per CT reconstruction acceleration techniques, algorithms and techniques have been proposed that allow for a reasonable image to be produced with as many as 50 projections [23]. In this case if a traditional reconstruction technique such as FBP is used the resulting image would be under-determined but by making use of techniques such as sparse sampling a full 3D image can be reconstructed from as little as 32 projections with an imaging time of 28 s. This is a speed improvement of approximately 30 times over a traditional reconstruction technique.

The drive to reduce imaging times is for two reasons. It reduces the amount of time an animal from a longitudinal study has to remain under anaesthetic and it reduces the chance of photo bleaching from overexposure to the bright light sources that are required for OPT imaging.

Optical Coherence Tomography (OCT) is a similar technique to OPT. It has lower resolution but a higher frame rate that approaches that of a typical video (8 fps) when used in zebrafish [24]. The technique typically only produces a single cross sectional 2D image rather than a 3D volume.

5 Optical Imaging

Optical imaging is the most established imaging modality with which to visualize the zebrafish central nervous system (CNS) (Table 2). Embedding methods for short and longer-term imaging have been established [39] and when employed in combination with optical imaging and genetically encoded markers the fundamental workings of the zebrafish brain have begun to be elucidated.

Fig. 3 (continued) commissure, *CantV* ventral anterior commissure, *CCeG* granular layer of cerebellum, *CCeGgl* ganglionic layer of cerebellum, *CCeM* molecular layer of cerebellum, *CZ* central zone of optic tectum, *DWZ* deep white zone of optic tectum, *LCa* medial caudal lobe of the cerebellum, *LFB* lateral forebrain bundle, *LLF* lateral longitudinal fascicle, *MFB* medial forebrain bundle, *MLF* medial longitudinal fascicle, *PGZ* periventricular grey zone, *SWGZ* superficial grey and white zone, *TMCa* anterior mesencephalo-cerebellar tract, *TTB* tecto-bulbar tract. Scale bar = 500 μm

Table 2 Neuroanatomical references available for the zebrafish

Database	Age	Imaging Modality	Data	Reference
Topological atlas of the zebrafish brain	Adult	Bright-field microscopy	Nissl sections and annotations	[104]
Atlas of early zebrafish development	2, 3, and 5 dpf	Bright-field microscopy	Histological sections of cellular and molecular markers. Annotation	[105]
FishNet	24 h to adult	Optical projection tomography	3D models	[102]
The zebrafish virtual atlas	48 h to adult	Bright-field microscopy	Hematoxylin and eosin histological slides, some annotation	http://zfatlas.psu.edu
Digital embryos	Up to 67 h post fertilization	Digital light sheet microscopy	Movies, figures, and Matlab data sets	[46, 49]
Virtual Brain Explorer (ViBE-Z)	2,3, and 4 dpf	Confocal microscopy	Digital 3D anatomical atlas, gene expression data, software	[106]
Zebrafish brain atlas	Larva	Confocal microscopy	Movies	www.zebrafishbrain.org
Z-space	6 dpf	Confocal microscopy	Digital reference atlas, Matlab data sets	[107]

Optical techniques for recording CNS activity are dependent on markers for neural activity. The simplicity of genetics and transgenesis in the zebrafish [40] has led to the creation of a large number of transgenic lines expressing a range of markers. Most prevalent are genetically engineered calcium indicators (GECIs), such as GCaMP that fluoresce when intracellular calcium is released [41]. Successive iterations of GCaMP have been created enabling single action potentials to be decoded from calcium signals [42]. Pan-neuronal GCaMP zebrafish lines are readily available and have been imaged to elegantly correlate tectal responses of free-swimming larval fish to prey items [43], examine whole-brain dynamics during fictive motor adaptation [44], examine functional development of visual circuits [45], and correlate taste categories with brain stem responses and behaviour [46]. While less frequently employed, optical imaging has also been performed in older zebrafish. Aoki et al. [25] imaged 1-month-old *HuC:IP* fish to identify a discrete area of the dorsal telencephalon essential for long-term memory retrieval.

Numerous other markers of neuronal activity also exist including: calcium-based activity markers that operate in alternate wavelengths including UV and red-shifted wavelengths [47, 48]; photoconvertible proteins such as the calcium-modulated photoactivatable ratiometric integrator (CaMPARI), which fluoresces in the green wavelength when high levels of Ca^{2+} are present and when exposed to UV light irreversibly convert to a red fluorophore [49] or photoactivatable GECIs (PA-GECIs), which enable visualization of both functional and structural activity [50]; and genetically encoded voltage sensors (GEVIs) including ArcLight [51], Arch [52], and ASAP1 [53]. To date, no studies have employed GEVIs to examine the zebrafish brain. However, recently a dual-function calcium and voltage reporter was used to study the zebrafish heard in vivo [54], and GEVIs studies in *Caenorhabditis elegans*, *Drosophila*, and mice [55–57] demonstrate the potential of this type of sensor.

The advent of light sheet imaging enables the in-vivo long term recording of development and neural activity from most of the central nervous system of a larval zebrafish [26]. Light-sheet microscopy (LSM) or selective plane illumination microscopy (SPIM) employs a thin plane of light to optically section a sample at subcellular resolution [35, 58]. A cylindrical lens is used to focus the excitation light into a sheet at the focal plane of the detection objective. In contrast to conventional fluorescent microscopy systems, including confocal, two-photon, or spinning disk microscopes, LSM minimizes specimen phototoxicity and fluorophore bleaching because only a thin portion of the sample is exposed to light at any one point in time during the acquisition. As only a sheet of the sample is illuminated, there is no out-of-focus light and acquisition speeds are significantly higher than line scanning microscopes. Early light-sheet microscopes consisted of a single illumination and detection objectives and suffered from artefacts induced by light scattering and absorption in dense tissue [58]. This resulted in significant image degradation due to a loss of excitation light and fluorescence on the sample side furthest from the illumination objective, and stripes and shadows. To overcome these artefacts the sample had to be rotated to acquire multiple views in a fashion similar to OPT, which were then realigned during image post-processing [58]. Multi-directional

SPIM (mSPIM) was developed to overcome these issues by using parallel illumination objectives and sequentially illuminating and imaging opposing sides of a sample [59]. The result was significant improvements in image quality due to uniform illumination, thinner optical sectioning and increased axial resolution.

In recent years, as LSM technology has steadily improved, it has become increasingly popular and was even named the 2014 Nature Method of the Year [60]. Some of the developments include digital scanned light sheet fluorescence microscopy (DSLM), which uses a laser scanner to rapidly move the laser light through the specimen thereby providing uniform intensity during imaging, reducing optical aberrations, increased illumination power, and enhanced contrast [35, 36]; two-photon light-sheet microscopy (2P-SPIM), which is excellent for imaging deeper into a sample [61–63]; multi-view light sheet microscopes with dual illumination and dual detection objectives [64–66]; lattice-light sheet microscopy, which minimizes photobleaching and phototoxicity while still acquiring data at the molecular level with high spatial and temporal resolution [67]; light sheet microscopes where the illuminations beams are orthogonal to each other [68]; and finally light sheet microscopes with altered light-sheet geometries such as the dual-view iSPIM [69], and swept confocally-aligned planar excitation (SCAPE) microscopy which utilizes an angled, swept light-sheet in a single objective [70]. The result is a multitude of LSM platforms with varying field-of-views, spatial and temporal resolutions, data sizes, costs, and ultimately functions. Due to the diversity of LSM configurations we suggest readers have a look at the following reviews [71, 72] for guidance.

Zebrafish have been extensively studied with LSM. The utility of DSLM was initially demonstrated in zebrafish embryos by tracking all nuclei locations and their movement up to 67 h post fertilization [35, 36, 73]. The 5 dpf zebrafish heart was visualized at the cellular level with a temporal resolution of 70 frames per second using 2P-SPIM [61]. Finally, LSM has also been applied to whole-brain imaging functional imaging of the 5–7 days post fertilization zebrafish brain [26, 68]. By imaging more than 80 % of the neurons in the larval zebrafish brain at cellular resolution, Ahrens et al [26] were able to identify two functionally defined networks, the hindbrain-oscillator and the hindbrain-spinal network.

6 Electron Microscopy

Transmission electron microscopy (TEM) is a microscopic technique that directs a broad beam of electrons through a thin sample (60–100 nm) before being focused onto an electronic sensor [74]. Examination of the zebrafish nervous system using TEM has primarily been performed on cellular structures at early stages of development to visualize processes such as differentiation of cell types or the formation of axon tracts. Due to sample size restrictions in TEM (1 $mm^2 \times 90$ nm) most zebrafish studies have focused on confined regions such as the retina [75], taste buds [76], lateral line, forebrain and midbrain [77–79], and hindbrain [80]. Although a single 281 gigapixel montage (1.5×0.6 mm^2 area and 1.6 nm resolution) of the entire zebrafish has been collected [81] no large scale volumetric circuit studies have been performed on the zebrafish brain in a similar fashion as studies of the central nervous system (CNS) of *Caenorhabditis elegans* [82]. This is due to the tedious nature of sectioning and the vast number of sections, required for such a dataset on a nano scale. This has limited circuit analyses in a more complex CNS such as the zebrafish.

Recently, new technology has revolutionized TEM and made it an ultrastructural imaging method capable of imaging large three-dimensional volumes in a relatively high-throughput manner [83]. First, automated tape-collecting ultra microtomes (ATUM) have been built that automatically cut (<30 nm thick) and then collect sections on carbon-coated tape before then being imaged on an electron microscope [84, 85]. While the resulting images must subsequently be stitched and registered into a three-dimensional volume, this technique keeps the sections on the tape permitting subsequent imaging. Second, new microscopes have been developed that place the tissue block inside a vacuum chamber with an SEM. The SEM images the face of the block before a section is removed either with a diamond knife as in serial block-face electron microscopy (SBEM) [86], or with a focused ion beam (FIB-SEM).

SBEM is a widely used technology because it can examine samples with a size up to 800 um^2 and a total volume of up to 1000 mm^3 while still maintaining a minimum slice thickness of ~25 nm and a maximum lateral resolution of ~5 nm^2. In contrast, FIB-SEM is more suited to high resolution automated serial imaging with a smaller field of view (<80 μm^2), but a maximum voxel resolution of ~3 nm^3. Moreover, FIB-SEM can target specific region of interest without damaging the rest of the block—enabling later resampling of different regions. In contrast to EM of sections collected by ATUM, both SBEM and FIB-SEM enable the collection of thousands of automatically cut and perfectly aligned images, however, the sections that created those images are destroyed. Finally, new commercial microscopes have also been developed that include up to 91 parallel electron beans enabling large field of views (up to 1 cm) with nanometer resolution [87]. Note that in most current EM techniques there is a potential loss of adjoining structure and acquisition times are prohibitively large for whole brain acquisition at high resolution.

To parallel technological advances in electron microscopes new staining methods and contrast agents have also been established. Whole-brain osmium-based staining protocols permit visualization of tissue ultrastructure throughout the brain including chemical synapses, spine necks, and small-caliber axons [88, 89]. Correlative light and electron microscopy (CLEM) contrast agents that combine localization data from fluorescent microscopy and ultrastructural information from TEM have also now exist [90–92]. Earlier methods relied on minimal fixation [93] and therefore results could be variable in larger samples where ultra-structural preservation is more difficult [94]. For detailed protocols see Schieber et al. 2010 [95]. Newer contrast agents have overcome this fixation issue by using transgenic methods. For example, variants of the photoconvertible Eos fluorescent protein, mEos4, were engineered that are compatible with the osmium fixation required for TEM [92]. In a second study, a modified soybean ascorbate peroxidase (APEX) was targeted to a GFP-binding protein by engineering a GFP binding peptide (GBP) to the APEX-tag, thereby enabling EM localization of any GFP-tagged protein [96].

Fig. 4 Reconstruction of zebrafish larvae olfactory bulb from serial blockface EM. From [97]

To further enhance the utility of this marker Ariotti et al. [96] also generated zebrafish strains with stable and inducible expression of APEX-GFP. While the zebrafish brain has not been examined yet, since these lines express APEX-GBP in all tissues they can be used in combination with any GFP-tagged line.

To date, no studies have employed these larger volume technologies or markers to map the 'zebrafish connectome', although one study has performed an initial analysis on the larval zebrafish olfactory bulb. Friedrich and Yaski [97] used transgenic zebrafish expressing a genetically encoded calcium indicator and multi-photon microscopy to measure primarily olfactory neuron responses to various odours. Subsequently, they performed SBEM on the same olfactory bulb and performed circuit reconstruction (Fig. 4). We anticipate this is just the beginning of CLEM studies.

7 Neuroimaging and Phenotyping

Neuroimaging plays a crucial role in phenotyping research and in studies of neurological diseases. By examining the neuroanatomy of animal models, morphological abnormalities can be identified and correlations made with behaviour. MRI and DWI are frequently used pre-clinically to identify morphological phenotypes in knockout models of neurological diseases. For example, in a mouse model of the γ–aminobutyric acid type A ($GABA_A$) $\gamma 2_{R43Q}$ mutation, MRI was used to establish that the volume of the dentate gyrus was 5 % larger in *R43Q* mice than controls [98]. MRI and DWI have also both been employed to identify autism-relevant phenotypes. Comparisons between the inbred mouse strain BTBR T+tf/J and two wild-type strains found significant differences in brain region volumes and white matter

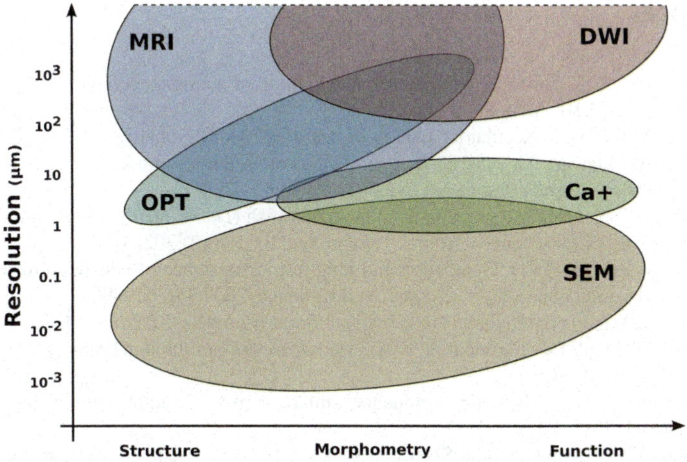

Fig. 5 Figure demonstrating the overlap of resolution and resulting data type for the techniques that are described in this chapter. With some modification some of these techniques can be made to work at higher or lower resolution but here we have shown their most typical use cases. e.g. large blockface SEM imaging projects have managed to acquire up to mm scale EM datasets but these are extremely time consuming to acquire and subsequently analyse for functional information

microstructure. These differences correlated with specific behaviours known to be associated with those regions [99]. Similar results were found in a knockout mouse model of X-linked Fragile X Mental Retardation 1 (FMR1) gene mutation [100]. Interestingly a zebrafish *fmr1* knockout model also exists [101–103], with a behavioural phenotype consistent with rodent models (however, a neuroimaging study is yet to be performed in this model).

In general, neuroimaging modalities for larval zebrafish show great promise for phenotyping (Fig. 5). For example, calcium imaging collects whole brain recordings at very high temporal resolutions while still permitting a zebrafish to behave relatively normally [26, 104]. These techniques have primarily been used to understand the fundamental workings of the zebrafish brain such as which neurons are responsible for escape behaviour [105], swimming speed [27], and swim posture [28], however similar imaging techniques could be applied to examining the seizure network in epileptic fish, or functional connectivity in autism models. This would be the first time these networks were examined across the entire brain yet still at the single cell scale! When coupled with microfluidics [106] and automated screening platforms [107, 108], precision medicine at a whole new level becomes possible.

Acknowledgements The authors acknowledge the support of the National Institute of Health (5R03NS077295-02) and the National Imaging Facility. JFPU was also supported by a University of Queensland Early Career Research Award.

References

1. Kalueff AV, et al. Towards a comprehensive catalog of zebrafish behavior 1.0 and beyond. Zebrafish. 2013;10(1):70–86.
2. Stewart AM, et al. Molecular psychiatry of zebrafish. Mol Psychiatry. 2015;20(1):2–17.
3. Kalueff AV, Stewart AM, Gerlai R. Zebrafish as an emerging model for studying complex brain disorders. Trends Pharmacol Sci. 2014;35(2):63–75.
4. Hill A, et al. Neurodevelopmental defects in zebrafish (*Danio rerio*) at environmentally relevant dioxin (TCDD) concentrations. Toxicol Sci. 2003;76(2):392–9.
5. Hinsch K, Zupanc GKH. Generation and long-term persistence of new neurons in the adult zebrafish brain: a quantitative analysis. Neuroscience. 2007;146:679–96.
6. Mueller T. What is the thalamus in zebrafish? Front Neurosci. 2012;6:1–14.
7. Ullmann JF, et al. Development of MRI-based atlases of non-human brains. J Comp Neurol. 2015;523(3):391–405.
8. Kabli S, et al. Magnetic resonance microscopy of adult zebrafish. Zebrafish. 2006;3(4):431–9.
9. Ullmann JFP, Cowin G, Collin SP. Magnetic resonance microscopy of the barramundi (*Lates calcarifer*) brain. J Morphol. 2010;271:1446–56.
10. Ullmann JFP, Cowin G, Collin SP. Quantitative assessment of brain volumes in fish: comparison of methodologies. Brain Behav Evol. 2010;76:261–70.
11. Ullmann JFP, et al. Magnetic resonance histology of the adult zebrafish brain: optimization of fixation and gadolinium contrast enhancement. NMR Biomed. 2010;23(4):341–6.
12. Ullmann JF, et al. A three-dimensional digital atlas of the zebrafish brain. Neuroimage. 2010;51(1):76–82.
13. Janke AL, Ullmann JF. Robust methods to create ex vivo minimum deformation atlases for brain mapping. Methods. 2015;73:18–26.
14. Ramirez IB, et al. Impaired neural development in a zebrafish model for Lowe syndrome. Hum Mol Genet. 2012;21(8):1744–59.
15. Mori S, et al. Three-dimensional tracking of axonal projections in the brain by magnetic resonance imaging. Ann Neurol. 1999;45(2):265–9.
16. Zhang J, van Zijl PCM, Mori S. Three-dimensional diffusion tensor magnetic resonance microimaging of adult mouse brain and hippocampus. Neuroimage. 2002;15:892–901.
17. Mori S, Zhang J. Principles of diffusion tensor imaging and its applications to basic neuroscience research. Neuron. 2006;51(5):527–39.
18. Shepherd TM, et al. Structural insights from high-resolution diffusion tensor imaging and tractography of the isolated rat hippocampus. Neuroimage. 2006;32(4):1499–509.
19. Freidlin RZ, et al. Diffusion tensor MR microscopy of adult zebrafish. In: Conference: proceedings of the international society for magnetic resonance in medicine, Kyoto Japan. 2004.
20. Ullmann JFP. Three-dimensional imaging of the teleost brain. PhD Thesis. Brisbane: School of Biomedical Sciences, The University of Queensland; 2010. p. 152.
21. Bassi A, Schmid B, Huisken J. Optical tomography complements light sheet microscopy for in toto imaging of zebrafish development. Development. 2015;142(5):1016–20.
22. Zhang J, Ge W, Yuan Z. In vivo three-dimensional characterization of the adult zebrafish brain using a 1325 nm spectral-domain optical coherence tomography system with the 27 frame/s video rate. Biomed Opt Express. 2015;6(10):3932–40.
23. Correia T, et al. Accelerated optical projection tomography applied to in vivo imaging of zebrafish. PLoS One. 2015;10(8), e0136213.
24. Rao KD, et al. Real-time in vivo imaging of adult zebrafish brain using optical coherence tomography. J Biophotonics. 2009;2:1–4.
25. Aoki T, et al. Imaging of neural ensemble for the retrieval of a learned behavioral program. Neuron. 2013;78(5):881–94.
26. Ahrens MB, et al. Whole-brain functional imaging at cellular resolution using light-sheet microscopy. Nat Methods. 2013;10(5):413–20.
27. Severi KE, et al. Neural control and modulation of swimming speed in the larval zebrafish. Neuron. 2014;83(3):692–707.

28. Thiele TR, Donovan JC, Baier H. Descending control of swim posture by a midbrain nucleus in zebrafish. Neuron. 2014;83(3):679–91.
29. Semmelhack JL, et al. A dedicated visual pathway for prey detection in larval zebrafish. eLife 2014;3:e04878.
30. Kubo F, et al. Functional architecture of an optic flow-responsive area that drives horizontal eye movements in zebrafish. Neuron. 2014;81(6):1344–59.
31. Bryson-Richardson RJ, et al. FishNet: an online database of zebrafish anatomy. BMC Biol. 2007;5(34):1–8.
32. Ullmann JF, et al. Enhanced characterization of the zebrafish brain as revealed by super-resolution track-density imaging. Brain Struct Funct. 2015;220(1):457–68.
33. Wullimann MF, Rupp B, Reichert H. Neuroanatomy of the zebrafish brain: a topological atlas. Basel: Birkhäuser Verlag; 1996. p. 144.
34. Mueller T, Wullimann MF. Atlas of early zebrafish brain development. Amsterdam: Elsevier; 2005.
35. Keller PJ, et al. Reconstruction of zebrafish early embryonic development by scanned light sheet microscopy. Science. 2008;322(5904):1065–9.
36. Keller PJ, et al. Fast, high-contrast imaging of animal development with scanned light sheet-based structured-illumination microscopy. Nat Methods. 2010;7(8):637–42.
37. Ronneberger O, et al. ViBE-Z: a framework for 3D virtual colocalization analysis in zebrafish larval brains. Nat Methods. 2012;9(7):735–42.
38. Randlett O, et al. Whole-brain activity mapping onto a zebrafish brain atlas. Nat Methods. 2015;12(11):1039–46.
39. Kaufmann A, et al. Multilayer mounting enables long-term imaging of zebrafish development in a light sheet microscope. Development. 2012;139(17):3242–7.
40. Scott EK, et al. Targeting neural circuitry in zebrafish using GAL4 enhancer trapping. Nat Methods. 2007;4(4):323–6.
41. Nakai J, Ohkura M, Imoto K. A high signal-to-noise Ca(2+) probe composed of a single green fluorescent protein. Nat Biotechnol. 2001;19(2):137–41.
42. Chen TW, et al. Ultrasensitive fluorescent proteins for imaging neuronal activity. Nature. 2013;499(7458):295–300.
43. Muto A, Kawakami K. Prey capture in zebrafish larvae serves as a model to study cognitive functions. Front Neural Circuits. 2013;7:110.
44. Ahrens MB, et al. Brain-wide neuronal dynamics during motor adaptation in zebrafish. Nature. 2012;485(7399):471–7.
45. Nikolaou N, Meyer MP. Lamination speeds the functional development of visual circuits. Neuron. 2015;88(5):999–1013.
46. Vendrell-Llopis N, Yaksi E. Evolutionary conserved brainstem circuits encode category, concentration and mixtures of taste. Sci Rep. 2015;5:17825.
47. Akerboom J, et al. Genetically encoded calcium indicators for multi-color neural activity imaging and combination with optogenetics. Front Mol Neurosci. 2013;6:2.
48. Zhao Y, et al. An expanded palette of genetically encoded Ca(2)(+) indicators. Science. 2011;333(6051):1888–91.
49. Fosque BF, et al. Neural circuits. Labeling of active neural circuits in vivo with designed calcium integrators. Science. 2015;347(6223):755–60.
50. Berlin S, et al. Photoactivatable genetically encoded calcium indicators for targeted neuronal imaging. Nat Methods. 2015;12(9):852–8.
51. Jin L, et al. Single action potentials and subthreshold electrical events imaged in neurons with a fluorescent protein voltage probe. Neuron. 2012;75(5):779–85.
52. Hochbaum DR, et al. All-optical electrophysiology in mammalian neurons using engineered microbial rhodopsins. Nat Methods. 2014;11(8):825–33.
53. St-Pierre F, et al. High-fidelity optical reporting of neuronal electrical activity with an ultra-fast fluorescent voltage sensor. Nat Neurosci. 2014;17(6):884–9.
54. Hou JH, et al. Simultaneous mapping of membrane voltage and calcium in zebrafish heart in vivo reveals chamber-specific developmental transitions in ionic currents. Front Physiol. 2014;5:344.

55. Cao G, et al. Genetically targeted optical electrophysiology in intact neural circuits. Cell. 2013;154(4):904–13.
56. Flytzanis NC, et al. Archaerhodopsin variants with enhanced voltage-sensitive fluorescence in mammalian and *Caenorhabditis elegans* neurons. Nat Commun. 2014;5:4894.
57. Gong Y, et al. High-speed recording of neural spikes in awake mice and flies with a fluorescent voltage sensor. Science. 2015;350:1361–1366.
58. Huisken J, et al. Optical sectioning deep inside live embryos by selective plane illumination microscopy. Science. 2004;305(5686):1007–9.
59. Huisken J, Stainier DYR. Even fluorescence excitation by multidirectional selective plane illumination microscopy (mSPIM). Opt Lett. 2007;32(17):2608–10.
60. de Souza N. Method of the year 2014. Nat Methods. 2015;12(1):1.
61. Truong TV, et al. Deep and fast live imaging with two-photon scanned light-sheet microscopy. Nat Methods. 2011;8(9):757–60.
62. Mahou P, et al. Multicolor two-photon light-sheet microscopy. Nat Methods. 2014;11(6):600–1.
63. Wolf S, et al. Whole-brain functional imaging with two-photon light-sheet microscopy. Nat Methods. 2015;12(5):379–80.
64. Chhetri RK, et al. Whole-animal functional and developmental imaging with isotropic spatial resolution. Nat Methods. 2015;12(12):1171–8.
65. Krzic U, et al. Multiview light-sheet microscope for rapid in toto imaging. Nat Methods. 2012;9(7):730–3.
66. Tomer R, et al. Quantitative high-speed imaging of entire developing embryos with simultaneous multiview light-sheet microscopy. Nat Methods. 2012;9(7):755–63.
67. Chen BC, et al. Lattice light-sheet microscopy: imaging molecules to embryos at high spatiotemporal resolution. Science. 2014;346(6208):1257998.
68. Vladimirov N, et al. Light-sheet functional imaging in fictively behaving zebrafish. Nat Methods. 2014;11(9):883–4.
69. Wu Y, et al. Spatially isotropic four-dimensional imaging with dual-view plane illumination microscopy. Nat Biotechnol. 2013;31(11):1032–8.
70. Bouchard MB, et al. Swept confocally-aligned planar excitation (SCAPE) microscopy for high speed volumetric imaging of behaving organisms. Nat Photonics. 2015;9(2):113–9.
71. Reynaud EG, et al. Guide to light-sheet microscopy for adventurous biologists. Nat Methods. 2015;12(1):30–4.
72. Keller PJ, Ahrens MB, Freeman J. Light-sheet imaging for systems neuroscience. Nat Methods. 2015;12(1):27–9.
73. Keller PJ, et al. Digital scanned laser light-sheet fluorescence microscopy (DSLM) of zebrafish and Drosophila embryonic development. Cold Spring Harb Protoc. 2011;2011(10):1235–43.
74. Ruska E, Knoll M. Das elektronenmikroskop. Z Phys. 1932;78:318–39.
75. Schmitt EA, Dowling JE. Early retinal development in the zebrafish, *Danio rerio*: light and electron microscopic analyses. J Comp Neurol. 1999;404(4):515–36.
76. Hansen A, Reutter K, Zeiske E. Taste bud development in the zebrafish, *Danio rerio*. Dev Dyn. 2002;223(4):483–96.
77. Wilson SW, et al. The development of a simple scaffold of axon tracts in the brain of the embryonic zebrafish, *Brachydanio rerio*. Development. 1990;108(1):121–45.
78. Hansen A, Zeiske E. Development of the olfactory organ in the zebrafish, *Brachydanio Rerio*. J Comp Neurol. 1993;333(2):289–300.
79. Lindsey BW, Darabie A, Tropepe V. The cellular composition of neurogenic periventricular zones in the adult zebrafish forebrain. J Comp Neurol. 2012;520(10):2275–316.
80. Kimmel CB, Sessions SK, Kimmel RJ. Morphogenesis and synaptogenesis of the zebrafish Mauthner neuron. J Comp Neurol. 1981;198(1):101–20.
81. Faas FG, et al. Virtual nanoscopy: generation of ultra-large high resolution electron microscopy maps. J Cell Biol. 2012;198(3):457–69.
82. White JG, et al. The structure of the nervous system of the nematode *Caenorhabditis elegans*. Philos Trans R Soc Lond B Biol Sci. 1986;314(1165):1–340.

83. Kasthuri N, et al. Saturated reconstruction of a volume of neocortex. Cell. 2015;162(3):648–61.
84. Kuwajima M, Mendenhall JM, Harris KM. Large-volume reconstruction of brain tissue from high-resolution serial section images acquired by SEM-based scanning transmission electron microscopy. Methods Mol Biol. 2013;950:253–73.
85. Hayworth KJ, et al. Imaging ATUM ultrathin section libraries with WaferMapper: a multi-scale approach to EM reconstruction of neural circuits. Front Neural Circuits. 2014;8:68.
86. Denk W, Horstmann H. Serial block-face scanning electron microscopy to reconstruct three-dimensional tissue nanostructure. PLoS Biol. 2004;2(11), e329.
87. Eberle AL, et al. High-resolution, high-throughput imaging with a multibeam scanning electron microscope. J Microsc. 2015;259(2):114–20.
88. Tapia JC, et al. High-contrast en bloc staining of neuronal tissue for field emission scanning electron microscopy. Nat Protoc. 2012;7(2):193–206.
89. Mikula S, Denk W. High-resolution whole-brain staining for electron microscopic circuit reconstruction. Nat Methods. 2015;12(6):541–6.
90. de Boer P, Hoogenboom JP, Giepmans BN. Correlated light and electron microscopy: ultrastructure lights up! Nat Methods. 2015;12(6):503–13.
91. Chang YW, et al. Correlated cryogenic photoactivated localization microscopy and cryo-electron tomography. Nat Methods. 2014;11(7):737–9.
92. Paez-Segala MG, et al. Fixation-resistant photoactivatable fluorescent proteins for CLEM. Nat Methods. 2015;12(3): 215–8, 4 p. following 218.
93. Nixon SJ, et al. A single method for cryofixation and correlative light, electron microscopy and tomography of zebrafish embryos. Traffic. 2009;10(2):131–6.
94. Watanabe S, et al. Protein localization in electron micrographs using fluorescence nanoscopy. Nat Methods. 2011;8(1):80–4.
95. Schieber NL, et al. Modern approaches for ultrastructural analysis of the zebrafish embryo. In: Müller-Reichert T, editor. Methods in cell biology. Amsterdam: Academic Press; 2010. p. 425–42.
96. Ariotti N, et al. Modular detection of GFP-labeled proteins for rapid screening by electron microscopy in cells and organisms. Dev Cell. 2015;35(4):513–25.
97. Friedrich RW, Genoud C, Wanner AA. Analyzing the structure and function of neuronal circuits in zebrafish. Front Neural Circuits. 2013;7:71.
98. Richards KL, et al. Hippocampal volume and cell density changes in a mouse model of human genetic epilepsy. Neurology. 2013;80(13):1240–6.
99. Ellegood J, et al. Neuroanatomical analysis of the BTBR mouse model of autism using magnetic resonance imaging and diffusion tensor imaging. Neuroimage. 2013;70:288–300.
100. Ellegood J, et al. Anatomical phenotyping in a mouse model of fragile X syndrome with magnetic resonance imaging. Neuroimage. 2010;53(3):1023–9.
101. Ng MC, Yang YL, Lu KT. Behavioral and synaptic circuit features in a zebrafish model of fragile X syndrome. PLoS One. 2013;8(3), e51456.
102. Kim L, et al. Anxiety, hyperactivity and stereotypy in a zebrafish model of fragile X syndrome and autism spectrum disorder. Prog Neuropsychopharmacol Biol Psychiatry. 2014;55:40–9.
103. den Broeder MJ, et al. Generation and characterization of FMR1 knockout zebrafish. PLoS One. 2009;4(11), e7910.
104. Ritter DA, Bhatt DH, Fetcho JR. In vivo imaging of zebrafish reveals differences in the spinal networks for escape and swimming movements. J Neurosci. 2001;21(22):8956–65.
105. Douglass AD, et al. Escape behavior elicited by single, channelrhodopsin-2-evoked spikes in zebrafish somatosensory neurons. Curr Biol. 2008;18(15):1133–7.
106. Candelier R, et al. A microfluidic device to study neuronal and motor responses to acute chemical stimuli in zebrafish. Sci Rep. 2015;5:12196.
107. Pardo-Martin C, et al. High-throughput in vivo vertebrate screening. Nat Methods. 2010;7(8):634–6.
108. Lin X, et al. High-throughput mapping of brain-wide activity in awake and drug-responsive vertebrates. Lab Chip. 2015;15(3):680–9.

Illustrated Zebrafish Neurobehavioral Glossary

Allan V. Kalueff

Abstract This Chapter contains the alphabetized Zebrafish Neurobehavioral Catalog (ZBC)—an illustrated glossary of known zebrafish behaviors (phenotypes), compiled by the International Zebrafish Neuroscience Research Consortium (ZNRC) and relevant to major behavioral domains in adult and larval zebrafish.

Keywords Zebrafish • Behavioral phenotypes • Experimental models and tests

1 Introduction

Describing zebrafish behaviors in the literature is often complicated by the lack of standardized terminology accepted universally. Therefore, providing necessary definitions and placing zebrafish behaviors in specific contexts become critical for improving our understanding of zebrafish phenotypes [1]. Using various behavioral terms in a laboratory-specific manner also represents a problem, especially since different terms can be used by different groups to describe the same zebrafish behaviors, and the same terms can describe completely different behaviors performed in distinct biological contexts (see [1] for details). Moreover, various behaviors in zebrafish may have similar or overlapping behavioral manifestations, with very subtle (but important and describable) differences [1]. For example, stress-related freezing behavior (i.e., lack of body movements except for eyes and slight fin undulation, with increased opercular movements) can be easily distinguished from sleep-like/resting behavior (defined as the lack of body movement, but

A.V. Kalueff (✉)
The International Zebrafish Neuroscience Research Consortium (ZNRC) and ZENEREI Research Center, Slidell, LA, USA

College of Food Science and Technology, Research Institute of Marine Drugs and Nutrition, Guangdong Ocean University, Zhanjiang, Guangdong, China

Institute of Translational Biomedicine, St. Petersburg State University, St. Petersburg, Russia

Institutes of Chemical Technology and Natural Sciences, Ural Federal University, Ekaterinburg, Russia
e-mail: avkalueff@gmail.com

A.V. Kalueff (ed.), *The rights and wrongs of zebrafish: Behavioral phenotyping of zebrafish*, DOI 10.1007/978-3-319-33774-6_14

occurring with reduced opercular activity) [1]. However, this distinction can only be clear if both behavioral phenotypes are properly named, described and quantified.

Generated by a network of active zebrafish laboratories working with both larval and adult models, the Zebrafish Neurobehavioral Catalog (ZBC) [1] represents a formalized consensus-based glossary standardizing all major terminology in this field. This chapter lists zebrafish behavioral phenotypes outlined in ZBC [1], reproduced here with permission from Mary Ann Liebert, Inc, the original publisher for this copyrighted material. In addition to listing the main behavioral terms observed in adult and larval zebrafish, the glossary also provides their synonyms as well as specific contexts (e.g., locomotor, social, given in italics) for terms that cover multiple distinct behavioral domains [1]. Where appropriate, behavioral parameters (indices) were also mentioned here as useful examples of measurable endpoints for the terms in this glossary [1]. For convenience of ZBC usage in the literature, the terms reflecting specific individual behaviors, are numbered from 1 to 190 to encourage citing their respective ZBC numbers in research articles, enabling better clarity, improved characterization and cross-study/cross-laboratory data interpretation [1].

In conclusion, the illustrated Glossary presented here offers a comprehensive standardized and up-to-date catalog of major zebrafish behavioral phenotypes currently known [1]. As a good example of a research community-driven neurophenotyping effort coordinated by the International Zebrafish Neuroscience Research Consortium (ZNRC), this comprehensive behavioral glossary enables a formalized and improved understanding of fish behavior [1]. While representing a continuous progress open to further modifications, revisions and updates by the members of zebrafish research community, the Glossary is expected to foster translational neuroscience and neurophenotyping research using this model species [1].

2 Zebrafish Neurobehavioral Catalog

1. *Abnormal body position:* Contortion of the body (e.g., droopy tail, Fig. 1 and bending); typically has a long-term nature (unlike short-term *twitches/ spasms*); caused by illness, genetic defects, toxic agents, or aging. In larval zebrafish, commonly associated with neurological phenotypes and/or neurodevelopmental abnormalities.

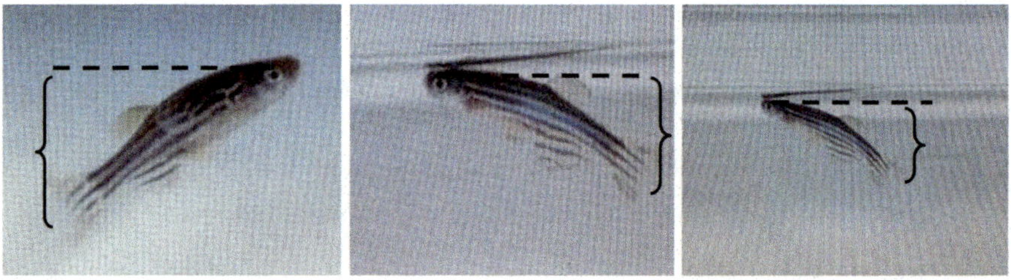

Fig. 1 Examples of zebrafish droopy tail—an aberrant motor phenotype differing from normal (horizontal) fish body position

Fig. 2 Zebrafish 'shoaling' as an example of fish aggregation behavior

2. *Aggregation behavior:* Seeking of conspecifics based on chemical, visual or other cues; exploratory approach that then leads to *shoaling* (Fig. 2), *schooling*, and/or *sexual aggregating behavior*, where applicable.
3. *Aggression:* Complex behaviors (including *approach, fin raise, undulating body movement, mouth opening behavior, body color change, biting, charging, chasing*, and *circling*) directed at conspecifics (or other objects) in adult zebrafish; may appear in the context of defending the territory (*territorial behavior*), protecting resources (e.g., females) and establishing dominance (see *Social interaction*). Related to *boldness phenotype*; can be affected by different pharmacological manipulations.
4. *Akinesia:* A slowness of swimming or loss of normal motor function; commonly observed in aged zebrafish or after exposure to selected compounds, such as dopamine-depleting drugs (e.g., reserpine), is often accompanied by *droopy tail*; can be assessed by a global reduction in distance traveled and/or swimming velocity (similar to *hypolocomotion*), also see *Ataxia* and *Motor incoordination*.
5. *Alarm reaction:* An adaptive escape reaction which serves as an anti-predatory response exhibited in the context of fear-inducing stimulation (e.g., chemical alarm cue or visual predator exposure). Typically characterized by increased speed of movement and rapid directional changes, a response set that is often referred to as *erratic movement* (also see *Zig-zagging*). Alarm reaction may also include *freezing* with frequent *opercular movements*, changes in *shoaling* (e.g., rapid 1–2 s decreases of shoal cohesion followed by longer-lasting, up to several minutes, increase of shoal cohesion) and *diving*.
6. *Anxiety (anxiety-like) behavior:* Complex behavior evoked by dangerous or potentially dangerous environment/stimuli. Includes reduced exploration, and typically manifests in *geotaxis* (*diving*), *thigmotaxis*, *scototaxis*, *freezing*, *opercular movements*, *body color change*, and *erratic movement* (*zig-zagging*). Anxiolytic drugs generally reduce anxiety-like behaviors, while anxiogenic agents potentiate these responses. Zebrafish anxiety-like behavior frequently

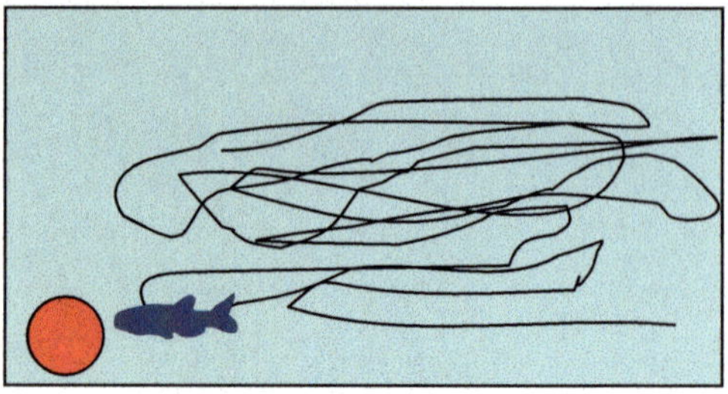

Fig. 3 Zebrafish approach behavior (as an example of boldness phenotype) in response to a novel object (*red ball*) exposure

overlaps with *fear-related behavior*, and future studies are needed to better characterize these two domains.

7. *Appetitive olfactory behavior* (also see *Olfactory response):* Increased rate of swimming and distance traveled with frequent directional changes (>90° turns) that serve to sample appetitive odor plumes (e.g., L-alanine, food extract). Once visual contact with food is established, *approach* and *nibbling* behaviors are displayed.
8. *Approach:* Display of presence, movement towards an object. *Sexual:* Abrupt swimming movement ('*present*') expressed independently of any male courtship behaviors; performed by females during *courtship. Exploratory:* approach to the novel object (part of *boldness* phenotype, Fig. 3), opposite of *avoidance. Appetitive:* can be part of *attraction* behavior (e.g., *food seeking*).
9. *Ataxia:* A general loss of normal body posture and/or coordination of movements (e.g., *laying on a side*, *swimming on a side or upside down*, *corkscrew swimming*); commonly observed as a result of neurotoxicity-induced *motor incoordination, akinesia, seizure behavior*, and/or *paralysis*.
10. *Attack (attacking):* Short bouts of fast swimming directed at an opponent, accompanied by the *mouth opening behavior* and *biting*; part of *aggression*-related behavior (differs from *strike* behavior by the presence of physical contact between fighting fish).
11. *Attraction:* Increased time spent nearby or movement towards an object (visual) or chemical stimulus (e.g., food extract); opposite to *Avoidance*.
12. *Avoidance:* Increased movement away and/or time spent away from an object or a stimulus (e.g., predator, bright light); opposite to *Attraction*.
13. *Background adaptation:* See *Camouflage response, Body color change*.
14. *Backward swimming:* Albeit rarely occurring in normal zebrafish, typically represents an aberrant motor behavior observed under some circumstances, such as following exposure to selected hallucinogenic drugs (e.g., lysergic acid diethylamide, LSD).
15. *Beat-and-glide:* An intermittent form of *swimming* characterized by tail beating followed by gliding; appears at ~4 dpf in larvae.
16. *Bend (bending):* Aberrant neurological phenotype involving swimming with the body in a laterally bent position; can be observed as part of *seizure* behavior (e.g., evoked by certain convulsant agents, such as caffeine). Short-lasting bouts of this behavior represent *twitch/spasm* behavior.

17. *Bite (biting):* Quick movement towards target, with mouth opening and closing, with physical contact. *Social:* zebrafish will often bite/nip (*nipping*) each other around the gill region or fins during 'fights'. *Predatory/food:* zebrafish can attempt to bite/consume any sufficiently small item moving through their field of vision at appropriate speeds; differs from non-aggressive *nibbling*.
18. *Boldness:* Behavior characterized by bold personality trait, typically manifested in increased *approach* towards novel objects (Fig. 3, also see *Risk-taking behavior*). Usually, bolder animals also present reduced *anxiety-like behavior, body color change*, and increased *exploratory activity*.
19. *Body color change (coloration response):* A general change in body pigmentation resulting in a darker or lighter appearance; can be a sign of anxiety, a natural response to lighting/environmental conditions (*camouflage response*), a part of social behavior (e.g., *display, fight* or *courtship*), as well as a result of stress/sickness, or drug-evoked dispersion (skin darkening) or aggregation (paling) of melanophores. Specific drugs (e.g., alcohol, ibogaine) evoke robust skin darkening in adult zebrafish, while some factors (e.g., cold exposure, pathogens) can evoke paling (e.g., *sickness behavior*). Coloration response can be assessed manually (by visual inspection) or using automated (luminescence-based) tools.
20. *Breeding (reproductive) behavior:* See *Spawning.*
21. *Buoyancy dysregulation:* Interference with the ability to control buoyancy. Characterized by an inability to remain at a constant elevation (sometimes in vertical or inclined/titles position) without exerting physical effort via swimming; most commonly caused by problems with the swim bladder or other peripheral systems; often manifests as *surfacing, vertical drifting, Cartesian diver behavior, inclined swimming, and tilting*.
22. *Burst-and-coast behavior:* Darting pattern specific to larval fish not yet able to perform continuous swimming. Fish move forward (burst) in a single motion and glide (coast) to a slow speed, or stop from which they burst forward again.
23. *Burst swim (swimming):* Fast forward swim with large bend angles, maximally at mid-body of larval zebrafish, appears at 2 dpf in larvae. Includes larger amplitude bending (large bend angles), faster speeds and greater yaw that during slow *swimming*; often associated with *escape* behaviors; pectoral fins are tucked against the body and not active).
24. *C-start (C-bend/turn, Mauthner reflex):* Quick *escape/startle response* in which the fish body first curves to form a C-shape, and then the fish propels itself away at an angle from its previous position using a fast swim. Exhibited by both adult and larval fish, and is regulated by Mauthner cells (also see *O-bend* and *S-start/bend*). In larval zebrafish, head stimulation generally elicits C-starts, while tail stimulation evokes both C- and S-starts (Fig. 4).
25. *Cannibalism:* Eating of dead or alive conspecifics (also see *Infanticide,* including egg cannibalism).
26. *Camouflage response (background adaptation),* also see: *Body color change*. A change in body pigmentation (resulting in a darker or lighter appearance) after being exposed to a darker or lighter background, respectively (Fig. 5). Occurs due to melanophore dispersion (skin darkening) or aggregation (paling). Part of

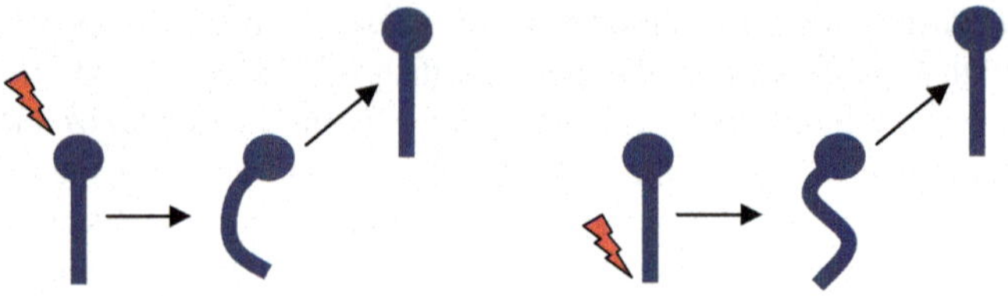

Fig. 4 Diagrams showing typical C-start (*left*) and S-start (*right*) in zebrafish locomotion

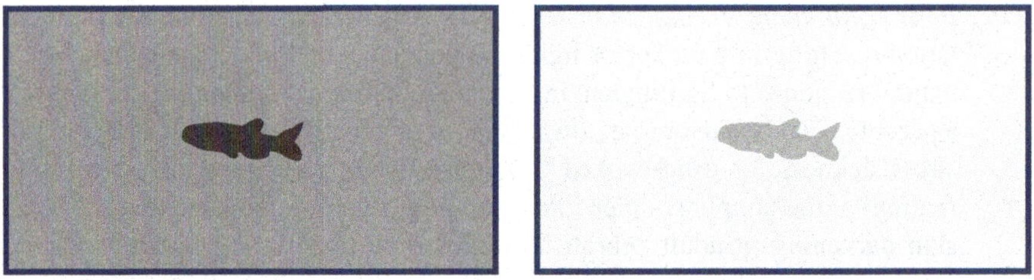

Fig. 5 An example of zebrafish camouflage response (background adaptation) to *dark* vs *light* backgrounds

body color change response; in fish phenotyping literature represents 'expanded melanophore phenotype' (e.g., lack of body color change may indicate deficits in light perception, leaving the larva in a dark adapted pigmentation state).

27. *Cartesian diver behavior:* An aberrant phenotype that involves alternating between passive *vertical drift* and *sinking*; induced in fish by some neuroactive substances, such as LSD.
28. *Charge (charging):* Movement towards a second fish, increasing acceleration, while second fish avoids the first. Establishes social dominance, and marks the resolution of a zebrafish '*fight*.'
29. *Chase (chasing):* See *Charge*.
30. *Chemotaxis (chemoattraction):* Movement to/preference towards specific chemical cues serving as chemoattractants for zebrafish. Chemically-mediated *attraction* behaviors are diverse and include *appetitive olfactory behavior* (elicited by L-alanine, food extract and others), chemically mediated *kin recognition* and *sexual aggregating behavior* (elicited by sex pheromones). Usually characterized by higher speed as fish follow the increasing concentrations of chemo-attractants and by slower speed when fish locate the signal source.
31. *Chewing:* While lacking an upper pharyngeal jaw, zebrafish can chew their food by grinding the teeth in their lower jaw against a chewing pad on the base of the skull.
32. *Circling (cycling, rotation):* Repetitive swimming in a circular direction (usually seen during seizures, neurological impairments, and following the selected drugs' action, Fig. 6). Normal behaviors with circling include *display (*circling plays a part in lateral display behavior) and *courtship (*circling can be seen in sexual behavior); can be quantified manually or using automated video-

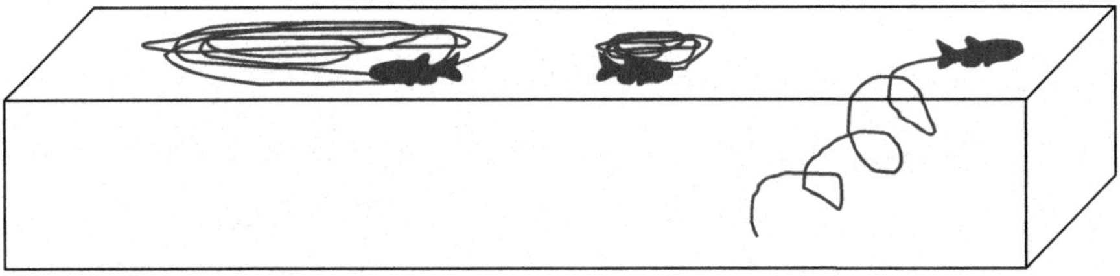

Fig. 6 Examples of zebrafish circling (*left*), tight circling (*center*) and cockscrew swimming (*right*)

tracking tools. Characteristic '*tight*' circling evoked by some treatments (e.g., glutamatergic antagonists) can be defined by their diameter, expressed in body length (e.g., two body lengths/~5 cm). Commonly used circling endpoints include the number of complete circles (360°) per trial, the number (%) of animals showing circles, and the direction of circling (left- or right-rotations); automated methods may also quantify turn angle and angular velocity. This behavior typically occurs in the same plane, thus differing from the spiral *corkscrew swimming* (occurring in 3D, Fig. 6)

33. *Coast (coasting):* Passive *sliding* without body/fin movements (i.e., after the fish stopped swimming actively), similar to *drifting* (also see *Creeping*, which involves slow swimming with only pectoral fin use).
34. *Coil (coiling):* Embryonic movement describing a full body contraction that brings the tip of the tail to the head (coils); can be spontaneous or evoked by touch. Involves single or alternating left-right bending of entire body; appears in embryo around 18 hpf, then gradually decreases in frequency.
35. *Color preference:* A natural preference/bias towards specific colors. For example, zebrafish will remain near some 'preferred' colors and keep away from those that induce an innate aversion (e.g., preferring black (*scototaxis*), or yellow, green, or red vs. blue), most likely associated with colors of natural threats, such as predators (this, however, may depend on context). The color of fish objects is also an important factor in *social interaction* (e.g., *shoaling* formation) in adult zebrafish.
36. *Coloration response:* See *Body color change.*
37. *Corkscrew swimming (spiraling, whirling):* Spiral swimming with an increased speed and in an uncoordinated direction; commonly observed as part of seizure phenotype (Fig. 6).
38. *Courtship:* Complex patterned behaviors that precede spawning. The male will *follow* or *chase* the female in a jerky swimming motion with his dorsal fin erect (*fin raise*), and attempt to *tail-nose touch* (this happens after the male makes visual contact, since first needs to chemically sense the readiness of female and display *sexual aggregating behavior*). If an immediate *spawning* attempt fails, the male may position himself just above the substrate with his body slightly angled downwards, and then will often display either *circling* (sometimes circling the female) or *zig-zagging*. The male will continually attempt to spawn with the female during this time. The female may *approach*,

Fig. 7 Representative courtship behaviors in adult zebrafish (*darker fish* is a male, *paler fish* is a female), from *left* to *right*: follow, tail-nose touch, circling (in the courtship context)

escort, present, and/or *lead.* If the male's advance is unwelcome, the female may *chase* the male away. Zebrafish courtship behavior can be quantified manually or using automated video-tracking systems, and characterized by the following endpoints: average distance between male and female; the number of contacts between male and female; time spent in spawning area by male and female; the number of entrances into spawning area by male and female; swimming distance and velocity inside and outside spawning area; total swimming activity and turning rate by male and female.

39. *Creeping:* Very slow *swimming* during which only the pectoral fins propel the fish forward; also see *coasting/drifting* (passive *sliding* without fin use).
40. *Cycling:* See *Circling* (Fig. 6 and 7).
41. *Dart (darting):* A single fast acceleration in one direction (e.g., as part of *escape* behavior) with the use of caudal fin. May be part of *dashing* or *erratic movement/zig-zagging* (associated with multiple darts, representing fast acceleration bouts in rapid succession, in which the direction of movement also changes in a seemingly stochastic manner between the darts). In some publications, darting was called *leaping* or *jumping* (which are presently defined in ZBC as separate, distinct behavior).
42. *Dashing:* A series of directed (propulsive) *darting* movements; commonly seen as an *escape response*.
43. *Depth preference:* Natural tendency to prefer depth over shallow water, Fig. 8 (note, however, that shallow water can trigger *breeding behavior*).
44. *Dispersion:* Rapid escape-like behaviors of multiple fish moving away from each other, before reuniting with the group; typically caused by the sudden exposure to a large (potentially dangerous) moving object, such as predator (also see *Shoaling*).
45. *Display:* Agonistic social behavior used to establish dominance/hierarchy, plays a role in *fighting behavior. Lateral* display: Two fish line up parallel to each other head to tail, raise their dorsal fins (*fin raise*), extend their caudal fins, darken in color (*body color change*), and swim in circles (*circling*), often ascending. *Frontal* display: Two fish *approach* each other from the front with the attempt of *nippling/biting.*
46. *Dive (diving, geotaxis):* Movement to/preference towards the bottom of the tank, often in response to threat. Generally, a very sensitive measure of *anxi-*

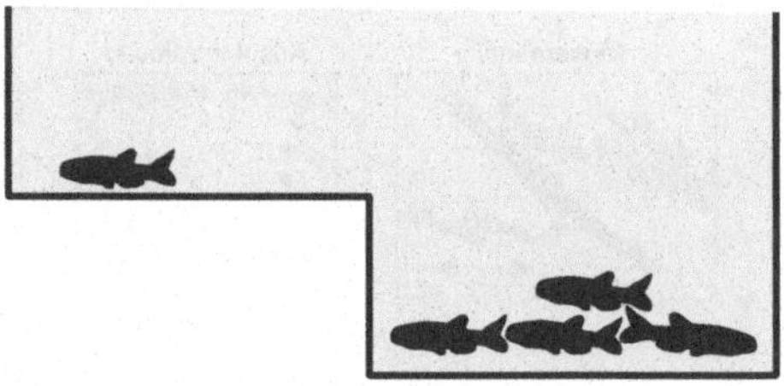

Fig. 8 The depth preference test in adult zebrafish

ety/avoidance behavior; can be quantified by latency to bottom, time in bottom, frequency of visits to the bottom, distance traveled in bottom, and also expressed as respective top:bottom ratios. Is commonly reduced during *habituation* or anxiolytic treatments, increased by sedative/anxiogenic drugs, and can be atypically reversed by some hallucinogenic drugs (e.g., ibogaine). Diving is an active, fast and *directed* zebrafish behavior (with body heading towards the bottom head first); differs from passive, more slow and undirected behaviors, such as *sinking* or *Cartesian diver behavior* (typically occurring in horizontal body position), or *resting behavior*.

47. *Dorsal light reflex (DLR):* A tilting of the body axis toward a light source, commonly observed in teleost fishes. Briefly, when illuminated horizontally, the fish inclines its dorsoventral axis and turns dorsal surface toward the light source, with its body tilt corrected antagonistically by the vestibular righting reflex (i.e., the body inclination increases with the illumination intensity but decreases with the gravity).
48. *Drift (drifting):* See *Sliding, Coasting*.
49. *Droopy (drooping) tail:* Motor phenotype associated with neurological deficits, *akinesia,* and global *hypolocomotion* (Fig. 1). Can be evoked by aging, motor impairments or genetic and pharmacological modulations (e.g., exposure to monoamine-depleting agent reserpine); extreme phenotypes may result in *inclined swimming*. Droopy tail is a long-lasting phenotype, and differs from *tail dip* (a short episode of *submissive behavior*); can also be part of a normal *resting/sleep behavior.*
50. *Epilepsy-like behavior:* See *Seizure behavior.*
51. *Erratic movement (erratic swimming/locomotion):* Complex behavior characterized by sharp changes in direction or velocity and repeated rapid *darting* (Fig. 9). Commonly observed in adult zebrafish, erratic movement is associated with multiple darts (fast acceleration bouts in rapid succession in which the direction of movement also changes in a seemingly stochastic manner between the rapid darts; also see *Zig-zagging*). Usually evoked by acute stressors (predator exposure, alarm cue release) or reflects a general baseline anxiety/fear state; commonly seen immediately before or after *freezing* bouts; part of the *alarm reaction*. Larval zebrafish can also display erratic movements, for

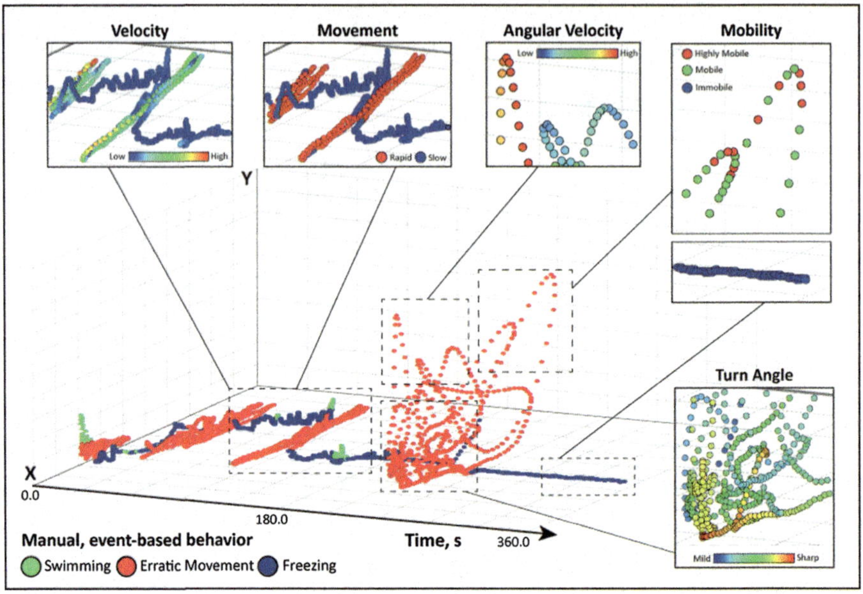

Fig. 9 Locomotor 3D-based dissection of adult zebrafish erratic behaviors (*bold inset*) in zebrafish (adapted from [2])

example, in response to sudden change in the light (e.g., exposed to a predator-looking shadow).

52. *Escape (startle response, tail thrash/ing):* A large body angular acceleration and displacement in response to a startling stimulus. The first stage is a bodily 'C-bend' (*C-start*), Fig. 4, followed by a contralateral bend and *tail beat(s)*. The initial acceleration is often followed by rapid *zig-zagging* near the bottom of the tank; in some cases, escape can lead to *jumping* behavior. In larval fish, involves *fast turning* followed by *burst swimming*.
53. *Escort (escorting):* Swimming alongside a male or remaining still while being courted (Fig. 10); performed by females during *courtship*.
54. *Exploratory activity:* A complex group of behaviors directed at exploration of novel environments. Related to, but not dependent on, locomotor activity and anxiety-related parameters (for example, the exploratory profile of zebrafish can be measured by quantifying the ratio of their activity in different horizontal sections and vertical areas of a tank).
55. *Fast turn (turning):* Escape-like turns in larval zebrafish, characterized by fast, large-angle turns that involve bending of the entire body with high angular velocity; takes 12 ms to turn head 180°, followed by a C-shaped counter-bend and vigorous swimming episode, as larvae swim away at a 90–180° angle. Associated with *escape* responses (e.g., in response to a stimulus), typically last 6–14 ms in larvae.

Fig. 10 Escorting behavior in adult male zebrafish

56. *Fear-like behavior:* See *Anxiety-like behavior* for details. Traditional clinical view of anxiety is that it is a state or response induced by potential (but not currently present) aversive stimuli, whereas fear is in direct response to the appearance or perception of such stimuli. Therefore, anxiety is more diffuse, and fear is more cue-oriented. Currently, it is unclear how exactly the two conditions translate into zebrafish behavior, although certain conditions (e.g., *alarm reaction*) are more relevant to fear, while others (e.g., withdrawal) seem to represent pathological anxiety-like state.
57. *Feeding:* Behaviors related to consumption of food (see *Biting, Chewing, Nibbling*); can include some specific types of food (e.g., *cannibalism, prey capture*).
58. *Fight (fighting):* Agonistic confrontation between two individuals often used to establish social dominance; comprises two distinct phases: the fish first assess each other by exhibiting *display*, *biting/nipping, flicking,* and *circling* behaviors, which continues until the first *chase/flee* occurs (Fig. 11). Next, the 'winner' (chaser) initiates all agonistic behaviors, while the 'loser' displays *fleeing*, *submission behavior,* or *freezing.*
59. *'Figure eight' swimming:* A specific swimming pattern observed in zebrafish following selected drug treatments (e.g., nicotine or ketamine); can also be part of natural *courtship* behavior (when male fish swims around female with raised fins, Fig. 12).
60. *Fin raise (fin extension/erection):* Raising the dorsal fin and/or extending the caudal fins; common in zebrafish during *aggression* and *courtship.*
61. *Flee (fleeing), flight behavior:* Accelerating movement away from another fish or stimulus.[67]
62. *Flick (flicking):* A specific agonistic behavior observed when two zebrafish swim towards each other, briefly touch mouths, and then simultaneously flick away in opposite directions; can be repeatedly displayed during agonistic interactions (*fight*).
63. *Flight behavior:* See *Flee.*
64. *Floating:* Passive swimming (typically near the water surface), differs from *surfacing* (typically a more active locomotion at the water), *drifting* (typically in the middle of the water layer) or *sleep/resting* (typically near/at the bottom); can be related to neurological impairments or *buoyancy dysregulation*.
65. *Food seeking:* A common form of zebrafish *foraging* behavior. Is triggered by hunger and can be suppressed by pathogenic conditions (e.g., *sickness behavior*) or by selected psychotropic drugs acting as appetite suppressants.

Fig. 11 Fighting behavior in adult male zebrafish (dominant winner is *darker*, loser fish is *paler*)

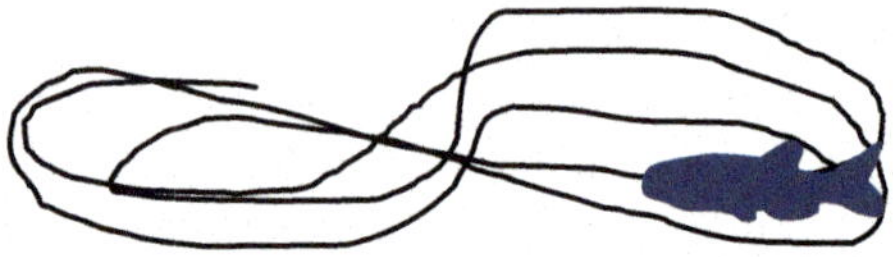

Fig. 12 Example of the zebrafish figure eight swimming (*top view*)

66. *Follow (following):* Behavior similar to *chase*, typically a nonaggressive movement towards (after) another fish; common during *courtship* and *social interaction.*
67. *Foraging:* Searching and/or probing movements typically in response to sensory cues (e.g., *food seeking*). Chemically induced foraging is characterized by frequent displays of directional turns as animal samples turbid chemical plumes (see *Appetitive olfactory behavior*).
68. *Freeze (freezing):* A complete cessation of movement (except for gills and eyes) by the fish while at the bottom of the tank. Generally, a result of high stress/anxiety or part of the *submissive behavior* (e.g., submissive immobile postures); can be quantified by assessing the latency, frequency, duration and location of freezing. *Opercular movements* (respiration/gill movements) are usually very frequent during stress-induced freezing. Freezing behavior differs from *immobility*, which is typically not associated with increased *opercular movements,* and usually caused by toxic/sedative agents (e.g., high ethanol concentrations), during which the animals also present *hypolocomotion* and *akinesia;* can also result in *sinking*.
69. *Fright:* See *Escape.*
70. *Geotaxis:* See *Dive/diving.*
71. *Jaw movements:* Stereotypic non-foraging *mouth opening behavior* observed following treatment with some drugs (e.g., hallucinogenic phencyclidine or convulsant agents, such as domoic acid).
72. *Habituation:* Tendency to show a robustly decreased response upon repeated exposure to a novel stimulus/environment. Includes inter-trial (inter-session) and intra-trial (intra-session) habituation. Over time, typically includes increased top exploration, reduced *diving* (Fig. 13), and unaltered *erratic movements.* Zebrafish habituation can be quantified by calculating the ratios of behavioral activity during the initial vs. latest trials, or by assessing the behavioral profile of fish across the trial(s).

73. *Head-butting:* Single or repeated pushing head against the vertical surface (e.g., glass, rock); commonly observed during *mirror stimulation response*, during *thrashing* behavior, or as the result of action of selected psychotropic/hallucinogenic drugs (e.g., LSD or ibogaine).
74. *Head shake movements:* A type of seizure/tremor-like behavior, in larval zebrafish often coupled with convulsions (typical for some convulsant drugs, e.g., domoic acid).
75. *Hide (hiding):* Attempt by the fish to conceal itself (e.g., under the stationary object/shelter).
76. *Homebase formation/behavior:* The tendency to establish a key 'safe' location which the fish spends more time in and repeatedly returns to after exploring a novel environment (Fig. 14). A natural form of *place preference* behavior; can be assessed by time spent, number of visits and distance traveled in homebase (vs. non-homebase) areas. May be sensitive to some pharmacological manipulations.
77. *Hyperactivity:* See *Hyperlocomotion, Hyperactivity burst.*
78. *Hyperactivity burst:* Episode of darting-like *erratic movements* with rapid turning and high velocity locomotion *within* a single behavioral bout; can be seen in both adult and larval fish, e.g., during high *anxiety* states or as *seizure behavior.*
79. *Hyperlocomotion (hyperactivity):* Abnormally fast swimming endured for an extended period of time; typically related to psychostimulant/convulsant action or *anxiety-like behavior*.
80. *Hypoactivity:* See *Hypolocomotion.*
81. *Hypolocomotion (hypoactivity):* Abnormally slow swimming for an extended period of time; typically related to sedation, neuromotor deficits, and *akinesia*.
82. *Immobility:* A complete cessation of movement (except for gills and eyes) at the bottom of the tank; differs from *freezing* and *resting* as not always associated with altered (respectively, increased or reduced) *opercular movements* (note, however, that immobility and freezing are often used as synonyms in

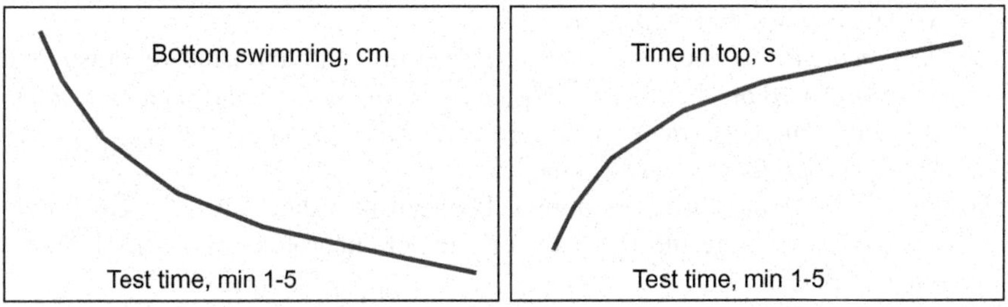

Fig. 13 A typical habituation curve for bottom swimming distance (*left*) and time spent in top (*right*) observed in the novel tank test

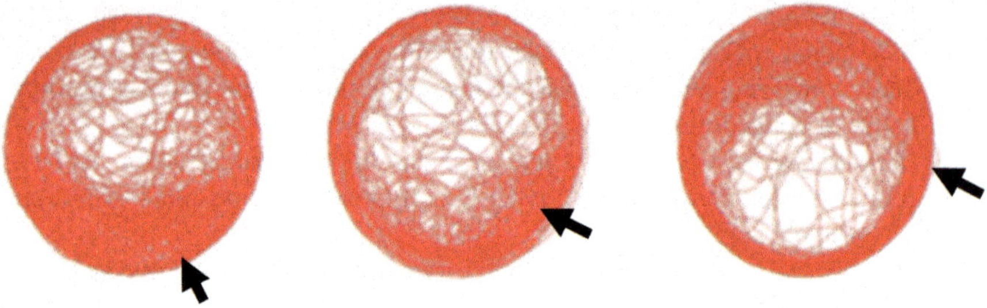

Fig. 14 Examples of zebrafish homebases (denoted by *arrows*) established by three representative zebrafish in the circular open field tanks

zebrafish literature). Can be caused by sedative agents (such as high ethanol concentrations), during which the animals may also present *hypolocomotion, akinesia*, or *paralysis.*

83. *Inclined swimming:* An aberrant phenotype (swimming with an angle relative to the water surface; *tilting*), commonly induced by neuroactive/neurotoxic substances (also see *vertical swimming* and *swimming upside down*). Can be related to *droopy tail, buoyancy dysregulation,* and triggered by motor deficits or aging.
84. *Infanticide:* Cannibalizing eggs (egg *cannibalism*) or larvae/fry.
85. *J-bend (J-turn), J-start:* Fine reorientation tuning in which the larva body slightly curves (~30°), with a characteristic bend at tail (Fig. 15).
86. *'Jittery' swimming:* A specific pattern of *swimming* characterized by multiple short 'jerky' movements with reduced smoothness of swimming trajectories; common for some seizure behavior, can be induced by selected consultants (e.g., RDX, strychnine).
87. *Jump (jumping):* A specific zebrafish behavior involving jumping out of water/tank (similar to *leaping*); usually caused by anxiogenic factors, as part of *escape* behavior or *alarm reaction* (e.g., can be triggered by predator or alarm cue exposure); also see *Terrestrial jump* (note, however, that in some publications, terms 'jumping' or 'leaping' are used to describe *darting* behavior).
88. *Kin preference:* The preference for kin vs. unrelated zebrafish, absent in larvae but particularly robust in juvenile 21-dpf zebrafish; based on chemical and visual cues (see *Kin recognition*).
89. *Kin recognition:* The ability to recognize kin (from unrelated zebrafish) and seek kin based on chemical and visual cues; involves *approach* and *attraction* and will ultimately lead to increased time spent near kin (*kin preference*; also see *Social preference/recognition*).
90. *Laying on a side:* Loss of normal body posture due to *ataxia*; commonly observed as a result of sedation and/or neurotoxicity-induced motor incoordination.

Fig. 15 Diagrams showing typical J-start (*left*) and O-start (*right*) in zebrafish locomotion

91. *Lead (leading):* Returning at least three times to one location in the tank; performed by females during *courtship.*
92. *Leap (leaping):* See *Jumping.*
93. *Lethargy:*Behavioral state indicative of chronic distress and/or illness (similar to a broader term *sickness behavior*) in adult zebrafish that includes decreased locomotor activity, reduced *escape* response, atypical body coloration and staying close to the bottom, with fins (especially dorsal) typically held close to the body. Differs from social *submissive behavior* by the chronic nature and independence of social context.
94. *Loop (looping):* Distinct circular swimming behavior in larvae around a virtual point outside of the larva's body (differs from *circling*); occurs as early as 5 dpf, common in mutants with visual feedback defects.
95. *Magnetic behavior:* Behavioral responsivity of zebrafish (e.g., preferred spatial orientation) to the magnetic fields.
96. *Mauthner reflex:* See *C-start/bend.*
97. *Meander (meandering):* Movement without a fixed direction or path; assessed as °/m (Fig. 16); can be increased during periods of high anxiety, especially during *erratic movement.*
98. *Mirror stimulation response:* Complex behaviors evoked in fish by mirror exposure; most likely linked to aggression; typically includes *approach, head-butting, biting* (the mirror), and *chasing* (own reflection, Fig. 17).
99. *Motor incoordination:* A general loss of normal coordination of body movements (e.g., *swimming on a side, corkscrew swimming*); commonly observed as a result of neurotoxicity or other neurological defects (also see *Akinesia, ataxia*).
100. *Mouth opening behavior:* Frequent mouth opening (different from *chewing* or *biting* behavior) which can be part of *aggression/attack, snapping* (e.g., when exposed to food odors) or a specific *stereotypic behavior* (e.g., *jaw movements)* observed following treatment with some drugs. Mouth opening rate is significantly reduced during *resting/sleep* states.
101. *Neophobia:* Avoidance of novel objects or food (e.g., food neophobia); can be assessed by measuring the latency and time spent/frequency of contacting the novel object.
102. *Nip (nipping):* See *Bite (biting).*

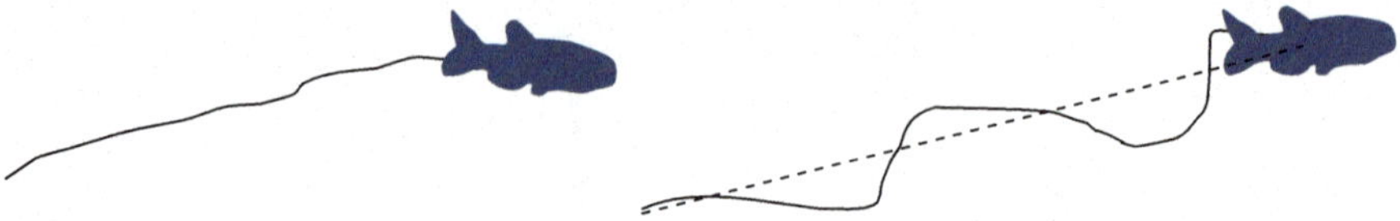

Fig. 16 Examples of normal, more straight swimming (*left*) vs. high-meandering locomotion (*right*)

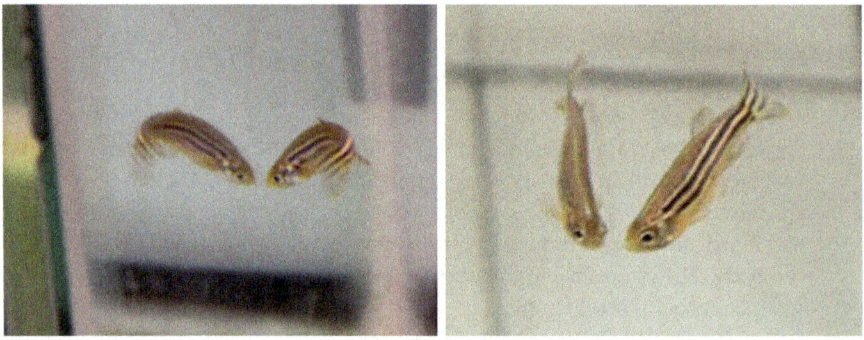

Fig. 17 Examples of zebrafish mirror-stimulation behaviors (photos courtesy of Matthew Singer, University of Idaho, ID, USA)

103. *Nibble (nibbling):* Nonaggressive *biting* on an object (usually, in *food seeking* or as part of object exploration).
104. *Nocifensive (pain-related) behavior:* Pain response to noxious stimuli, often can be experimentally induced in fish by chemical, thermal, or electrical stimulation; characterized by increased *swimming*, *escape,* and *tail-beating* responses (phasic stimuli induce tail beating and escape, while tonic stimuli induce *rubbing*, *tail-beating*, and increased *opercular movements*). While early views questioned pain responses in fish, mounting evidence indicates the presence of pain and pain-related behavior in zebrafish.
105. *O-bend (O-start, O-turn):* Orientation movement in which the larval zebrafish body curves to change the orientation (~180°) of swimming. In contrast to *C-bend*, this response is slower and independent of Mauthner cells; commonly elicited by dark flashes (Fig. 15).
106. *Olfactory response (olfactory behavior):* Complex, odorant-evoked behavioral activity (also see *Chemotaxis*). Common changes in behavior include altered swimming speed and distance traveled (*appetitive olfactory behavior*), *avoidance* (e.g., *alarm response*) *and attraction* (i.e., during *foraging, spawning* or *kin recognition*).
107. *Opercular movements:* Respiration/gill movements of zebrafish, can be visualized using slow-mode video-recording; bi-directionally modulated by various psychotropic drugs, are markedly increased during distress (e.g., during

stress-induced *freezing,* where opercular beat rate can serve as an additional index of anxiety) and reduced during *resting/sleep.*

108. *Optokinetic response/reflex (OKR):* Stereotyped tracking eye movements triggered by moving objects across the visual field. Has two components: a smooth *pursuit* movement following the object, and a fast saccadic movement resetting the eyes after the object has left the visual field.
109. *Optomotor response/reflex* (*OMR*)*:* Locomotion induced by a repetitive moving stimulus presentation (e.g., rotating drum), as zebrafish will generally swim in the same direction as the moving pattern.
110. *Oscillations of locomotor activity:* Sinusoidal aspect of zebrafish locomotion/swimming, with alternating high- and lower-velocity phases (usually, with the frequency of 4–5 min) when exposed to novel environments.
111. *Oviposition:* Release of eggs by the female during *spawning.*
112. *Pain-related behavior:* See *Nocifensive behavior.*
113. *Parallel (paralleling):* Behavior during *spawning*, when the male swims alongside the female, in contact but slightly behind it, with head approximately leveling the female's operculum.
114. *Paralysis:* A complete cessation of all movement, including eyes, gills/operculum and fins (similar to *ataxia*, but with more severe/global motor impairment, often with an abnormal posture, such as laying on the side, floating upside down or standing vertically). Usually caused by selected neuroparalyzing agents or genetic neurological mutations.
115. *Photokinesis (phototaxis):* General movement in response to light, including *positive* (light seeking/dark avoidance, *scotophobia*) and *negative* (light avoidance, dark preference, *scototaxis*). Zebrafish display sensitivity to visible light (positive phototaxis in larval fish; light avoidance in adult fish), and negative phototaxis to ultraviolet (UV) light (*UV avoidance*).
116. *Photomotor response (PMR):* A stereotypic series of motor behaviors in embryonic zebrafish in response to light stimulation, as zebrafish show motor excitation (lasting 5–7 s) with vigorous shaking, followed by a refractory phase, during which basal locomotion is suppressed and animals do not respond to another light pulse.
117. *Phototaxis:* See positive and negative *photokinesis (scotophobia* and *scototaxis).*
118. *Piping:* Gulping air at the water surface; can be indicative of distress (e.g., hypoxia or toxicity), but may also be seen during depth-related adjustment of swim bladder volume (during *diving).*
119. *Place preference:* The tendency to establish a preferred location in which the fish spends more time. Can be induced by drugs (e.g., in conditioned place preference paradigms, CPP), repeated administration of food/food odors, social reward, or be based on natural behaviors (e.g., *homebase formation*, see Fig. 4) or preferences (e.g., *depth preference, scototaxis, thigmotaxis*).
120. *Polarization:* Behavioral characteristic of adult zebrafish reflecting the degree to which members of the group are moving in the same direction (Fig. 18); is high in zebrafish schools (see *Schooling*) and reduced in shoals (see *Shoaling*).

Fig. 18 Polarization in zebrafish groups of approximately the same size (*left* to *right*: high to low)

121. *Prey capture (capturing):* A complex behavior of larval zebrafish; consists of identifying the prey (e.g., paramecium) visually or using chemosensation, tracking it with a series of *routine turns*, forward slow *swim* and/or *J-turns,* followed by capture and ingestion (*feeding* behavior). The initial bends have low amplitude and are prominent at far-caudal locations; later bends originate more rostrally, have higher amplitude and are accompanied with increased *tail-beat* frequency.
122. *Predator inspection:* An exploratory/*boldness*-related *risk-taking behavior* in fish associated with either increased or decreased tendency to *approach* a predator, potentially to gather information about the identity, precise location and/or current motivational state of the predator. Commonly observed in fish when *shoaling*, as they leave the shoal, swim towards the predator, and then return to the group.
123. *Predatory attack:* Adult and larval zebrafish may attempt to bite/consume any sufficiently small item moving through their field of vision at appropriate speeds (see *Biting*); in larval zebrafish, develops at ~4 dpf, and manifests as *prey capture* behavior.
124. *Present (presenting):* Halting and exposing side in front of a male or swimming up and down in front of male; performed by females during courtship.
125. *Quiver (quivering):* High frequency, low amplitude tail oscillation by a male while aligned against the side of a female; occurs during *spawning.*
126. *Reflection chase (chasing):* A behavior that includes chasing own reflection (e.g., in the observation tank or as part of the *mirror stimulation response);* can also be triggered by selected psychoactive (e.g., hallucinogenic) drugs.
127. *Reproductive (breeding) behavior:* see *Spawning.*
128. *Rest behavior (resting): Sleep*-related behavior in adult and larval zebrafish. Typical rest behavior in larval zebrafish includes floating with head down, or staying horizontal, often close to the bottom of the tank (*immobility, hypolocomotion).*
129. *Retreat (retreating):* A social behavior relevant to dominance in zebrafish, generally involves a submissive fish swimming rapidly away from the opponent (e.g., from a dominant fish) in response to an *attack* (e.g., after a *strike, bite, chase, or charge*), part of *fleeing* behavior.
130. *Rheotaxis:* A common behavior in aquatic species, includes turning towards a current and a tendency to swim upstream; displayed by both larval and adult zebrafish. In experimental setting, manifests in *avoidance* by zebrafish of the sucking source (e.g., sucking pump or standpipe).

131. *Risk-taking behavior:* Propensity of zebrafish to engage in dangerous situations (part of their *boldness* phenotype); commonly occurs when a prey fish approaches/inspects a predator (see *Approach, Predator inspection*), and/or when a shoal member leaves the group (see *Shoaling*).
132. *Rotation behavior:* See *Circling, Cycling.*
133. *Routine turn(ing); R-turn(ing):* A slow spontaneous turn (20–30 ms) with a large bend angle (~60°) resulting in reorientation of the larva before forward *swimming, prey capture;* lacks the large counter-bend (shown in *escape turns),* with only a small portion of the tail bending; has a slow angular velocity with relatively slow turning angles.
134. *Rub (rubbing):* A characteristic aberrant zebrafish behavior involving rubbing body sides on the sides of the tank (or the surface of other objects); typically caused by pathogenic conditions (e.g., skin disease and/or parasitic infection).
135. *Scoot swim (swimming):* See *Slow swim.*
136. *Scotophobia:* A natural preference for light (or avoidance of dark) lighting/environment, commonly observed in larval fish; Fig. 19; usually is replaced with *scototaxis* in adult fish (also see *Photokinesis*).
137. *Scototaxis:* A natural preference for dark (or avoidance of bright) lighting/environment in adult zebrafish (Fig. 19). Generally, a measure of anxiety (reduced by anxiolytic drugs and increased by anxiogenic agents). Note that larval zebrafish display opposite behavior (*scotophobia*). Can be quantified in the light–dark box tests (by assessing the latency, time spent in light or dark, distance traveled, the number of visits, the average duration of a visit; and by the respective behaviors' light:dark ratios); also see *Photokinesis.*
138. *School (schooling):* Formation of a relatively polarized group (school) in which multiple fish swim together, in a coordinated/synchronous fashion; part of *aggregation behavior* that increases *foraging* efficiency as well as the ability to detect and/or avoid predators. Schools may disperse into shoals (see *Shoaling*) which show *reduced polarization* (Fig. 20).
139. *Seizure (seizure-like/epilepsy-like) behavior:* Involuntary, rapid movements of body (usually, as a result of pathology, such as epilepsy) observed in both larval and adult zebrafish; include *ataxia, corkscrew (spiral) swimming, hyperactivity, circling, spasms, weavering, head shake movements, tremor,* and/or *jittery locomotion.* Severe cases include death. Can be quantified manually (using seizure scale) or applying automated video-tracking tools (by assessing the velocity and distance traveled).
140. *Sexual aggregating behavior:* Instinctive response to chemical cues released from females during ovulation; males display *attraction, approach* and *courtship* behaviors once female is identified.
141. *Shoal (shoaling):* Formation of a relatively nonpolarized group (shoal) of adult zebrafish, held together by social pressures (i.e., not by individual attraction to an external stimulus); part of *aggregation behavior.* Anxiety/fear causes the shoal to 'tighten' (the fish swim closer together) and potentially form a school (see *Schooling,* Fig 20). Hunger/habituation causes the shoal to

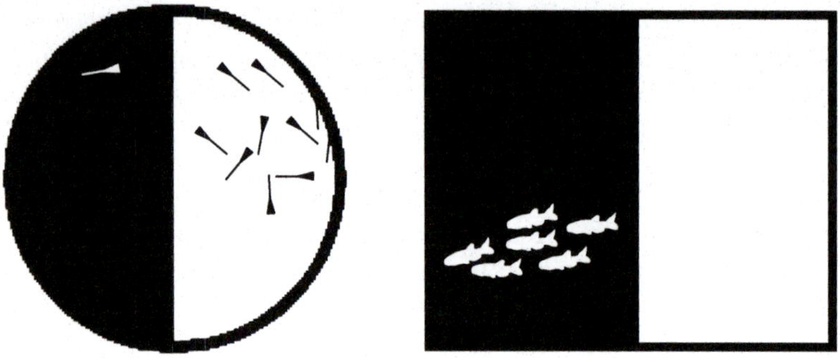

Fig. 19 A typical scotophobic response in larval (*left*) and scototaxic response in adult (*right*) zebrafish

Fig. 20 A typical zebrafish school (*left*) and shoal (*right*). (*Insets*: the respective top view photos; images courtesy of Dr. Noam Miller, Wilfrid Laurier University, ON, Canada)

become looser and less organized. Zebrafish shoaling has an oscillating dynamic, and this behavior can be quantified manually or using automated video-tracking systems, assessing several endpoints, including the average inter-fish distance; shoal area size; proximity (time each member of the shoal spent within a specified distance from each other); nearest and farthest neighbor distances; time spent in shoal; time spent away from shoal; number of animals leaving the shoal (also relevant to *risk-taking behavior*) and *polarization* (reflecting the uniformity of heading).

142. *Shyness:* A reduced *exploratory activity*, reduced general activity in a novel environment and/or in response to stimuli, or reduced risk-taking behavior (opposite to *boldness*, Fig. 21).
143. *Sink (sinking):* Freezing behavior during which the fish remains immobile (except for the eyes and gills) but changes its position in the water column (moving from top to bottom) without moving any of its fins (also see *Cartesian diver behavior*).
144. *Sickness behavior:* A broad cluster of behaviors indicative of illness (or pain) that include *hypoactivity*, inhibited exploration, *feeding* or *food seeking*, pale *body color*, and *lethargy* (with fins typically held close to the body).
145. *Sleep (sleep-like behavior):* Activity characterized in zebrafish by *rest behavior*, including reversible *immobility/hypolocomotion* (Fig. 22), elevated

arousal threshold, reduced respiratory rate (e.g., *opercular movements*) and mouth opening frequency, and a compensatory rebound in response to sleep deprivation. In adult zebrafish, includes brief periods of inactivity, often with a drooping caudal fin (see *Droopy tail*), alternated with active periods of *swimming;* can be easily reversed by startling stimuli, such as tapping, sound, or weak electric field.

146. *Slide (sliding):* See *Coasting.*
147. *Slow swim (slow/scoot swimming):* Larval zebrafish slow *swimming* (scoots) characterized by small bend angles with bend location near the tail. Maximal bending occurs close to the tail; low degree of bending and tail beat frequency; yaw angles are <3°; pectoral fins are active and alternate right to left between adduction and abduction.
148. *Snap (snapping):* Reflexive opening and closing of mouth during exposure to high concentrations of appetitive stimuli (e.g., L-alanine, food extract); signals initiation of ingestive phase during *feeding.*
149. *Spasm (twitch, twitching):* Spontaneous, rapid movements of body (usually, as a result of neurological/neurotoxic impairment, such as *seizure*).

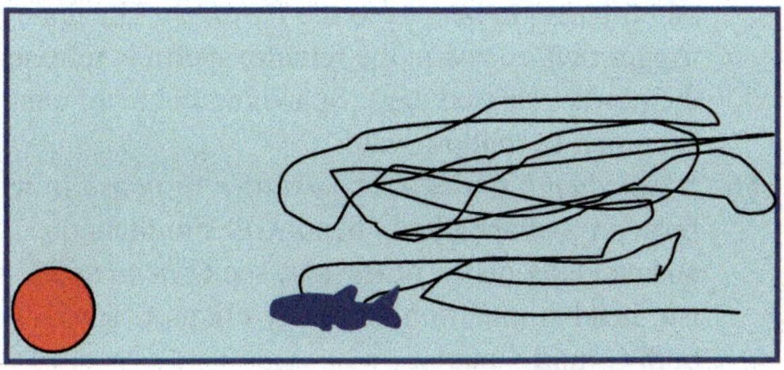

Fig. 21 Zebrafish shyness behavior in response to a novel object (*red ball*)

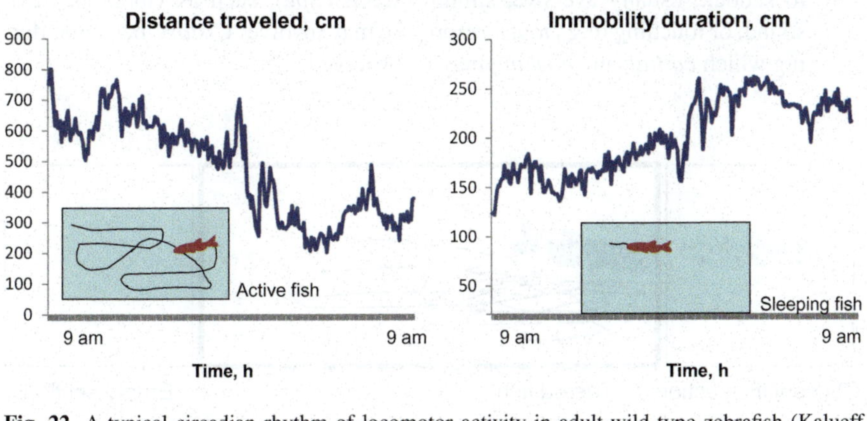

Fig. 22 A typical circadian rhythm of locomotor activity in adult wild type zebrafish (Kalueff et al., unpublished data, 2014–2015); insets: a typical active (*left*) vs sleeping (*right*) zebrafish

150. *Spatiotemporal stability:* The ability to withstand changes in environmental characteristics during exploration of a novel environment, primarily by scaling locomotor activity (e.g., distance traveled) to the size of the environment, but retaining the temporal budgeting of the activity; can be evaluated by temporal distribution of locomotion and position in a test tank (e.g., distance traveled, transitions and time spent in each area).
151. *Social interaction:* Normal social behavior of zebrafish, represents a reciprocal change in zebrafish behavior influenced by the presence or actions of other conspecifics. Some examples include *fighting/aggression, shoaling/schooling, courtship* and *spawning*; can also manifest in *approach/boldness* (social investigation), *social recognition*, and *social preference*.
152. *Social preference:* A natural tendency to spend time close to conspecifics; can be observed as part of *shoaling* behavior, *kin recognition, social recognition,* or preference of the 'conspecific' vs. 'empty' compartments of the tank (Fig. 23).
153. *Social recognition:* The ability of zebrafish to recognize familiar from unfamiliar zebrafish.
154. *Spiraling:* See *Corkscrew swimming, whirling* (Fig. 6).
155. *Spawning (breeding/reproductive behavior):* During breeding, the male zebrafish approaches the female and curves body around, positioning his genital pore next to hers (also see *Parallel*). The male then quivers in an attempt to trigger *oviposition* in the female; sperm is released simultaneously to fertilize the newly released eggs. Spawning behavior can be promoted by exposure of zebrafish to shallow water.
156. *S-start/bend:* Quick *escape/startle response* in which the fish body curves to form an S-shaped body bend with simultaneous activity rostrally on one side, and caudally on the other (also see *O-bend* and *C-start/bend*). In larval zebrafish, head stimulation generally elicits C-starts, while tail stimulation evokes both C- and S-starts.
157. *Startle response:* An evolutionarily conserved, adaptive behavior in response to sudden, usually aversive, stimuli (see *Escape*), such as vibration, light, sound, or touching (e.g., *touch response*); may involve '*C-start*' behavior, during which *coiling* and *dashing* may be observed.

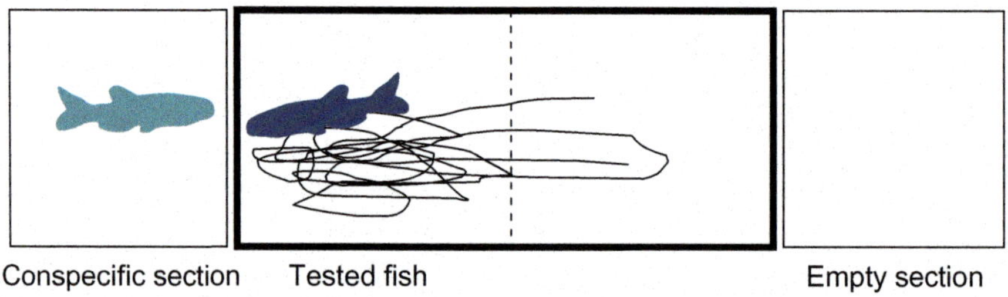

Fig. 23 Conspecific preference in zebrafish

158. *Stereotypic behaviors:* A pattern of rigid, repetitive behaviors other than *swimming* (see *Stereotypic locomotion/swimming*), evoked in zebrafish under some conditions, e.g., stereotypic *mouth opening behavior* following treatment with some hallucinogenic drugs (e.g., phencyclidine).
159. *Stereotypic locomotion (stereotypic swimming):* A pattern of rigid, repetitive behaviors (e.g., swimming from corner to corner) evoked in zebrafish under some conditions (e.g., treatment with psychostimulants like nicotine, Fig. 24, and caffeine, or hallucinogens like ibogaine or phencyclidine).
160. *Strike (striking):* An *aggression*-related behavior, observed in zebrafish when the fish swims rapidly toward the opponent, but without physical contacts between them. Differs from *approach* by a generally much higher velocity and its aggressive (rather than investigatory) nature (also see *Attack,* which occurs *with* physical contact).
161. *Struggle (struggling):* A behavior observed in larval zebrafish, characterized by longer alternating motor bursts at lower frequencies than *swimming*. During struggling, motor bursts propagate in inverse directions along the body, compared with swimming behavior.
162. *Submissive behavior:* A social behavior following aggressive confrontations between zebrafish. Submissive fish stays immobile (with fins retracted), typically near the bottom or near the surface of the aquaria (also see *Freezing*), with the caudal part of the body oriented downward (also see *Tail dip*, representing the initial form of submissive behavior).
163. *Surface touching:* See *Surfacing*.
164. *Surfacing (surface touching):* Dwelling of fish at the surface of the water; is usually evoked by selected neuroactive drugs, mainly serotonergic agonists and glutamatergic antagonists, may also be related to *buoyancy dysregulation*.

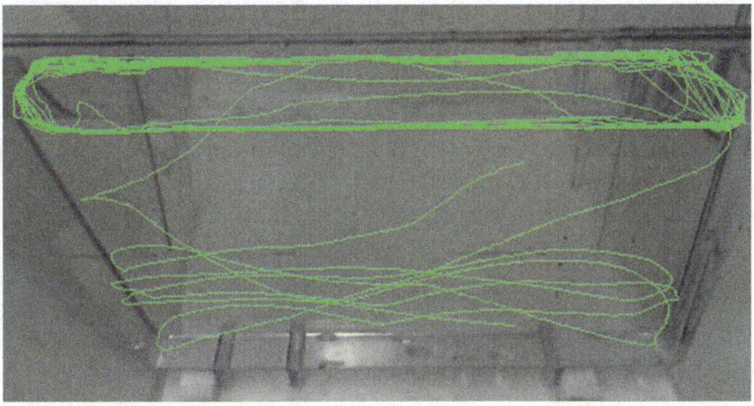

Fig. 24 Example of characteristic stereotypic 'peripheral' corner-to-corner top swimming evoked by acute exposure to 10 mg/L nicotine in adult zebrafish

165. *Swim (swimming):* Simple zebrafish locomotion; can be categorized by its duration as 'prolonged,' which may be maintained for minutes, or as 'sustained,' which may be maintained for hours. In larvae, includes *slow swim* (point of maximal body bending occurs close to the tail) and *burst swim* (maximal body bending occurs near the mid-body; maintained over seconds or less, includes larger-amplitude bending, faster speed, and greater yaw (vs. slow swim), also see *Beat-and-glide)*. Selected forms of aberrant swimming include *swimming on a side, upside down, vertical swimming, and Cartesian diver behavior.*
166. *Swimming on a side:* Loss of normal body posture due to *ataxia*; commonly observed as a result of sedation and/or neurotoxicity-induced motor incoordination.
167. *Swimming upside down:* An aberrant phenotype (swimming in an upside down position), commonly induced by neuroactive/neurotoxic substances as part of ataxia; also see *Inclined* and *Vertical swimming*.
168. *Tail dip (tail dipping):* An agonistic behavior during *fighting*, when one fish slightly drops its tail for a short period of time, to signal its submission and end the confrontation (differs from *droopy tail*, which a long-lasting behavior; see *Submissive behavior*).
169. *Tail beat (beating, slapping):* Characterized by repeating episodes of rhythmic, rostro-caudally progressing peripheral nerve discharges that are alternated between the two sides of the body. Viewed from above, tail beating is physically apparent in side-to-side sweeping of the tail (that can be measured as tail beat amplitude).
170. *Tail-nose touch(ing):* Touching the side or tail of another fish with the nose or head. Performed by males during *social interaction,* especially *courtship.*
171. *Terrestrial jump(ing):* A coordinated *leap* (using tail-flip) in response to placement on a damp surface (e.g., after *jumping* out of water); can be quantified by amplitude (height) and frequency of the leaps (Fig. 25).
172. *Territorial behavior:* Monopolization and aggressive defense of a defined area in a habitat/tank (e.g., *spawning* sites are a commonly defended territory in zebrafish). Trespassers into the territory may be chased or bitten by the dominant fish, or can '*fight*' to challenge its dominance.
173. *Thigmotaxis:* A preference for staying in close proximity to the edge/side (and avoiding the central open areas); generally, can serve as a measure of zebrafish anxiety (Fig. 26).
174. *Thrashing:* Forceful swimming with the use of the caudal fin while physically in contact with the side or bottom wall of the tank; swimming against the glass wall of the tank that appears as if the fish is trying to swim through the glass barrier. This behavior can be thrashing *towards* an appetitive cue (e.g., food) or thrashing *away* from an aversive stimulus (e.g., representing an *escape*); can be similar to *head-butting* behavior (e.g., during the mirror stimulation test).

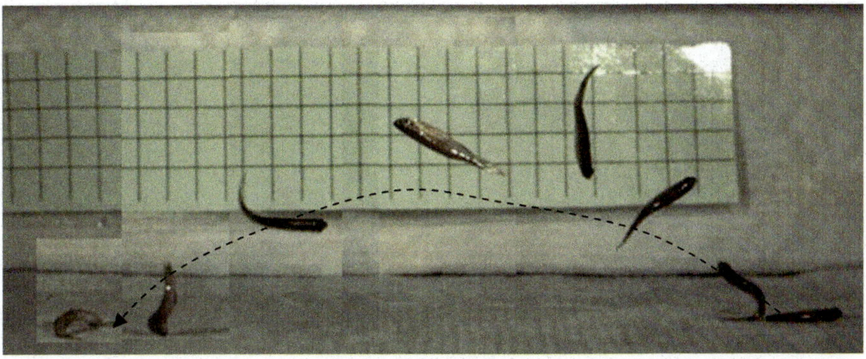

Fig. 25 Zebrafish terrestrial jump (a series of snapshots from video available in public domain on Youtube https://www.youtube.com/watch?v=bsSyXla6qeU)

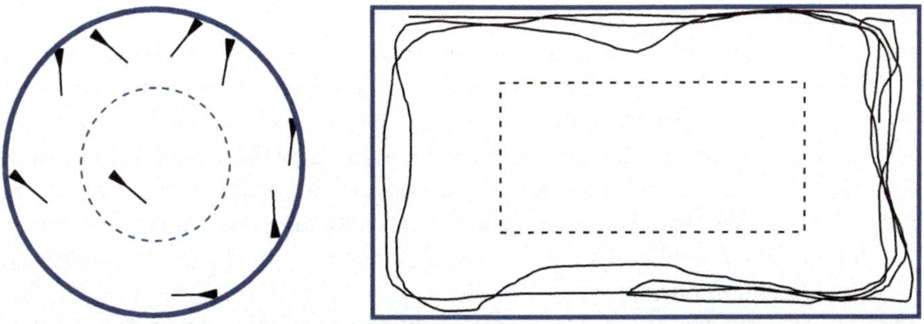

Fig. 26 Thigmotaxis in larval (*left*) and adult (*right*) zebrafish; the arena center is denoted by the *dashed line* (*top view*)

175. *Tilt (tilting):* Deviation from the horizontal position; is often seen during *inclined swimming.*
176. *Top dwelling:* Dwelling in the top half of the tank; can include aberrant swimming very close to water surface (*surfacing*).
177. *Touch response: Startle behavior*-related phenotype evoked by the touch in an embryo, which responds with fast *coiling* of the trunk bending over the head; appears around 21 hpf.
178. *Trance-like swim(ming):* A slow swimming motion induced in fish by specific psychotropic drugs (e.g., chronic fluoxetine or hallucinogens, such as LSD or salvinorin A); often includes a swimming pattern characterized by a slow (albeit active) bout of horizontal swimming for 1–2 s, followed by a similarly short passive gliding motion (Fig. 27).
179. *Tremor behavior:* Specific shivering-like behavior, most typically evoked in adult or larval zebrafish by selected neurotoxic/convulsant drugs, such as domoic acid (also see *Seizures* and *Weavering*); particularly visible in the tail area.

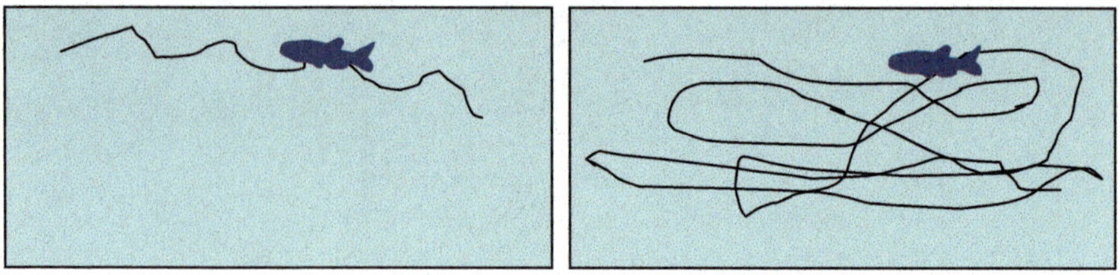

Fig. 27 Examples of trance-like swimming (*left*) vs. normal control fish locomotion

180. *Turn (turning):* A simple change in swimming direction observed in both adult or larval zebrafish. Larval zebrafish show several specific forms of turning behavior, including *slow routine turns* (lacking the large counterbend of escape turns) and *fast turns* (fast, large angle turns characteristic of *escape* responses).
181. *Twitch (twitching):* See *Spasm.*
182. *Undulating body movement:* A wave-like or snake-like motion; part of *aggression*-related behaviors, and occurs mainly at the beginning of the *fight,* especially between two equal opponents. This behavior is common in fish species, and is likely related to the use of lateral line by fish to size-up the opponent by the waves generated by its movement. Although less visible in zebrafish (compared to other fishes), their undulating body movement can be observed using high-resolution videos (sometimes this behavior is followed by forceful *tail beats*).
183. *UV avoidance:* Avoidance of UV light (negative *photokinesis/phototaxis*) reported in zebrafish larvae as early as 4 dpf.
184. *Vertical drift (drifting):* An aberrant phenotype that involves passive floating vertically (i.e., passive vertical motion from bottom to top, Fig. 28); commonly induced by neuroactive/neurotoxic substances or agents related to *buoyancy dysregulation* (also see *Vertical swimming*); opposite to *sinking* (also see *Cartesian diver behavior*).
185. *Vertical swim (swimming):* An aberrant fish phenotype that involves swimming vertically (typically heads up at the surface, Fig. 28); commonly induced by neuroactive (e.g., high doses of LSD) or neurotoxic substances, or agents related to *buoyancy dysregulation* (also see *Inclined swimming* and *Swimming upside down*).
186. *Vestibulo-ocular reflex (VOR):* Compensatory eye movements in zebrafish in response to linear/angular acceleration as well as changes in head position with respect to gravity, in order to stabilize the retinal image. These movements are evoked through the semicircular canals of the vestibular apparatus in zebrafish.
187. *Weaver (weavering):* Aberrant tremor/shaking-like phenotype, typically evoked by selected neurotoxic/convulsant agents (e.g., strychnine); similar to *tremor* and *head shake movements* (also see *Seizure behavior*).
188. *Whirl (whirling):* See *Corkscrew swimming, spiraling.*

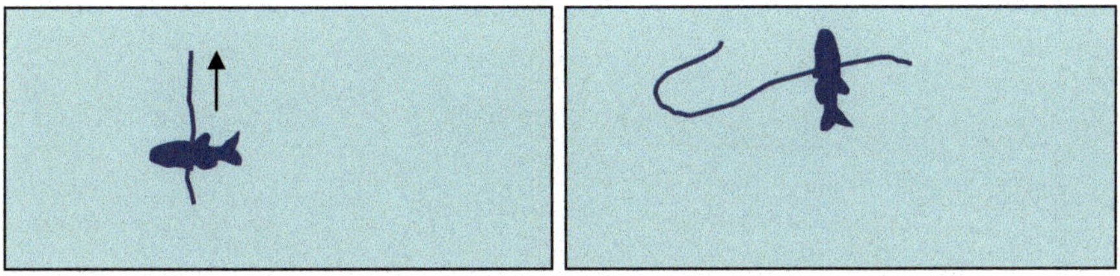

Fig. 28 Zebrafish vertical drifting (*right*) and vertical swimming (*left*) phenotypes

189. *Withdrawal-related behavior:* A characteristic behavioral syndrome observed in zebrafish following discontinuation of drugs of abuse (e.g., ethanol, morphine); typically characterized by elevated *anxiety/fear-like behavior.*
190. *Zig-zagging (zigzagging): Sexual:* a tail sweep and circle along a female's body resembling a 'figure eight'; typically performed by males during *courtship. Stress-induced: erratic movement* with multiple *darts,* during which the direction of movement changes in a seemingly alternating (zig-zag-like) manner between the darts.

References

1. Kalueff AV, Gebhardt M, Stewart AM, Cachat JM, Brimmer M, Chawla JS, Craddock C, Kyzar EJ, Roth A, Landsman S, Gaikwad S, Robinson K, Baatrup E, Tierney K, Shamchuk A, Norton W, Miller N, Nicolson T, Braubach O, Gilman CP, Pittman J, Rosemberg DB, Gerlai R, Echevarria D, Lamb E, Neuhauss SC, Weng W, Bally-Cuif L, Schneider H. Towards a comprehensive catalog of zebrafish behavior 1.0 and beyond. Zebrafish. 2013;10(1):70–86.
2. Cachat JM, Stewart A, Utterback E, Hart P, Gaikwad S, Wong K, Kyzar E, Wu N, Kalueff AV. Three-dimensional neurophenotyping of adult zebrafish behavior. PLoS One. 2011;6(3), e17597.

Index

A.V. Kalueff (ed.), *The rights and wrongs of zebrafish: Behavioral phenotyping of zebrafish*, DOI 10.1007/978-3-319-33774-6

Z

CPSIA information can be obtained
at www.ICGtesting.com
Printed in the USA
LVOW01*1927230217
525239LV00002B/29/P

9 783319 337739